jmp®

Version 9

Using JMP

"The real voyage of discovery consists not in seeking new landscapes, but in having new eyes."

Marcel Proust

JMP, A Business Unit of SAS
SAS Campus Drive
Cary, NC 27513

9.0.2

The correct bibliographic citation for this manual is as follows: SAS Institute Inc. 2010. *Using JMP 9*. Cary, NC: SAS Institute Inc.

Using JMP® 9

ISBN 978-1-60764-599-3

SAS Institute Inc., SAS Campus Drive, Cary, North Carolina 27513.

1st printing, September 2010

2nd printing, January 2011

3rd printing, June 2011

Contents
Using JMP

6 Output Reports

7 Reshaping Data

8 Summarizing Data

C JMP Preferences

D Formula Functions Reference

Index

Credits and Acknowledgments

Origin

JMP was developed by SAS Institute Inc., Cary, NC. JMP is not a part of the SAS System, though portions of JMP were adapted from routines in the SAS System, particularly for linear algebra and probability calculations. Version 1 of JMP went into production in October 1989.

Credits

JMP was conceived and started by John Sall. Design and development were done by John Sall, Chung-Wei Ng, Michael Hecht, Richard Potter, Brian Corcoran, Annie Dudley Zangi, Bradley Jones, Craige Hales, Chris Gotwalt, Paul Nelson, Xan Gregg, Jianfeng Ding, Eric Hill, John Schroedl, Laura Lancaster, Scott McQuiggan, Melinda Thielbar, Clay Barker, Peng Liu, Dave Barbour, Jeff Polzin, John Ponte, and Steve Amerige.

In the SAS Institute Technical Support division, Duane Hayes, Wendy Murphrey, Rosemary Lucas, Win LeDinh, Bobby Riggs, Glen Grimme, Sue Walsh, Mike Stockstill, Kathleen Kiernan, and Liz Edwards provide technical support.

Nicole Jones, Kyoko Keener, Hui Di, Joseph Morgan, Wenjun Bao, Fang Chen, Susan Shao, Michael Crotty, Jong-Seok Lee, Tonya Mauldin, Audrey Ventura, Ani Eloyan, Bo Meng, and Sequola McNeill provide ongoing quality assurance. Additional testing and technical support are provided by Noriki Inoue, Kyoko Takenaka, Yusuke Ono, Masakazu Okada, and Naohiro Masukawa from SAS Japan.

Bob Hickey and Jim Borek are the release engineers.

The JMP books were written by Ann Lehman, Lee Creighton, John Sall, Bradley Jones, Erin Vang, Melanie Drake, Meredith Blackwelder, Diane Perhac, Jonathan Gatlin, Susan Conaghan, and Sheila Loring, with contributions from Annie Dudley Zangi and Brian Corcoran. Creative services and production was done by SAS Publications. Melanie Drake implemented the Help system.

Jon Weisz and Jeff Perkinson provided project management. Also thanks to Lou Valente, Ian Cox, Mark Bailey, and Malcolm Moore for technical advice.

Thanks also to Georges Guirguis, Warren Sarle, Gordon Johnston, Duane Hayes, Russell Wolfinger, Randall Tobias, Robert N. Rodriguez, Ying So, Warren Kuhfeld, George MacKensie, Bob Lucas, Warren Kuhfeld, Mike Leonard, and Padraic Neville for statistical R&D support. Thanks are also due to Doug Melzer, Bryan Wolfe, Vincent DelGobbo, Biff Beers, Russell Gonsalves, Mitchel Soltys, Dave Mackie, and Stephanie Smith, who helped us get started with SAS Foundation Services from JMP.

Acknowledgments

We owe special gratitude to the people that encouraged us to start JMP, to the alpha and beta testers of JMP, and to the reviewers of the documentation. In particular we thank Michael Benson, Howard Yetter (d),

Andy Mauromoustakos, Al Best, Stan Young, Robert Muenchen, Lenore Herzenberg, Ramon Leon, Tom Lange, Homer Hegedus, Skip Weed, Michael Emptage, Pat Spagan, Paul Wenz, Mike Bowen, Lori Gates, Georgia Morgan, David Tanaka, Zoe Jewell, Sky Alibhai, David Coleman, Linda Blazek, Michael Friendly, Joe Hockman, Frank Shen, J.H. Goodman, David Iklé, Barry Hembree, Dan Obermiller, Jeff Sweeney, Lynn Vanatta, and Kris Ghosh.

Also, we thank Dick DeVeaux, Gray McQuarrie, Robert Stine, George Fraction, Avigdor Cahaner, José Ramirez, Gudmunder Axelsson, Al Fulmer, Cary Tuckfield, Ron Thisted, Nancy McDermott, Veronica Czitrom, Tom Johnson, Cy Wegman, Paul Dwyer, DaRon Huffaker, Kevin Norwood, Mike Thompson, Jack Reese, Francois Mainville, and John Wass.

We also thank the following individuals for expert advice in their statistical specialties: R. Hocking and P. Spector for advice on effective hypotheses; Robert Mee for screening design generators; Roselinde Kessels for advice on choice experiments; Greg Piepel, Peter Goos, J. Stuart Hunter, Dennis Lin, Doug Montgomery, and Chris Nachtsheim for advice on design of experiments; Jason Hsu for advice on multiple comparisons methods (not all of which we were able to incorporate in JMP); Ralph O'Brien for advice on homogeneity of variance tests; Ralph O'Brien and S. Paul Wright for advice on statistical power; Keith Muller for advice in multivariate methods, Harry Martz, Wayne Nelson, Ramon Leon, Dave Trindade, Paul Tobias, and William Q. Meeker for advice on reliability plots; Lijian Yang and J.S. Marron for bivariate smoothing design; George Milliken and Yurii Bulavski for development of mixed models; Will Potts and Cathy Maahs-Fladung for data mining; Clay Thompson for advice on contour plotting algorithms; and Tom Little, Damon Stoddard, Blanton Godfrey, Tim Clapp, and Joe Ficalora for advice in the area of Six Sigma; and Josef Schmee and Alan Bowman for advice on simulation and tolerance design.

For sample data, thanks to Patrice Strahle for Pareto examples, the Texas air control board for the pollution data, and David Coleman for the pollen (eureka) data.

Translations

Trish O'Grady coordinates localization. Special thanks to Noriki Inoue, Kyoko Takenaka, Masakazu Okada, Naohiro Masukawa, and Yusuke Ono (SAS Japan); and Professor Toshiro Haga (retired, Tokyo University of Science) and Professor Hirohiko Asano (Tokyo Metropolitan University) for reviewing our Japanese translation; François Bergeret for reviewing the French translation; Bertram Schäfer and David Meintrup (consultants, StatCon) for reviewing the German translation; Patrizia Omodei, Maria Scaccabarozzi, and Letizia Bazzani (SAS Italy) for reviewing the Italian translation; RuiQi Qiao, Rula Li, and Molly Li for reviewing Simplified Chinese translation (SAS R&D Beijing); Finally, thanks to all the members of our outstanding translation and engineering teams.

Past Support

Many people were important in the evolution of JMP. Special thanks to David DeLong, Mary Cole, Kristin Nauta, Aaron Walker, Ike Walker, Eric Gjertsen, Dave Tilley, Ruth Lee, Annette Sanders, Tim Christensen, Eric Wasserman, Charles Soper, Wenjie Bao, and Junji Kishimoto. Thanks to SAS Institute quality assurance by Jeanne Martin, Fouad Younan, and Frank Lassiter. Additional testing for Versions 3 and 4 was done by Li Yang, Brenda Sun, Katrina Hauser, and Andrea Ritter.

Also thanks to Jenny Kendall, John Hansen, Eddie Routten, David Schlotzhauer, and James Mulherin. Thanks to Steve Shack, Greg Weier, and Maura Stokes for testing JMP Version 1.

Thanks for support from Charles Shipp, Harold Gugel (d), Jim Winters, Matthew Lay, Tim Rey, Rubin Gabriel, Brian Ruff, William Lisowski, David Morganstein, Tom Esposito, Susan West, Chris Fehily, Dan Chilko, Jim Shook, Ken Bodner, Rick Blahunka, Dana C. Aultman, and William Fehlner.

Technology License Notices

Chapter 1

Preliminaries

Introducing JMP

Using JMP statistical software, you can interact with your graphs and data to do the following:

Discover Using graphics, you can see patterns and relationships in your data, and find data that does not fit identifiable patterns.

Interact Using JMP interactive features, you can follow up on clues and try different approaches. The more approaches you try, the more you likely you are to discover trends in your data.

Understand Using graphics, you can see how the data and the model work together to produce the statistics. Because JMP is a progressively disclosed system, you learn statistics methods in the right context.

Here are just a few of the things that you can do with JMP:

- Interact with data tables and reports.
- Compute values using the Formula Editor.
- Design experiments.
- Use scripting features to automate frequently used processes.
- Open SAS data sets, run stored processes, and submit SAS code.

Figure 1.1 Interacting with JMP

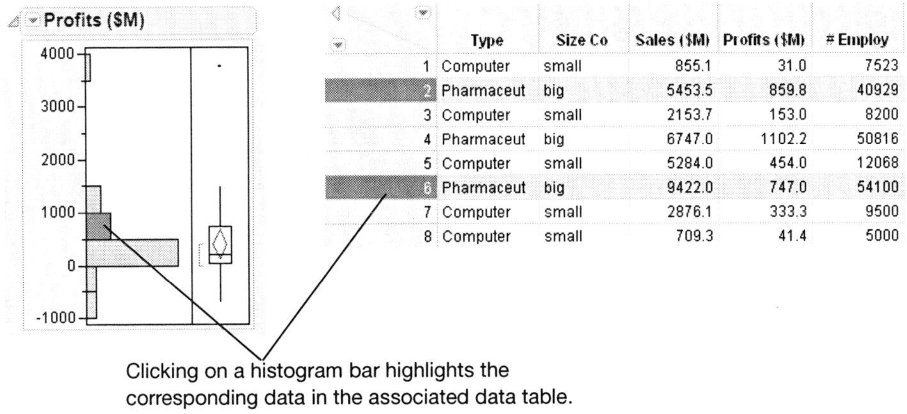

Contents

Prerequisites

Before you begin using JMP, note the following information:

- You can use many JMP features, such as data manipulation, graphs, and scripting features, without any statistical knowledge.
- A basic understanding of central statistical concepts, such as mean and variation, is recommended.
- Analytical features require statistical knowledge appropriate for the feature.

JMP Terminology

- You can enter, view, edit, and manipulate data using *data tables*. In a data table, each variable is a *column*, and each observation is a *row*.
- You can access a *platform* from the **Analyze** and **Graph** menus. Platforms contain interactive windows that you can use to analyze data and work with graphs.
- Platforms use these windows:
  - *Launch windows* where you set up and run your analysis.
  - *Report windows* showing the output of your analysis.
- Report windows normally contain the following items:
  - A graph of some type (such as a scatterplot or a chart).
  - Specific *reports* that you can show or hide using the *disclosure button* .
  - Platform *options* that are located within *red triangle menus* .

Learning about JMP

JMP provides numerous resources to help you learn about the software. Most of them can be found within the **Help** menu. You can also access context-sensitive Help from within JMP.

Note: For further information about all of the options in the **Help** menu, see "Use JMP Help," p. 6.

About JMP Documentation

You can view the JMP documentation suite by selecting **Help > Books**.

Table 1.1 describes documents in the JMP documentation suite and the purpose of each document.

Table 1.1 About JMP Documentation

Document	Who Should Read This Document	What this Document Covers
Discovering JMP	If you are not familiar with JMP, start here.	Introduces you to JMP and gets you started using JMP
Using JMP	If you want to understand JMP data tables and how to perform basic operations in JMP, start here.	• General JMP concepts and features that span across all of JMP • Material in these JMP Starter categories: File, Tables, and SAS
Basic Analysis and Graphing	If you want to perform basic analysis and graphing functions.	• These Analyze platforms: – Distribution – Fit Y by X – Matched Pairs • Many basic graphing platforms • Material in these JMP Starter categories: Basic and Graph

Table 1.1 About JMP Documentation *(Continued)*

Document	Who Should Read This Document	What this Document Covers
Modeling and Multivariate Methods	If you want to perform modeling or multivariate methods	• These Analyze platforms: – Fit Model – Screening – Nonlinear – Neural – Gaussian Process – Partition – Time Series – Categorical – Choice – Multivariate – Cluster – Principal Components – Discriminant – PLS (Partial Least Squares) – Item Analysis • These Graph platforms: – Profilers – Surface Plot • Material in these JMP Starter categories: Model, Multivariate, and Surface

Table 1.1 About JMP Documentation *(Continued)*

Document	Who Should Read This Document	What this Document Covers
Quality and Reliability Methods	If you want to perform quality control or reliability engineering	• Life Distribution • Fit Life by X • Recurrence Analysis • Degradation • Survival • Fit Parametric Survival • Fit Proportional Hazards • These Graph platforms: – Variability/Gauge Chart – Control Charts – Capability – Pareto Plot – Diagram (Ishikawa) • Material in these JMP Starter window categories: Reliability, Measure, and Control
Design of Experiments	If you want to design experiments	• Everything related to the **DOE** menu • Material in this JMP Starter window category: DOE
Scripting Guide	If you want to use the JMP Scripting Language (JSL)	A reference guide for using JSL commands

In addition, the *New Features* document is available at http://www.jmp.com/support/downloads/documentation.shtml.

Note: The **Books** menu also contains two reference cards: The *Menu Card* describes JMP menus, and the *Quick Reference* describes JMP keyboard shortcuts. You can print these for ease of use.

Use JMP Help

You can access JMP Help in two ways:

- Access the context-sensitive Help by selecting the Help tool from the **Tools** menu. Place the Help tool anywhere in a data table or report window to see the Help for that area.
- Within a window, click on the **Help** button.

Search and view JMP Help using the **Help > Contents**, **Search**, and **Index** options.

Use Tutorials

You can access JMP tutorials by selecting **Help > Tutorials**. The first item on the **Tutorials** menu is **Tutorials Directory**. This opens a new window with all the tutorials grouped by category.

If you are not familiar with JMP, then start with the **Beginners Tutorial**. It steps you through the JMP interface and explains the basics of using JMP.

The rest of the tutorials help you with specific aspects of JMP, such as creating a pie chart, using Graph Builder, and so on.

Access Sample Data Tables

All of the examples in the JMP documentation suite use sample data. To access JMP's sample data tables, select **Help > Sample Data**. From here, you can do the following:

- Open the sample data directory.
- Open an alphabetized list of all sample data tables.
- Find a sample data table within a category.

Alternatively, the sample data tables are installed in the following directory:

On Windows: C:\Program Files\SAS\JMP\9\Support Files <language>\Sample Data

On Macintosh: \Library\Application Support\JMP\9\<language>\Sample Data

Learn About Statistical and JSL Terms

The **Help** menu contains the following indexes:

Table 1.2 Descriptions of Help Menu Indexes

Statistics Index	Provides definitions of statistical terms.
JSL Functions	Provides definitions of JSL functions.
Object Scripting	Provides a list of JSL scriptable objects and the messages that can be sent to those objects.
DisplayBox Scripting	Provides a list of the JSL objects that comprise a JMP report.

For more details about these indexes, see "The Help Menu," p. 440 in the "Main Menu" chapter.

Learn JMP Tips and Tricks

When you first start JMP, you see the Tip of the Day window.

To turn off the Tip of the Day, clear the **Show tips at startup** check box. To view it again, select **Help > Tip of the Day**. Or, you can turn it off using the Preferences window. See "General," p. 445 in the "JMP Preferences" chapter for details.

You can use the *Quick Reference* to learn more advanced commands in JMP. View this document by selecting **Help > Books > Quick Reference**.

Access Resources on the Web

To access JMP resources on the Web, select **Help > JMP.com** or **Help > JMP User Community**.

The **JMP.com** option takes you to the JMP Web site, and the **JMP User Community** option takes you to JMP online user forums.

Conventions

The following conventions help you relate written material to information that you see on your screen.

- Most open data table names that are used in examples appear in Helvetica font (Animals or Animals.jmp) in this document. References to variable names in data tables and items in reports also appear in Helvetica according to the way they appear on the screen or in the documentation.

Note: Special information, warnings, and limitations are noted as such in **boldface**.

- Reference to menu names (**File** menu) or menu items (**Save** option) appear in **Helvetica bold**.
- Words or phrases that are important or have definitions specific to JMP are in *italics* the first time they occur in the text. For example, the word *platform* is in italics the first time you see it. Most words in italics are defined when they are first used unless clear in the context of use.

Common Features Throughout Platforms

The following features are common to multiple areas of JMP:

- "Launch Window Features," p. 8
- "Script Menus," p. 11
- "Automatic Recalc Feature," p. 13

Launch Window Features

Each analysis platform prompts you with a launch window. Table 1.3 describes three panes that all launch windows have in common.

Table 1.3 Descriptions of Common Panes in Launch Windows

Select Columns	Lists all of the variables in your current data table. For details about the red triangle menu, see "Columns Filter Menu," p. 10.
Cast Selected Columns into Roles	Moves selected columns into roles (such as **Y**, **X**, and so on.) You cast a column into the role of a variable (like an actor is cast into a role). See "Cast Selected Columns into Roles Buttons," p. 9. This pane does not exist in the Graph Builder platform.
Action	**OK** performs the analysis.
	Cancel stops the analysis and quits the launch window.
	Remove deletes any selected variables from a role.
	Recall populates the launch window with the last analysis that you performed.
	Help takes you to the Help for the launch window.

Cast Selected Columns into Roles Buttons

Table 1.4 describes buttons that appear frequently throughout launch windows. Buttons that are specific to certain platforms are described in the chapter for the platform.

Table 1.4 Descriptions of Common Buttons in Launch Windows

Y	Identifies a column as a response or dependent variable whose distribution is to be studied.
X	Identifies a column as an independent, classification, or explanatory variable whose values divide the rows into sample groups.
Weight	Identifies the data table column whose variables assign weight (such as importance or influence) to the data.
Freq	Identifies the data table column whose values assign a frequency to each row. This option is useful when a frequency is assigned to each row in summarized data.
By	Identifies a column that creates a report consisting of separate analyses for each level of the variable.

Columns Filter Menu

In most of the platform launch windows there is a Column Filter menu. This menu appears as a red triangle within the Select Columns panel.

Figure 1.2 Example of the Columns Filter Menu

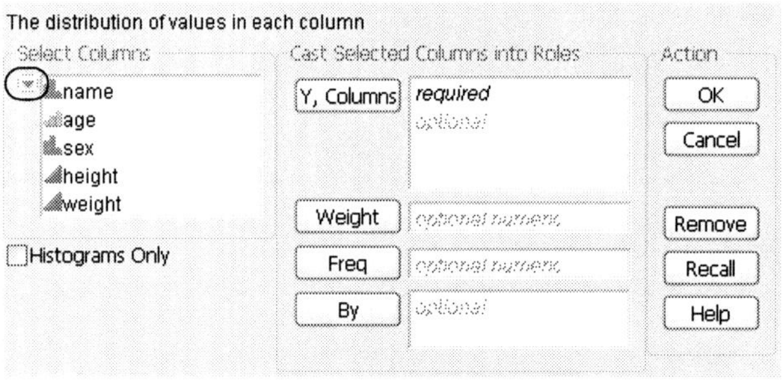

Table 1.5 Options within the Columns Filter Menu

Reset	Resets the columns to its original list.
Sort by Name	Sorts the columns in alphabetical order by name.
Continuous	Shows or hides columns whose modeling type is continuous.
Ordinal	Shows or hides columns whose modeling type is ordinal.
Nominal	Shows or hides columns whose modeling type is nominal.
Numeric	Shows or hides columns whose data type is numeric.
Character	Shows or hides columns whose data type is character.
Match case	(Only applicable to the Name options below) Makes your search case-sensitive.
Name Contains	Searches for column names containing specified text. To remove the text box, select **Reset**.
Name Starts With	Searches for column names that begin with specified text. To remove the text box, select **Reset**.
Name Ends With	Searches for column names that end with specified text. To remove the text box, select **Reset**.

Script Menus

The red triangle menu at the top level of every JMP report contains a **Script** menu.

Figure 1.3 The Script menu

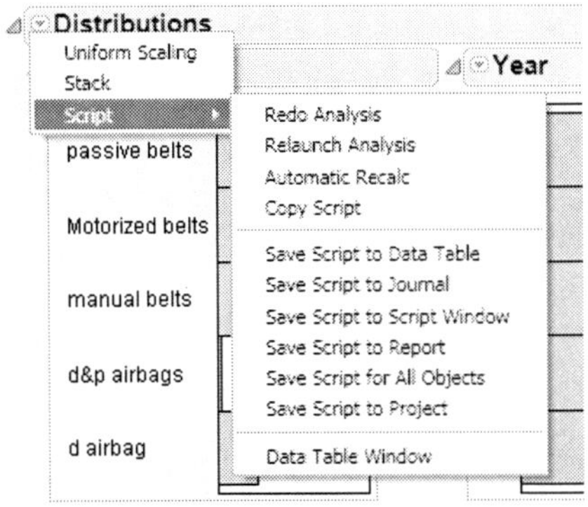

Most of these options are the same throughout JMP. A few platforms add extra options that are described in the specific platform chapters. Table 1.6 lists the **Script** menu options that are common to all platforms.

Table 1.6 Descriptions of Script Options

Redo Analysis	If the values in the data table that was used to produce the report have changed, this option duplicates the analysis based on the new data. The new analysis appears in a new report window.
Relaunch Analysis	Opens the platform Launch window and recalls the settings used to create the report.
Automatic Recalc	Automatically updates analyses and graphics when data table values change. See "Automatic Recalc Feature," p. 13.
Copy Script	Places the script that reproduces the report on the clipboard so that it can be pasted elsewhere.
Save Script to Data Table	Saves the script to the data table that was used to produce the report.
Save Script to Journal	Saves a button that runs the script in a journal. The script is added to the current journal.
Save Script to Script Window	Opens a script editor window and adds the script to it. If you have already saved a script to a script window, additional scripts are added to the bottom of the same script window.
Save Script to Report	Adds the script to the top of the report window.

Table 1.6 Descriptions of Script Options *(Continued)*

Save Script for All Objects	If you have By groups or similar multiple reports, a script for each object is saved to the script window. Otherwise, this option is the same as **Save Script to Script Window**.
Save Script to Project	Saves the script in a project. If you have a project open that contains the report, the script is added to that project. If you do not have a project that contains the report, a new project is created and the script is added to it.
Data Table Window	Brings the data table that was used to create the report to the front and makes it the active window.

If you have specified a **By** variable in the platform launch window, the **Script All By-Groups** menu also appears.

Figure 1.4 Example of the Script All By-Groups Menu

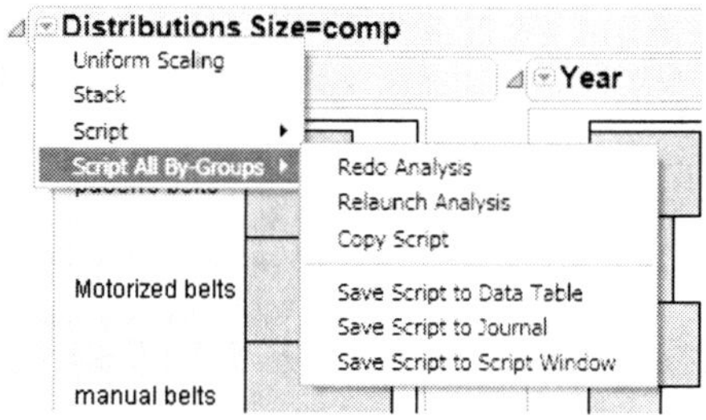

If you specified a **By** variable, the script options in Table 1.7 apply to a report for a single level of a **By** variable. The **Script All By-Groups** options apply to the reports for all the levels of the **By** variable.

Table 1.7 Descriptions of Script All By-Groups Options

Redo Analysis	If the values in the data table that was used to produce the reports have changed, this option duplicates the analysis based on the new data and produces new reports.
Relaunch Analysis	Opens the platform Launch window and recalls the settings used to create the reports.
Copy Script	Places the script that reproduces the reports on the clipboard so that it can be pasted elsewhere.
Save Script to Data Table	Saves the script to the data table that was used to produce the reports.

Table 1.7 Descriptions of Script All By-Groups Options *(Continued)*

Save Script to Journal	Saves a button that runs the script in a journal. The script is added to the current journal.
Save Script to Script Window	Opens a script editor window and adds the script to it. If you have already saved a script to a script window, additional scripts are added to the bottom of the same script window.

Automatic Recalc Feature

The Automatic Recalc feature immediately reflects changes that you make to the data table in the corresponding report window. You can make any of the following data table changes:

- exclude or unexclude data table rows
- delete or add data table rows

This powerful feature immediately reflects these changes to the corresponding analyses, statistics, and graphs that are located in a report window.

To turn on **Automatic Recalc** for a report window, click on the platform red triangle menu and select **Script > Automatic Recalc**. See Figure 1.3. To turn it off, deselect the same option. You can also turn on **Automatic Recalc** using JSL.

Note the following:

- By default, **Automatic Recalc** is turned off for platforms in the **Analyze** menu and turned on for platforms in the **Graph** menu. The exceptions are the **Capability**, **Variability/Gauge Chart**, and **Control Chart > Run Chart** platforms.
- For some platforms, the **Automatic Recalc** feature is not appropriate, and therefore is not supported. These platforms include the following: DOE, Profilers, Choice, Partition, Nonlinear, Neural, Partial Least Squares, Fit Model (REML, GLM, Log Variance), Gaussian Process, Item Analysis, Fit Proportional Hazard, and Control Charts (except Run Chart).

Chapter 2

Getting Started

Data Tables and Scripts

The first thing you typically do in JMP is open or create a JMP data table:

- To import an existing file into JMP, either click the **Open Data Table** button in the JMP Starter or select **File > Open**. See "Open Existing JMP Files," p. 24, for details.
- To manually type data into JMP, either click the **New Data Table** button in the JMP Starter or select **File > New > Data Table**. This creates an empty JMP data table, and you can start typing in your data. See "Create New Data Tables," p. 23, for details.

This chapter tells you more about getting started with the JMP environment, creating new data tables, importing data into new tables, exporting data to other formats, and getting further assistance.

Figure 2.1 JMP Data Table

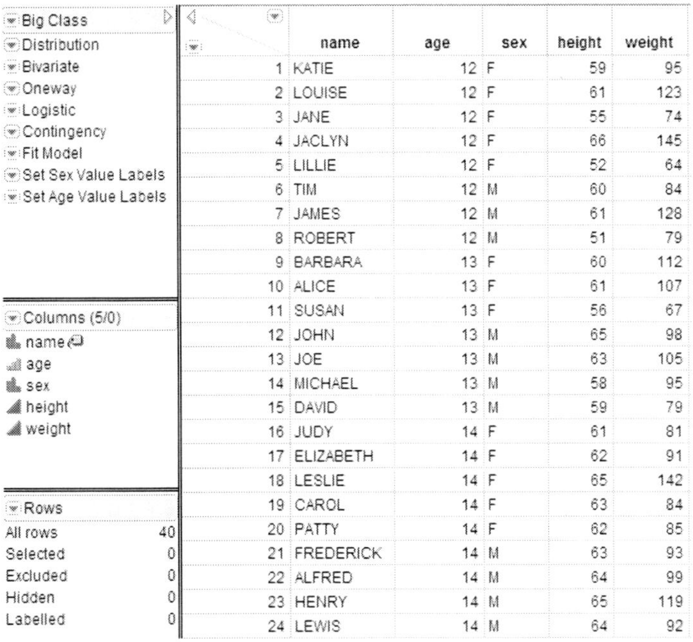

	name	age	sex	height	weight
1	KATIE	12	F	59	95
2	LOUISE	12	F	61	123
3	JANE	12	F	55	74
4	JACLYN	12	F	66	145
5	LILLIE	12	F	52	64
6	TIM	12	M	60	84
7	JAMES	12	M	61	128
8	ROBERT	12	M	51	79
9	BARBARA	13	F	60	112
10	ALICE	13	F	61	107
11	SUSAN	13	F	56	67
12	JOHN	13	M	65	98
13	JOE	13	M	63	105
14	MICHAEL	13	M	58	95
15	DAVID	13	M	59	79
16	JUDY	14	F	61	81
17	ELIZABETH	14	F	62	91
18	LESLIE	14	F	65	142
19	CAROL	14	F	63	84
20	PATTY	14	F	62	85
21	FREDERICK	14	M	63	93
22	ALFRED	14	M	64	99
23	HENRY	14	M	65	119
24	LEWIS	14	M	64	92

Contents

Before You Start

Before you begin using JMP, familiarize yourself with its initial windows: the Tip of the Day window, JMP Home Window, and JMP Starter window.

The Tip of the Day Window

When JMP opens, you see the Tip of the Day window. This window provides tips about using JMP that you might not know. Some tips are basic introductory information, and others give hidden power features that you should learn after getting comfortable with the basics.

The Tip of the Day window has the following features:

- **Show tips at startup** When checked, displays the **Tip of the Day** window each time you start JMP. This option is also accessible in **File** (**JMP** on the Macintosh) **> Preferences > General > Show the Tip of the Day at startup.**
- **Enter Beginner's Tutorial** Click to start the beginner's tutorial. This tutorial introduces beginners to JMP and how its commands and controls work. To access the other tutorials, select **Help > Tutorials.**
- **Previous Tip** Returns to the previous tip in the **Tip of the Day** window.
- **Next Tip** Advances to the next tip in the **Tip of the Day** window.

The Tip of the Day folder is installed with JMP. It contains HTML files and images that are used for producing the Tip of the Day window.

Note: Add your own tips by naming them tip*XX*.htm, where *XX* is the next unused number in a tip filename.

To open the Tip of the Day window after closing it, select **Help > Tip of the Day.**

The JMP Home Window (Windows Only)

When you open JMP on Windows, the JMP Home Window is displayed behind the Tip of the Day window. The JMP Home Window provides access to the following features:

- JMP menus and toolbars. See the "Personalizing JMP" chapter for details.
- *Recent Files*: Shows a list of files you recently opened. See "Work with Recent Files," p. 18 for details. (These files are also displayed in the **File > Recent Files** menu.)
- *Window List*: Shows a list of open JMP windows. See "Work with Open Windows," p. 20 for details. (You select **View > Window List** to display the list in a new window.)

Figure 2.2 Features of the JMP Home Window

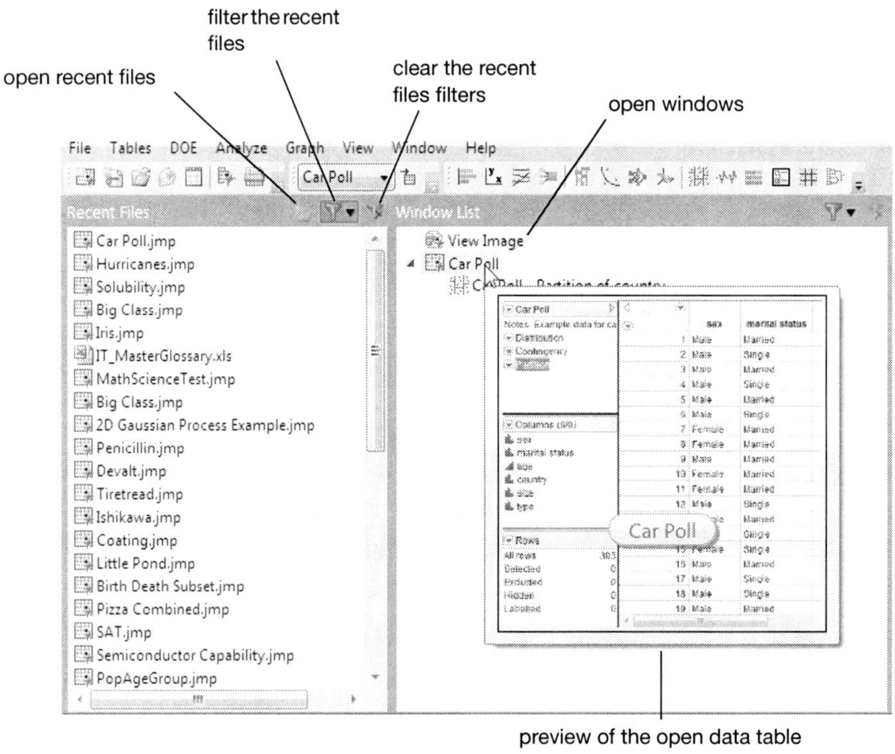

Note: By default, the JMP Home Window appears when you open JMP. You can set the preferences to display the JMP Starter or Window List instead. (Select **File > Preferences > General** and select an option from the **Initial JMP Window** list.) To display the Home Window at any time, select **View > Home Window**. Alternatively, click the **JMP Home Window** button in the lower right corner of most JMP windows. To display the Window List at any time, select **View > Window List**.

Work with Recent Files

The Recent Files list provides quick access to files that you recently opened in one of the following ways: by selecting **File > Open**, clicking the **Open** toolbar button, or dragging and dropping the file into the Home Window. You can double-click one or more files to open them, filter the list to display only certain types of files, and clear the list.

When JMP cannot find the file that you recently opened, the filename is grayed out in the list. Right-click the filename and select **Remove From List** to clear it from the list.

To Open Recent Files

1. To open one file, do one of the following:

- Double-click the filename.
- Right-click the filename, then select **Open**. (This option does not appear for non-JMP files, such as text files.)
- Select the filename, then select the **Open** button on the Recent Files bar.

2. To open several files, select the filenames and do one of the following:
   - Press ENTER.
   - Right-click, then select **Open**.
   - Select the **Open** button on the Recent Files bar.
3. To open all files, click in the Recent Files window and do one of the following:
   - Hold down the CTRL key, press A, and then press ENTER.
   - Hold down the CTRL key, press A, right-click, and then select **Open**.
   - Hold down the CTRL key, press A, and then select the **Open** button on the Recent Files bar.

Note: When you double-click a non-JMP file (such as a text file) in the Recent Files list, JMP applies your import preferences to arrange the data. This also occurs when you click the **Open** button on the Recent Files bar. You can instead select the import method or open the file in a text editing window when you right-click a file in the list. (Your import preference is bolded in the right-click menu.) For details about JMP preferences, see "Text Data Files," p. 456 in the "JMP Preferences" chapter. For details about importing text files, see "Import Data," p. 25.

To Filter the Recent Files List and Clear the Filters

1. Click the  button on the Recent Files bar.
2. Select the type of files that you want to display. The file types vary depending on which file types you have recently opened. Some common file types are described in the following table.

Table 2.1 Descriptions of File and Report Filtering Options

All Files	Shows all file types in the list.
Data Tables	Shows only data tables in the list.
Excel Files	Shows only Excel files in the list.
Other Data Files	Shows data file types other than JMP tables or Excel files (such as .sav files) in the list.
All Data Files	Shows all data tables, Excel files, and other data files in the list.
Scripts	Shows only scripts in the list.

3. To show several types of files (for example, data tables *and* Excel files), repeat the preceding steps and select a different file type.

4. To remove the filter for a specific file type, click the button on the Recent Files bar, and then deselect that file type.
5. To remove filters so that all files are displayed in the list, click the button on the Recent Files bar. (If this button is grayed out, no filters are selected.) Alternatively, select **All Files** from the **Filter** drop-down menu.

To Clear the Recent Files List

- To clear one file, right-click the file and select **Remove from List**.
- To clear the entire list, press CTRL+A, right-click, and then select **Remove from List**.
- To clear selected files, select those files, right-click, and then select **Remove from List**.
- To clear one or more files, select them, and then press DELETE.

Work with Open Windows

The Window List displays a list of open JMP windows, such as windows for data tables, scripts, and tutorials. The list also lets you quickly hide and rearrange JMP windows. For example, you can display a data table and its open reports side-by-side or tile some types of JMP windows. The tiled windows are all displayed at once to fill the screen. If you are using JMP on your primary monitor and have an extended desktop, the tiled windows are displayed on the second monitor if necessary.

The Window List is displayed in the JMP Home Window. However, you can open the list at any time by selecting **View > Window List**. The list appears in a new window.

Note: To deselect the selected window names, right-click and select **Clear Selections**.

To Display Open Windows

- To display one window in the Window List, double-click the window name. Alternatively, right-click the window name and select **View**.
- To display several windows in the Window List, select the window names, right-click, then select **View**.

To Rearrange Open Windows

- To display a data table and its reports side-by-side, select the files, right-click them in the Window List, then select **Arrange**.
- To display several windows at once, hold down the CTRL key, select the window names, right-click, and select **Arrange**.
- To display all windows at once, right-click in the Window List and select **Select All**. Right-click again and select **Arrange**.
- To display windows behind other JMP windows, right-click the window names, then select **Move To Back**.

To Hide and Unhide Open Windows

- To hide windows, right-click the window names, then select **Hide**.

The names of the hidden windows are grayed out in the Window List. The hidden windows are also no longer displayed on the Windows task bar.

- To unhide windows, right-click the grayed-out window names, then select **Unhide**.

To Filter the Window List and Clear the Filters

1. Click the button on the Window List bar.
2. Select the window type that you want to show.

   Only windows of the selected type appear.
3. To show several types of windows (for example, data tables *and* scripts), repeat the preceding steps and select a different window type.
4. To remove the filter for a specific window type, click the button on the Window List bar, and then deselect that window type.
5. To remove filters so that all window names are displayed in the list, click the button on the Window List bar. (If this button is grayed out, no filters are selected.) Alternatively, select **All Windows** from the **Filter** drop-down menu.

To Display the Window List in Maximized Windows

The Window List can be automatically displayed inside maximized data table, script, journal, and report windows. Then the list is always available when you are working in a maximized window. See “Windows Specific,” p. 459 in the “JMP Preferences” chapter for details.

To Close the Windows

1. Select the windows that you want to close.
2. Right-click and select **Close**.

   If changes in the windows need to be saved, you are prompted to do one of the following: save all changes, save no changes, display each modified file and then decide whether to save the changes, or cancel to close no windows.

The JMP Starter Window

The JMP Starter Window displays most of the commands found in the main menu and toolbars. For example, you can open data tables and create graphs through the JMP Starter, main menu, or toolbars.

You might find the JMP Starter helpful if you are not familiar with JMP or data analysis, because the Starter describes each option and report. See “Overview of the JMP Starter Window,” p. 393, for details.

To open and close the JMP Starter, select **View** (**Window** on the Macintosh) **> JMP Starter**.

To display the JMP Starter automatically when you open JMP, do one of the following:

- (Windows) Select **File > Preferences**, and then select **JMP Starter** from the Initial JMP Window list.
- (Macintosh) Select **JMP > Preferences**, and then select **Initial Starter Window**.

Figure 2.3 The JMP Starter

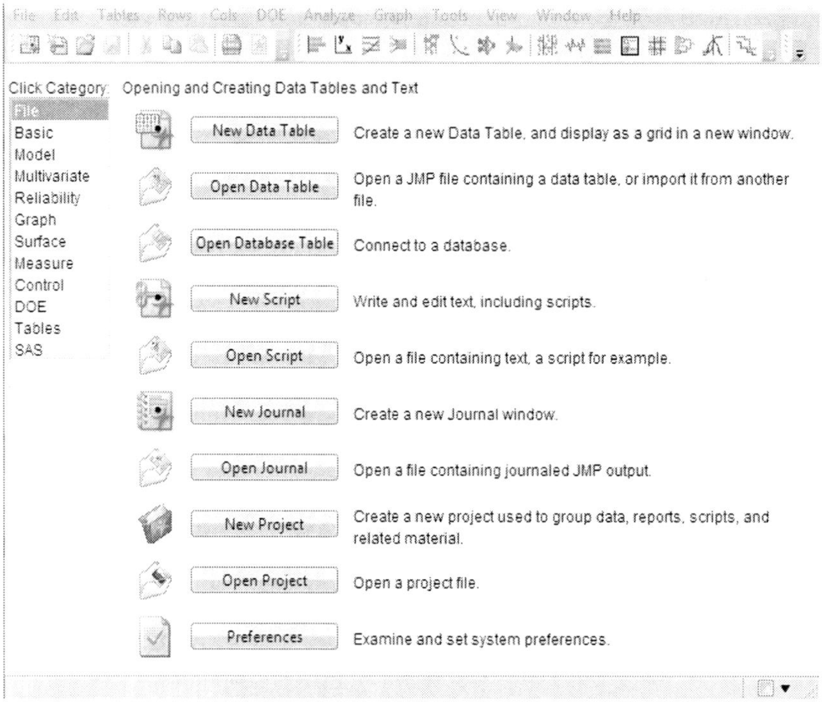

Display and Arrange Open Windows

In JMP, you typically have several windows open at once (for example, data tables, reports, and the JMP Home Window). JMP provides several ways to arrange and display these open windows.

- Select options from the **Window** menu. See "The Window Menu," p. 437 for details.
- (Windows only) In the Window List, select options from the right-click menu. See "Work with Open Windows," p. 20 for details.
- (Windows only) To display the JMP Home Window, click the **JMP Home Window** button  in the lower right corner of most windows, or hold down CTRL and press 1.
- (Windows only) To display all open windows side-by-side, do one of the following:
  - Select **Window > Arrange > Select All for Arrange** and then select **Window > Arrange > Arrange Selected Windows.**
  - Select **Window > Arrange > Select All for Arrange.** In one of the windows, click the **Arrange Menu** arrow next to the check box () in the bottom right corner, and then select **Arrange Selected Windows.**

Note: If you have extended monitors, windows that do not fit on the primary monitor are displayed on the extended monitors.

- (Windows only) To display specific windows side-by-side, do one of the following:
  - Select the check box in the bottom right corner of each window that you want to arrange. Then select **Window > Arrange > Arrange Selected Windows** in one of the windows.
  - Select the check box in the bottom right corner of each window that you want to arrange. In one of the selected windows, click the **Arrange Menu** arrow next to the check box ( ) in the bottom right corner, and then select **Arrange Selected Windows.**
  - In each window that you want to arrange, select **Window > Arrange > Select for Arrange.** Then select **Window > Arrange > Arrange Selected Windows** in one of the windows.
- (Windows only) To display the data table for a report, click the **View Associated Data** button in the lower right corner of the report.
- (Windows only) To show an associated report when you are viewing a data table, double-click the thumbnail preview of the report in the bottom pane. (Hover the cursor over the thumbnail to enlarge the preview as shown in the following figure.)

Figure 2.4 Thumbnail Previews of Open Reports

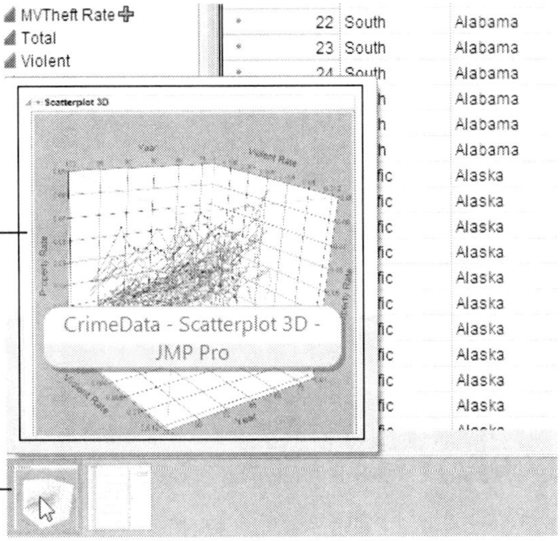

Create New Data Tables

To analyze data, you must first create a data table.

To create a new data table:

1. Select **File > New > Data Table**. This shows an empty data table with no rows and one numeric column, labeled Column 1.
2. Move the cursor onto a cell.
3. Click in the cell. The cursor appears as a line in the cell, as shown in Figure 2.5.

Figure 2.5 A New Data Table

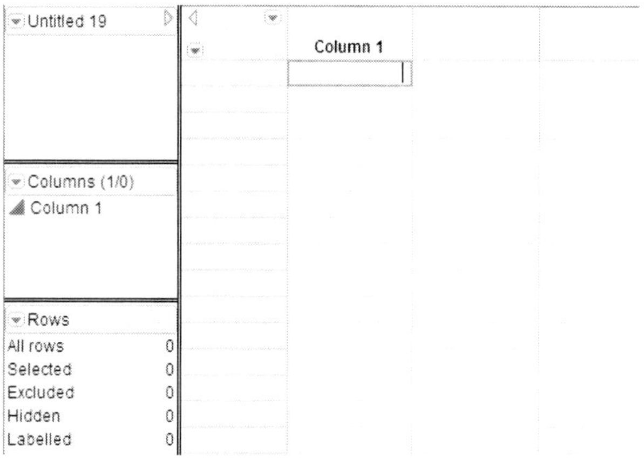

4. Enter a value.

There are several ways to fill a table with values:

- Create new rows and columns and type or paste data into the data grid. (See "Adding and Deleting Rows," p. 63.)
- Construct a formula to calculate column values. (See "Creating a Formula," p. 303.)
- Import data from another application. (See "Import Data," p. 25.)
- Copy values from another application and paste them into the table.
- Use a measuring instrument to read external measures. (See "Reading in Real-Time Data (Windows Only)," p. 388 in the "External Data and Analytical Sources" chapter for details.)
- Drag columns from one table to another.

See "Entering, Editing, and Managing Data," p. 45, for details about how to format, edit, and work with data tables.

Open Existing JMP Files

If you want to import a file that is a JMP data table (.jmp), script (.jsl), journal (.jrn), or report (.jrp):

1. Select **File > Open**.

2. Select the file type from the window that appears.
3. (Optional, Windows only) Select the **Select this filter the next time this dialog is invoked** option to open the same file type every time. This option is available for any file filter, including text import options. See “Open Text Files,” p. 27.

File Type	Operating System	Instructions for Opening
JMP data table	Windows	From the list next to **File name** (or the **Files of type** list on Windows XP), select **JMP Data Tables.** The highlighted table’s notes and number of rows and columns appear at the bottom of the window. Select the check box beside **Select Columns** to display only certain columns in the data table when opened.
	Macintosh	The highlighted table’s notes and number of rows and columns appear at the bottom of the window. Select the check box beside **Select columns before opening** to display only certain columns in the data table when opened.
JMP script	Windows	1. From the list next to **File name** (or the **Files of type** list on Windows XP), select **JMP Files** or **JMP Scripts.** 2. Select a script. The **Run this script only** option appears at the bottom of the window. Check this option to automatically begin running the script.
	Macintosh	After selecting a script, the **Run after opening** option appears at the bottom of the window. Check it to automatically begin running the script.
JMP journal, report, or project	Windows	From the list next to **File name** (or the **Files of type** list on Windows XP), select **JMP Journals**, **JMP Reports**, or **JMP Projects.**
	Macintosh	No further action is necessary.

3. Select the name of the file that you would like to open.
4. Click **Open.**

Note: JMP can open JMP files that are version 3 and later.

Import Data

You can import many file formats into JMP and save them as data tables. JMP opens many files by default. Other files, such as Microsoft Excel 2007 files, require specific Open Database Connectivity (ODBC) drivers.

Supported by default:

- Comma-separated (.csv)
- .dat files that consist of text
- ESRI shapefiles (.shp)
- Flow Cytometry versions 2.0 and 3.0 (.fcs)
- HTML (.htm, .html)
- Microsoft Excel 1997–2003 (.xls)
- Minitab (.mtw, .mtp, but not .mpj)
- Plain text (.txt)
- SAS transport (.xpt, .stx)
- SAS versions 6–9 on Macintosh (.sas7bdat, .ssd, .ssd01, .saseb$data)
- SAS versions 6–9 on Windows (.sd2, .sd5, .sd7, .sas7bdat)
- SPSS files (.sav)
- Tab-separated (.tsv)

Note: On Windows, you can open SAS data sets directly through the **File > Open** command. Another option is connecting to a SAS server by selecting **File > SAS > Browse Data**. On Macintosh, you must connect through a SAS server first. See "Open SAS Data Sets," p. 39 for details on opening SAS data sets directly. See "Opening SAS Data Sets through a SAS Server," p. 365 in the "External Data and Analytical Sources" chapter for details on connecting through a SAS server.

Require ODBC drivers:

- Database (dBASE) (.dbf, .ndx, .mdx) is supported with a V3+ compliant ODBC driver.
- Microsoft Access Database (.mdb) is supported with a V3+ compliant ODBC driver.
- Microsoft Excel 2007+ (.xlsm, .xlsx, .xlsb) is supported with at least a version 3 compliant ODBC driver installed on the machine. 64-bit JMP requires a 64-bit ODBC driver.

See "Connecting to Databases," p. 380 in the "External Data and Analytical Sources" chapter for details for working with databases.

Your computer's available memory affects data import. Very large files might load slowly or not at all. Consider splitting up large files before importing them. In JMP, you can then join or concatenate the tables. For more information, see "Concatenating Data Tables," p. 254 in the "Reshaping Data" chapter and "Joining Data Tables," p. 256 in the "Reshaping Data" chapter.

Note: You can open R code (.R) and SAS program files (.sas) in JMP, but the text opens in a Script window, not in a data table.

Open Text Files

You can open text files with the extensions .txt, .csv, and .tsv, and the text is converted to a data table. Files with the .dat extension that consist of text are also supported. Text files can be delimited using almost any character, or they can be fixed-width files.

In the JMP preferences, you change the import settings so that JMP determines the best way to structure and format the data table. Alternatively, you can manually select the import settings as you open the file. Another option is opening the file in the Script editor, editing the content, and then importing the content. This option is helpful when you need to add text delimiters or modify the text.

Note: When your text file contains columns of missing data, select **Treat empty columns as numeric** in the Preferences to import the columns as numeric rather than character. A period, unicode dot, *NaN*, or a blank string are possible missing value indicators. This option is deselected by default.

Table 2.2 lists the steps involved in opening text files directly.

Table 2.2 Opening Text Files

	Automatically Determining Data Arrangement	Manually Specifying Data Arrangement
Windows	1. Select **File > Open.** 2. From the list next to **File name** (or the **Files of type** list on Windows XP), select **Text Files.** 3. To use the import rules from the preferences, select **Data, using Text Import preferences.** (See "Text Data Files," p. 456 in the "JMP Preferences" chapter.) To have text import use its best guess to arrange the data, select the **Data, using best guess** option. (Optional) Select **Select this filter the next time this dialog is invoked** to apply the filter that you chose by default. 4. Select the file that you want to open. 5. Click **Open.** **Tip**: The JMP Home window provides a shortcut to the above steps if you recently opened the file. Right-click the file in the Recent Files list and select **Import (Preferences)** or **Import (Best Guess).** (Your import preference is bolded in the right-click menu.)	1. Select **File > Open.** 2. From the list next to **File name** (or the **Files of type** list on Windows XP), select **Text Files.** 3. Select **Data with Preview** next to **Open** at the bottom of the window. 4. Select the file that you want to open. 5. Click **Open.** 6. Complete the Text Import window. See "Text Import Preview Options," p. 29, for details. 7. Click **Import.** **Tip**: The JMP Home window provides a shortcut to the above steps if you recently opened the file. Right-click the file in the Recent Files list and select **Import (Preview).**
Macintosh	1. Select **File > Open.** 2. Select the file that you want to open. 3. From the **Open As** field, select **Data (Best Guess)** or **Data (Using Preferences).** 4. Click **Open.**	1. Select **File > Open.** 2. Select the file that you want to open. From the **Open As** field, select **Data (Using Preview).** 3. Click **Open.** 4. Complete the Text Import Preview window. See "Text Import Preview Options," p. 29, for details. 5. Click **Import.**

Note: On Windows, JMP can open text files in your computer's default text editor. Select **File > Open**, and then select **All Files (*.*)** from the **File name** list (or the **Files of type** list on Windows XP). Select the text file, and then select **Use default program to open. Uncheck to open as text.**

For details about importing text from a Script window, see "Import Text from the Script Window," p. 34.

Text Import Preview Options

When you open a text file that JMP supports, JMP can show a preview of the text before opening the file as a data table. This option lets you manually arrange and format the data. For example, you can specify the end-of-line character or strip quotation marks.

JMP detects the file's structure and shows options for importing text with either delimiters or fixed width fields. If JMP chooses the wrong file structure, click the **Delimited fields** or **Fixed width fields** radio button to import the data as the correct format. (For example, the fixed width window might appear when your file is actually delimited.)

The text import preview options are shown in the following figures:

Figure 2.6 Text Import Preview for Fixed Width Files

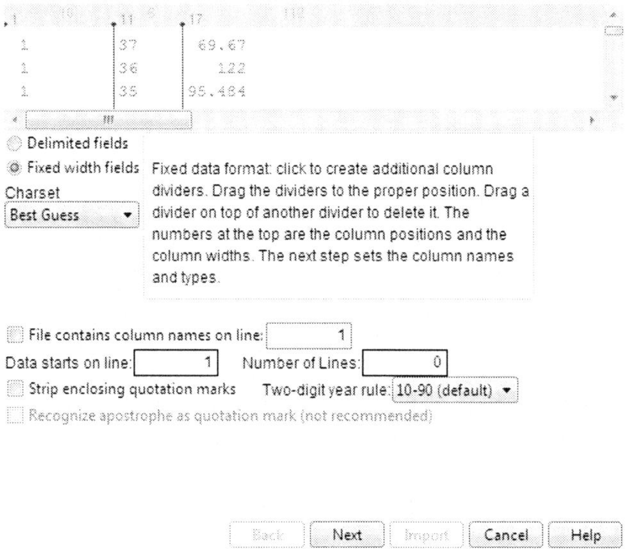

Figure 2.7 Text Import Preview for Delimited Files

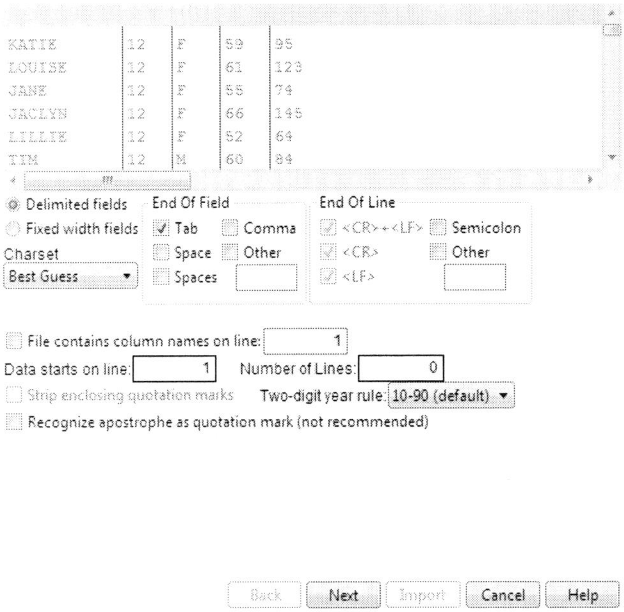

Charset Select the character set used in the imported file, or let JMP detect the character set. If incorrect characters are displayed in the imported file, open the file again and select another character set.

End of Field (Available only in the Delimited Import window) Select the check boxes beside the character that marks the end of a field. Alternatively, select the check box beside **Other** and enter a character if the appropriate character is not listed.

End of Line (Available only in the Delimited Import window) Select the check boxes beside the character that marks the end of a line (row). Alternatively, select the check box beside **Other** and enter a character if the appropriate character is not listed. Note that when JMP finds double quotation marks, the delimiter rules change to look for an end double quotation mark. Other text delimiters, including spaces embedded within the quotes, are ignored and treated as part of the text string.

File contains column names on line Tell JMP where to find data to use as column names. For example, if the column names in your text file are on line (row) 3, select this option and type 3 in the check box. Otherwise, JMP uses the data in the first line of the imported file as the column name in the JMP data table or takes the first line as data.

Data starts on line Specify the number of the first line that contains data.

Number of Lines Specify the number of lines (rows) that you want to import.

Strip enclosing quotation marks Available only on fixed-width imports. Select this check box when you want JMP to remove quotation marks that enclose data in the text file.

Two-digit year rule Specify how year numbers are displayed.

- **10-90 (default)** Refer to Table 2.3 to see how the 10-90 rule works in JMP.
- **19xx** JMP adds 19 before dates in the file that have only two digits that indicate the year.
- **20xx** JMP adds 20 before dates in the file that have only two digits that indicate the year.
- **Custom** Lets you open and implement a JSL script that customizes how JMP handles dates.

Recognize apostrophe as quotation mark (not recommended) (Available only in the Delimited Import window). Use this option only if your data comes from a nonstandard source that places apostrophes around data fields rather than quotation marks.

Table 2.3 The 10-90 Rule for Importing Two-Digit Years

Last Two Digits of Current Year (as Shown on Your Computer)	Two-Digit Date in Text File	JMP Display Format
Between 11 and 89	not applicable	The current century is used as the first two digits.
Between 00 and 10	Less than 90	The current century is used as the first two digits.
	Between 90 and 99	The previous century is used as the first two digits.
Between 90 and 00	Between 0 and 10	The next century is used as the first two digits.
	Between 11 and 99	The current century is used as the first two digits.

When you are finished selecting the settings, click **Next**. The next window shows each column's modeling type. To change the default modeling types, do one of the following:

- Click on the data type icon to change the data type from numeric ( ) to character ( ). Clicking the icon cycles between the data type and exclude ( ). Exclude means that the column is not imported.
- To change a numeric column's data format, select the format from the red triangle menu.
- Click on the column heading to modify the text.

The top of the Text Import window shows a preview of the text file as it appears when imported into a JMP data file. Click the **Import** button to import the data.

Figure 2.8 Text Import Preview Window with Column Options

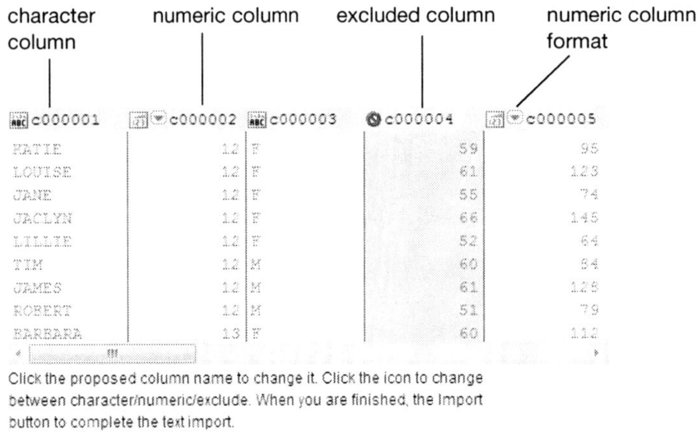

6. When you are finished, click **Import** to complete the text import.

Open a Text File in a Text Editing Window

You can open a text file in a Script window, where you edit the text. Then you can import the text as a data table. This feature is helpful when you want to reformat the text before importing it as a data table. For example, you might need to insert the correct delimiters or modify the text.

Another option is opening a JMP add-in definition (.**def**) file as text and then editing it in a Script window.

To open a text file in a text editing window (Windows):

Files that you recently opened are listed in the JMP Home window. For most files, right-click the text file and select **Open as Plain Text** to open the file in a text editing window. JMP add-in definition files cannot be opened as plain text from the JMP Home window.

When you are opening the file for the first time, follow these steps:

1. Select **File > Open**.
2. Do one of the following:
   – (Windows Vista and Windows 7) To open a JMP add-in definition file as text, select **All JMP Files** or **JMP Add-In Files** from the list next to **File name**. Click the **Open** button arrow, and then select **Open as Plain Text**. The file opens in a Script window. Skip the remaining steps.
   – To open other text files, select **Text Files** from the list next to **File name** (or the **Files of type** list on Windows XP).

Figure 2.9 Select Text Files

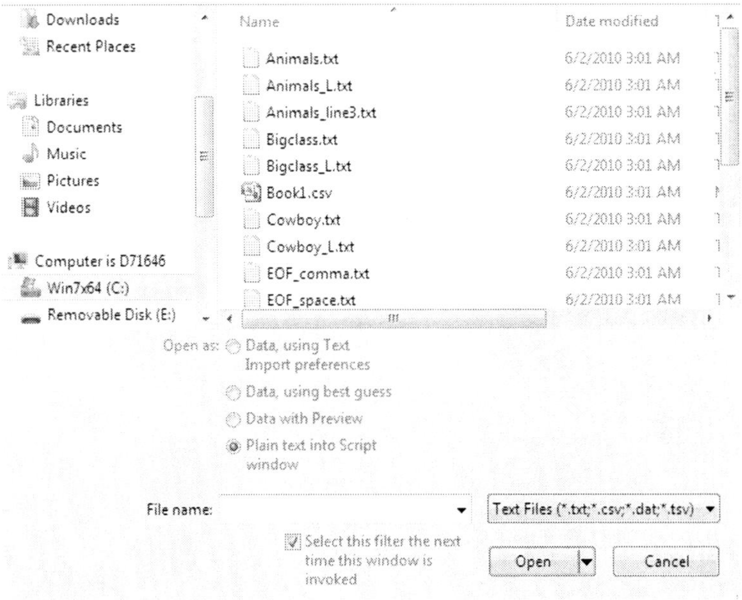

3. (Optional) To set the default option file type to **Text Files**, select the check box beside **Select this filter the next time this dialog is invoked**.
4. Select the file.
5. Select **Plain text into Script window** next to **Open as.**
6. Click **Open**.

   The text appears in a Script window.

To open a text file in a text editing window (Macintosh):

1. Select **File > Open**.

Figure 2.10 Opening a Text Document on the Macintosh

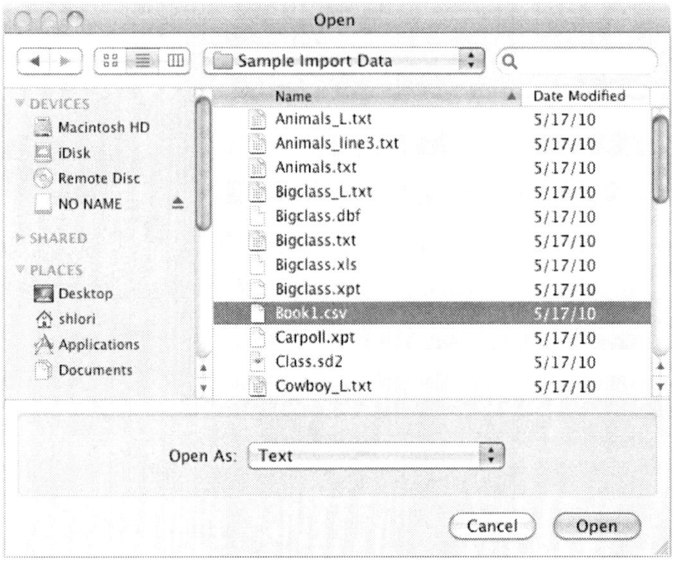

2. Select the file.
3. Select **Text** from the **Open As** list.
4. Click **Open**.

   The text appears in a Script window.

For details about converting the text to a data table, follow step 3 in "Import Text from the Script Window," p. 34.

Import Text from the Script Window

You can import text from the Script window as a data table. The text can be in a table format (for example, from a Microsoft Word document or web page) or in plain text format. This feature is helpful when you want to reformat the text before importing it as a data table. For example, you might need to insert the correct delimiters or modify the text.

JMP uses the import settings in the preferences to determine how to structure and format the text. Some options include removing quotation marks around text and specifying the rows that contain column headings and data. See "Text Data Files," p. 456 in the "JMP Preferences" chapter for details.

Note: You can also import an entire web page as a data table. See "Open Remote Files and Web Pages," p. 35 for details.

This section describes how to import text that you paste into the Script window. For details about opening a text file in the Script window, see "Open a Text File in a Text Editing Window," p. 32.

To import text from the Script window, proceed as follow:

1. Open a new Script window in JMP by selecting **File > New > Script** (Windows) or **File > New > New Script** (Macintosh).
2. Copy and paste the text into the Script window.
3. Do one of the following:
   - To import all text from the Script window, select **File > Import as Data** (Windows) or **Edit > Import as Data** (Macintosh).
   - To import specific text, select the text, and then select **File > Import as Data** (Windows) or **Edit > Import as Data** (Macintosh).

   The text is imported into a JMP data table.

Open Remote Files and Web Pages

You import data from Internet sites, intranet sites, FTP sites, or other computers by selecting **File > Internet Open**. The file paths begin with the Uniform Resource Locators (URLs) *http, ftp*, or *file*; a drive letter; or the path to a network drive (relative or absolute).

You also use this feature to import a Web page on a local or network drive as a data table. When you open a Web page by selecting **File > Open** instead, the page opens in a browser, not as a data table.

SAS stored process reports open in an HTML view. From that view, you can import the file as a data table.

JMP inserts the location of the original data as a note in the data table. For example, the path *Z:\example.html* is included as a note when you import example.html from the Z drive.

Open a Remote File

1. Select **File > Internet Open**. The window shown in Figure 2.11 appears.

Figure 2.11 Internet Open Window

2. Enter the URL.
3. Click the list under **Open As**, and select the option that specifies how you would like JMP to display the imported data:

   Data Imports the file as a data table.

   Web page Opens the Web page in a browser. (Select this option to import data generated by web page scripts and server-side requests.)

Text Opens the file in a JMP Script window. In an HTML file, the HTML tags of a .html file are displayed.

4. Click **OK**.

   One of the following occurs:

   – The file opens as you specified.
   – If you imported a web page as data, a window appears that lists all tables on the page. Select the table or tables that you want to open, and then click **OK**. Each table opens in a new data table.
   – If you imported a web page as a web page, select **File > Import Data as Data Table** in the browser. Select the table or tables that you want to open, and then click **OK**. Each table opens in a new data table.
   – If the file is on an FTP server, the window in Figure 2.12 appears.

Figure 2.12 FTP Login Window

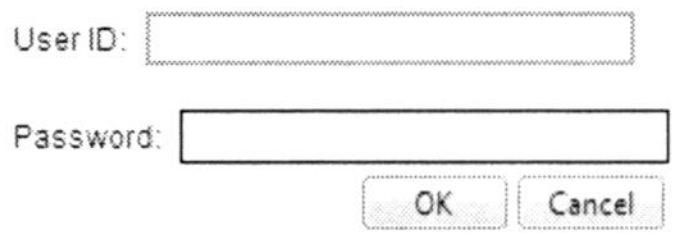

For an anonymous account, click **OK**. For an authenticated login, enter your user ID and password. The file then opens as you specified.

Note: Some anonymous FTP servers require a user ID. If the data table does not open, try typing either **ftp** or **anonymous** in the **User ID** text box. Leave the **Password** text box empty and click **OK**.

Open a SAS Stored Process Report as a Data Table

SAS stored processes open reports in an HTML view by default. To open a report as a data table, follow these steps:

1. In the HTML view, select **File > Import Table as Data Table**.

   A window appears that lists the tables found in the Web page.

2. Select the table or tables that you want to import.
3. Click **OK**.

   Each table is opened as a new data table.

See “Running Stored Processes,” p. 371 in the “External Data and Analytical Sources” chapterfor more information about stored process reports.

Open SPSS Files

JMP opens SPSS files as data tables and maintains several SPSS features:

- General numeric and character data with minimal formatting are supported.
- SPSS date, datetime, and time formats are supported.
- By default, labels are converted to column headings. When you select this option, and the data contains no labels, the columns are named *Column 1*, *Column 2*, and so on.

  You also have the option of selecting the conversion method for column headings when opening an SPSS file. The method that you select then overrides the preferences.

  To change the default conversion method, select **File > Preferences** (or **JMP > Preferences** on Macintosh). On the General page, deselect **Use SPSS labels for column names during import**. Variable names are then imported automatically as column headings.
- The value labels that you defined in the SPSS file are saved as *Value Labels* column properties. The value label then appears in each data table cell instead of the original value. For details about Value Label properties, see "Using Value Labels," p. 151 in the "Properties and Characteristics of Data" chapter.

SPSS can assign certain values in a variable to be treated as missing for analyses. For example, the value *64* could be regarded as missing for a **Height** variable. Then, the calculation of the distribution of height would ignore values of *64*. When you import SPSS into JMP, these values are included in the **Missing Value Codes** column property for the appropriate variable.

At the time of publication, custom currency formats selected in an SPSS file are not maintained on import. In addition, JMP does not read SPSS data that contains double-byte characters, such as non-Unicode Japanese characters.

Note: As with importing other files, you might experience a delay when opening and saving large SPSS files.

Open an SPSS File (Windows)

1. Select **File > Open**.
2. From the list next to **File name** (or the **Files of type** list on Windows XP), select **SPSS Data Files (*.sav)**.
3. Select the SPSS file.
4. (Optional) To specify the column headings, select one of the following **Set JMP column names from** options:
   - **SPSS Labels** (or **Use Label Names for Column Names** on Windows XP) creates column headings from SPSS labels.
   - **SPSS Variable Names** (or **Use Variable Names for Column Names** on Windows XP) creates column headings from variable names.
5. Click **Open**.

   JMP opens the file as a data table.

Open an SPSS File (Macintosh)

1. Select **File > Open**.

2. Select the SPSS file.
3. (Optional) To specify the column headings, do one of the following
   - Deselect **Use SPSS Labels as Headings** to convert variable names to column headings.
   - Select **Use SPSS Labels as Headings** to convert labels to column headings.
4. Click **Open**.

   JMP opens the file as a data table.

Open Excel Files

When you open an Excel file in JMP, the file is automatically converted to a data table. JMP can automatically convert the first row into column headings. If you do not select this option, the columns are named *Column 1*, *Column 2*, and so on.

JMP can also automatically open each worksheet in the spreadsheet as a separate data table. You select this option in the preferences or when you open a spreadsheet. See "General," p. 445 in the "JMP Preferences" chapter for details about Excel preferences.

JMP also opens Excel files from Web sites that do not require you to log in. On Windows, follow the procedure in this section. On Macintosh, use the **File > Internet Open** command. See "Open Remote Files and Web Pages," p. 35 for more information.

Note: In addition to supporting Microsoft Excel 97-2003 files, you can open Excel 2007+ files if at least a version 3 compliant ODBC driver is installed on the machine. In 64-bit JMP, a 64-bit ODBC driver is required.

ODBC, however, does not fully support some Excel 2007+ formatting, such as currency symbols and thousands separators. You might need to assign such column properties manually in the JMP data table. See "Assigning Properties to Columns," p. 145 in the "Properties and Characteristics of Data" chapter for details.

To open an Excel File (Windows):

1. Select **File > Open**.
2. Do one of the following:
   - To open an Excel 97-2003 file, select **Excel 97-2003 Files (*.xls)** from the list next to **File name** (or the **Files of type** list on Windows XP).
   - To open an Excel 2007+ file, select **Excel 2007 (*.xlsm, *.xlsx, *.xlsb)** from the list next to **File name** (or the **Files of type** list on Windows XP).
3. Select the file or enter the URL.
4. (Optional) To convert text in the first row to column headings, select **Always** next to **Always enforce Excel Row 1 as labels** (or **Should Row 1 be Labels?** on Windows XP). If you do not want to import specific worksheets, click **Open**.
5. (Optional) To open specific worksheets, do one of the following:

- (Windows 7 and Vista) Click the **Open** button arrow, select **Open Selected Worksheets**, select one ore more worksheets, and then click **OK**. You can also click **Select All** if you change your mind and want to import all worksheets.
- (Windows XP) Click **Allow individual worksheet selection**, click **Open**, select the worksheets, and then click **OK**.

The spreadsheet is opened as you indicated.

If you tried to open an Excel 2007+ file and do not have the correct ODBC driver installed, an error appears.

To open an Excel file (Macintosh):

1. Select **File > Open**.
2. Select the file. (If the filename is grayed out, the required ODBC driver is not installed, so the file type is not supported.)
3. (Optional) To convert text in the first row to column headings, select **Use Excel Labels as Headings**.
4. (Optional) To open specific worksheets, select **Select Individual Excel Worksheets**.
5. Click **Open**.

   If you chose to open specific worksheets, select those worksheets from the list, and then click **OK**. You can also click **Select All** if you change your mind and want to import all worksheets.

   The spreadsheet is opened as you indicated.

Open SAS Data Sets

SAS data sets are saved in one of many SAS formats:

- Windows formats are .sd2, .sd5, .sd7, .sas7bdat.
- Macintosh formats are .sas7bdat, .ssd, .ssd01, .saseb$data.

When you open a data set in JMP, the file opens as a data table. Columns names are assigned as follows:

- On Windows, JMP uses SAS labels as JMP data table column names by default. You can override the default and request that JMP uses SAS variable names as JMP column names. See step 4 below.
- On the Macintosh, JMP uses SAS variable names as table column names by default. After the data is imported as a JMP data table, you can view the SAS labels by double-clicking the column name. The SAS label appears as a Notes property. You can also replace the table column names with the SAS labels by running the script JMP automatically created when it opened the data table. The script is located in the tables panel under the data table name.

Note: On Windows, you can open SAS data sets directly through the **File > Open** command, which is covered in this section. Another option is connecting to a SAS server by selecting **File > SAS > Browse Data**. On Macintosh, you must connect through a SAS server first. See "Opening SAS Data Sets through a SAS Server," p. 365 in the "External Data and Analytical Sources" chapter for details.

To open a SAS data set:

1. Select **File > Open**.
2. On Windows, select **SAS Data Sets** or **Data Files** from the list next to **File name** (or the **Files of type** list on Windows XP), as shown in Figure 2.13.

Note: SAS variable names and formats are preserved and can be saved after changes are made to the SAS data set. See "Save as a SAS Data Set (Windows Only)," p. 110 in the "Saving Tables, Reports, and Sessions" chapter.

3. Select the file.

Figure 2.13 Open SAS Data Set

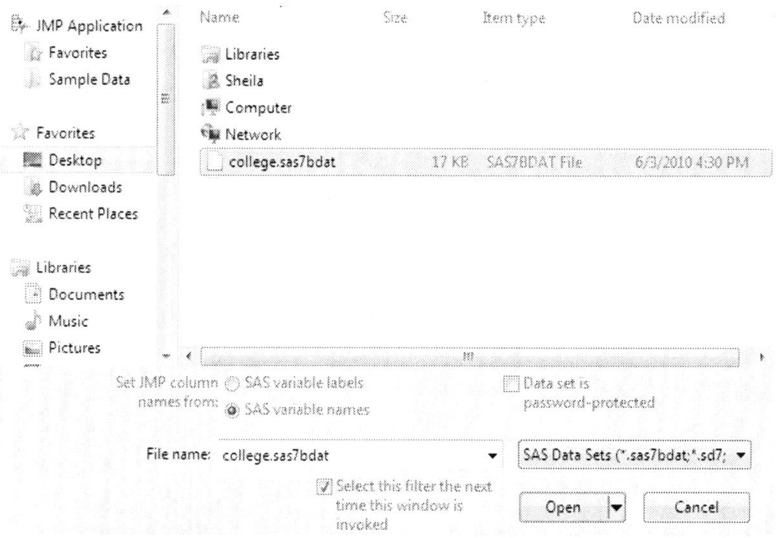

4. (Optional on Windows) Select any of the following options:
   - **Select this filter the next time this dialog is invoked** Sets the default file type choice to the option that you select next to the **File name** list (or in the **Files of Type** list on Windows XP). If selected, the default file type will be **SAS Data Sets** the next time you reach this window.
   - **SAS variable labels** (Windows Vista and Windows 7 only) Uses the SAS variable labels (instead of variable names) as the column names in the JMP data table.
   - **SAS variable names** (or **Use SAS Variable Names for Column Names** on Windows XP) Uses the SAS variable names (instead of the labels) as the column names in the JMP data table.
   - **Data set is password-protected** (or **Dataset is Password Protected** on Windows XP) Select this check box if you know the file is password protected. See "Opening Password-Protected Data Sets," p. 371 in the "External Data and Analytical Sources" chapter for details.

   (Optional) Select any of the following for a SAS Transport (.xpt) file:

- **Select member** Lets you enter the name of a specific member, or table, for JMP to open.
- **Open all members** Opens all members, or tables, in the transport file.
- **Save all members** Saves the file as a JMP file as soon as you open it. The file is saved to the same directory where the SAS transport file was opened.
- **Select Columns** Tells JMP to open only certain columns from the transport file. When you click **Open**, JMP provides a list of columns with checks next to the columns that it intends to open. Add or remove checks to modify the list.

5. Click **Open**.

Note: If you are importing date variables from a SAS file, JMP looks for a SAS date format and translates it to a JMP date column.

Create SAS Transport Files in SAS

JMP can open SAS transport files that were saved using the SAS XPORT engine. For example, below is sample SAS code that creates a transport file called test.

Note: `misc` and `work` are SAS libref names.

```
data test;
input name $ age weight;
cards;
Susan 12 72
Melanie 10 68
Jonathan 11 77
Sheila 13 67
;
libname misc xport 'C:/test.xpt';
proc copy in=work out=misc;
run;
```

E-mail Tables and Reports (Windows Only)

If you are using a Windows operating system, JMP gives you a one-click method of e-mailing a data table and report.

To send e-mail from JMP:

1. Open a data table or create a report.
2. Select **File > Send**. The open window is submitted to your default e-mail application. The file is attached to the e-mail.

Encrypt and Decrypt Scripts

If you want to protect a JMP Scripting Language (JSL) file, you can encrypt it so only someone who knows the password can view it; you can also require a password to run it. This is useful in situations when you want to implement controlled sharing of a script.

This section describes how to encrypt and decrypt scripts in the JMP Script window. For details on encrypting scripts in data tables, see the *Scripting Guide* book.

To encrypt a script:

1. Open the script that you want to encrypt.
2. Select **Edit** > **Encrypt Script**.
3. Assign password(s) to encrypt the files:
   - To encrypt a script so that a user can run it without a password, but needs a password to view it, enter only a decrypt password.
   - To encrypt a script so that a user must enter one password to run it and another password to view it, enter both a run and a decrypt password.

Note: The password must consist of single-byte characters; using a text Input Method Editor (IME) does not work.

4. Click **OK**.
5. If you entered only a decrypt password, click **Yes** to confirm that you do not want to assign a run password.

   The encrypted script opens in a new window.

Figure 2.14 Example of Encrypted Script

```
//-e6.0.2
JUN;OFMFCXAMQ;VWIHMR<ZMVXLSPT>>XESHVSKUDVOMRQB@NXV;
```

6. Save the encrypted script.

To decrypt a JSL script:

1. Open the encrypted script in JMP.
2. Select **Edit > Decrypt Script**.
3. Enter the decrypt password and click **OK**.

   The decrypted script opens in a new window.

To run an encrypted JSL script:

Note: You should examine a script from an unknown source before running it. To examine an encrypted script, you need the decrypt password.

---

1. Open the encrypted script in JMP.
2. Select **Edit > Run Script**.
3. Enter the run password and click **OK**.

   The script runs.

   If the script references a data table, you are prompted to open the data table, and then the script runs.

   If the script requires an empty data table, you must create the table and then run the encrypted script.

Note that entering the run password runs the script, but does not show the script: you must supply the decrypt password to actually view the script.

Chapter **3**

Entering, Editing, and Managing Data

Preparing for Analyses

After you import data into JMP or create a new data table, you need to format your data and the table so it will be ready for analysis.

This chapter contains information that helps you:

- Understand parts of a JMP data table
- Specify data types and modeling types
- Select your data's format
- Edit data tables
- Lock data tables
- Select rows and columns

Contents

Elements of JMP Data Tables

JMP data are organized into rows and columns referred to as the *data table*. A data table has two parts, as shown in Figure 3.1: the *data table panels* at left and the *data grid* at right.

Figure 3.1 Parts of a Data Table

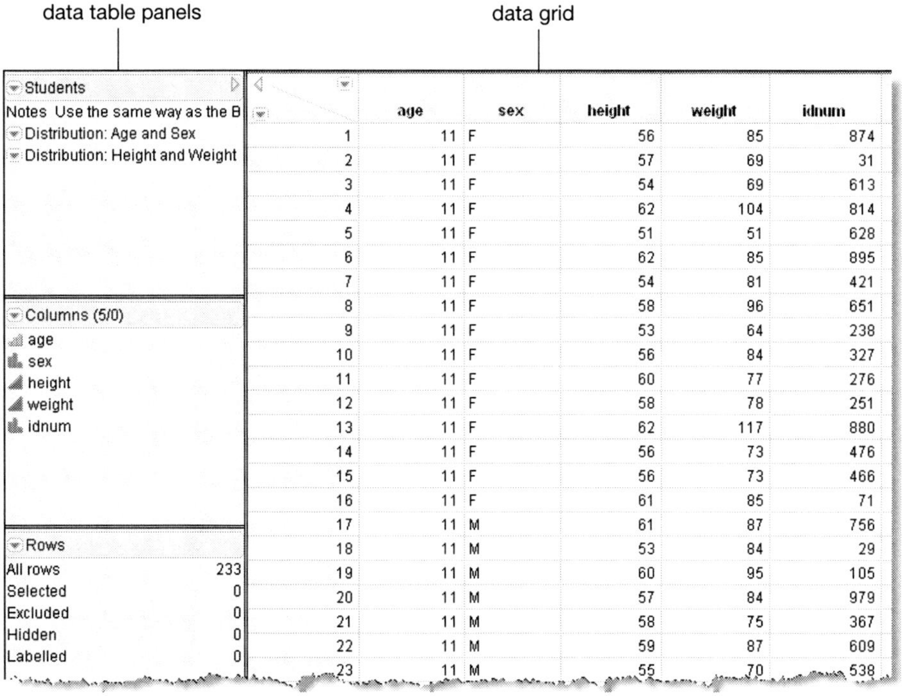

	age	sex	height	weight	idnum
1	11	F	56	85	874
2	11	F	57	69	31
3	11	F	54	69	613
4	11	F	62	104	814
5	11	F	51	51	628
6	11	F	62	85	895
7	11	F	54	81	421
8	11	F	58	96	651
9	11	F	53	64	238
10	11	F	56	84	327
11	11	F	60	77	276
12	11	F	58	78	251
13	11	F	62	117	880
14	11	F	56	73	476
15	11	F	56	73	466
16	11	F	61	85	71
17	11	M	61	87	756
18	11	M	53	84	29
19	11	M	60	95	105
20	11	M	57	84	979
21	11	M	58	75	367
22	11	M	59	87	609
23	11	M	55	70	538

The data table has the following characteristics:

- Column names can contain any keyboard character, including spaces, and can be up to 255 characters long.
- The maximum length of the data table's name depends on your computer's operating system.
- Change the default size and font for names and values selecting **File > Preferences > Fonts.** (On the Macintosh, select **JMP > Preferences > Fonts.**)
- Column names automatically wrap in the column name area to accommodate the column width that you specify.
- Move column boundaries and enlarge the column to view long values. Adjust widths of all selected columns at once by pressing the ALT key as you drag the double arrow cursor on any of the selected column boundaries.

- There is no limit to the number of rows or columns in a data table. However, the table must fit in your computer's memory.

The Data Table Panels

There are three data table panels—the table panel, the columns panel, and the rows panel. The data table panels are arranged to the left of the data grid, as shown in Figure 3.2. These panels contain information about the table and its contents. Each have interactive components, as shown in Figure 3.2.

Figure 3.2 Interacting with the Data Table Panels

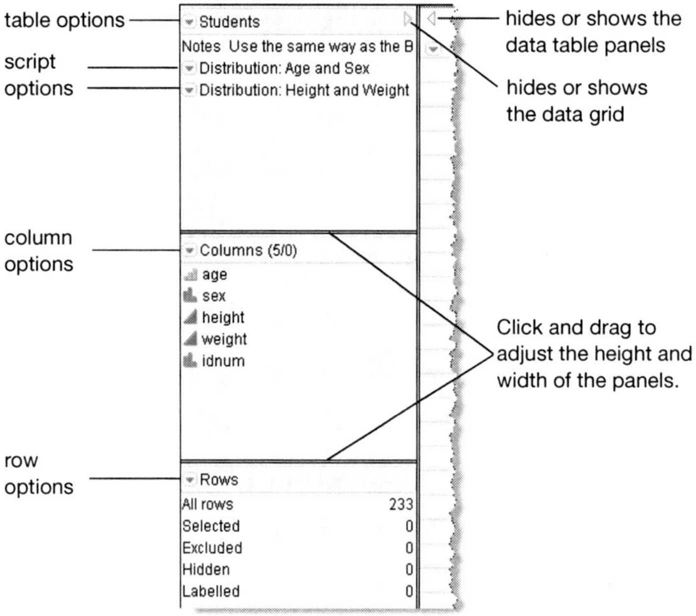

The Table Panel

The table panel contains the following elements:

- Data table name
- Red triangle menus containing table and script options
- Table variables
- Table scripts

Figure 3.3 Example of a Table Panel

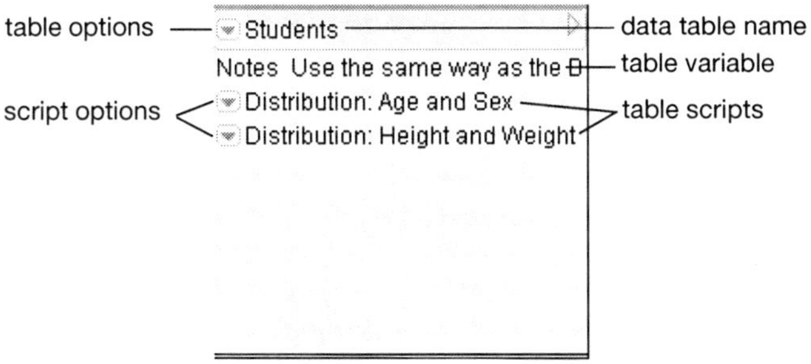

Table Options

Clicking on the red triangle icon next to the data table name gives you these options:

Tables Gives you the same options as selecting the **Tables** command from the main menu, which is discussed in detail in "Reshaping Data," p. 229.

New Table Variable Lets you create a new table variable, which can be text or any other constant character value that you always want to be available in the data table. Table variables are mostly used to document tables. You can also access them when you are using the formula editor so they can be incorporated into formulas or JSL scripts. JMP sometimes automatically creates table variables, such as:

- If you open a data table created by an earlier release of JMP and the table has table notes in it, a table variable called Notes appears with the note's text. (See "Adding Table Variables," p. 81, for an example.)
- If you create a design table with the Design of Experiments (DOE) commands, JMP creates a table variable named Design with the name of the design type as its value.

New Script Lets you create a JSL script to save with the data table. You can also access scripts when you are using the formula editor so they can be incorporated into formulas that calculate column values. (See "Creating a Formula," p. 303, for details.) After selecting this command, name the script and type in the value (the JSL commands). After you click **OK**, the new script is listed in the table panel and you can click its red triangle icon to run, edit, or delete it. See "Creating Scripts," p. 84, for details.

Suppress Formula Eval Turns off the feature that automatically evaluates formulas. You can turn off evaluation and build sections of a formula, and then turn evaluation on to test the formula.

Lock Data Table Locks the data table so values cannot be edited or added. You can still run analyses, assign characteristics, add rows and columns, and so on. See "Locking Tables," p. 80, for details.

Copy Table Script Copies the script that recreates the table. Paste the script into a new script and run it to recreate the table.

Script Options

Clicking on the red triangle icon next to the script name gives you these options:

Run Script Runs the script.

Edit Lets you edit the script.

Delete Lets you delete the script.

Copy Copies the script. You can then paste it into the table panel of another data table.

Additional Options

In addition to the red triangle options, there are other actions you can take in the table panel:

- Click twice on a data table or script name to edit it.
- Double-click a table variable or script name to edit the name and content.
- Click and drag on a table variable or script to rearrange it.

The Columns Panel

The columns panel (Figure 3.4) contains a list of columns found in the data table, each column's modeling type, and any attributes assigned to the columns.

Figure 3.4 Example of a Columns Panel

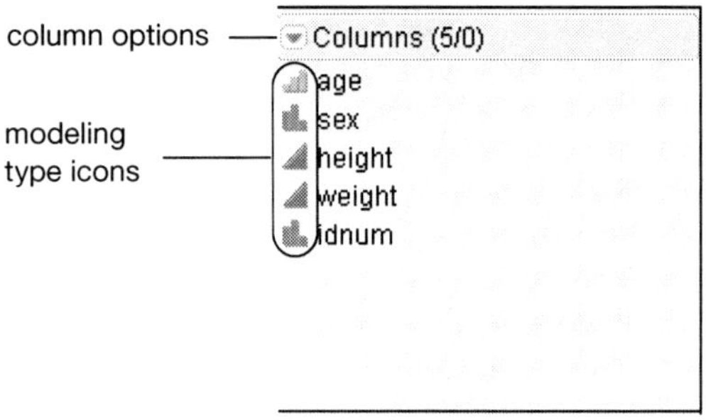

To access column options, click on the red triangle icon to the left of the word **Columns**, as shown in Figure 3.4. These options give you the same options as selecting the **Cols** command from the main menu, which is discussed in "The Cols Menu," p. 425.

To change the name of a column, click the column name in the column panel, and then click the name again.

Modeling Type Icons

Icons to the left of each column name indicate the column's modeling type, which can be one of the following types:

- Continuous
- Ordinal
- Nominal

Modeling types tell JMP how to use the column values in an analysis. To change the modeling type, click on the icon and select a different type. See "About Modeling Types," p. 58, for details.

Icons Representing Column Characteristics and Properties

Icons to the right of each column name, as shown in Figure 3.5, indicate characteristics and properties the columns contain. See "Assigning Characteristics to Rows and Columns," p. 135 and "Assigning Properties to Columns," p. 145, for details.

Figure 3.5 Elements Indicating Column Characteristics and Properties

Note: Italics indicate that the column is locked into place. When you scroll horizontally, the column remains visible.

Table 3.1 Description of Icons Representing Column Characteristics and Properties

	Can be X, Y, W, or F. Indicates that the column has been assigned a preselected role. In analyses, the column always appears as the *x*, *y*, *weight*, or *frequency* role.
	Indicates that points on plots corresponding to the column will be identified by the value instead of the row number.
	Indicates that the column will be excluded from the windows.
	Indicates that the column is hidden in the data grid.
*	Indicates that the column contains one or more properties. Click to reveal a list of properties the column contains.
	Indicates that the values in the column result from a formula. When formula evaluation is suppressed, part of the icon becomes gray. (See "Using Formula Editor Options," p. 316, for details.) Double-click to view and edit the formula.
	Indicates that the range check or the list check option is turned on. Click to view and edit the range or list.

Note: For detailed descriptions of these icons, see the "Properties and Characteristics of Data" chapter.

The Rows Panel

The rows panel (Figure 3.6) shows the number of total rows, selected (or highlighted) rows, excluded rows, hidden rows, and labeled rows.

To access row options, click the red triangle icon to the left of the word **Rows**, as shown in Figure 3.6. These options give you the same options as selecting the **Rows** command from the main menu, which is discussed in "The Rows Menu," p. 423.

Figure 3.6 Example of a Rows Panel

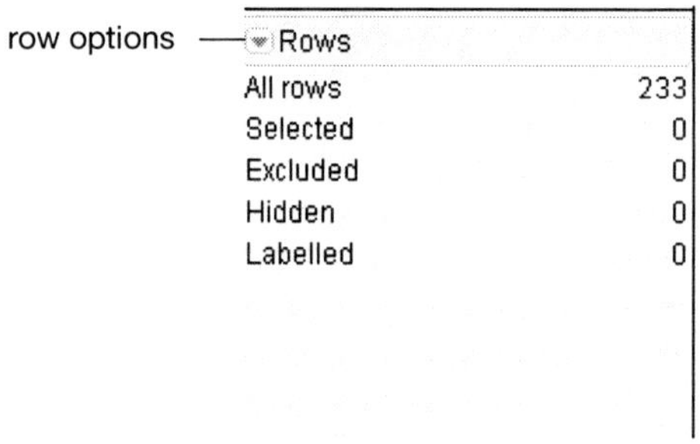

You can also right-click the categories in the rows panel (**All rows**, **Selected**, **Excluded**, **Hidden**, or **Labeled**) to select those rows, to clear the selection, or to create a *data view.*

Note: The menu is dimmed if it is not applicable. For example, if you have no rows selected, and you right-click on **Selected**, the menu is dimmed.

Creating a data view provides a convenient way to see a linked subset of the main data table. For example, if several rows are marked hidden, but you want to open a window that shows you only the hidden rows, right-click **Hidden** in the rows panel and select **Data View**.

Figure 3.7 Creating a Data View from the Rows Panel

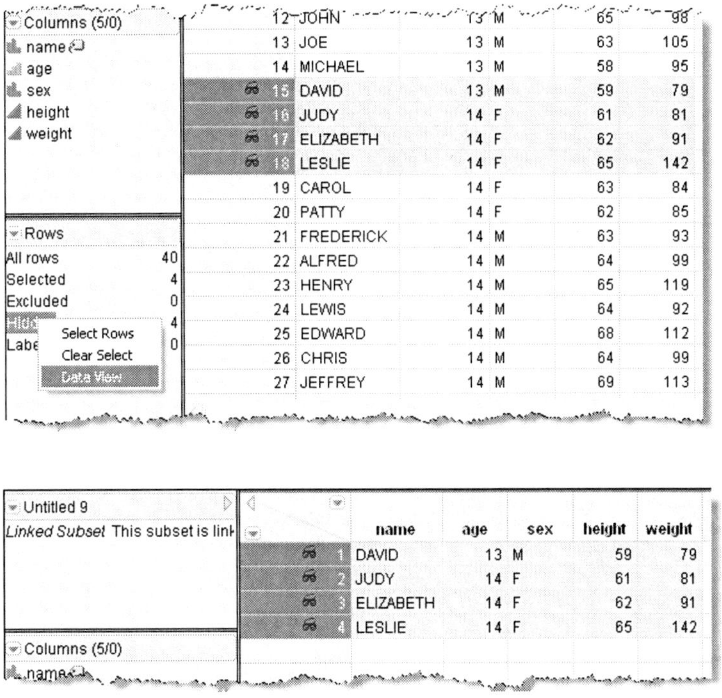

You should continue to do most of your editing in the main data table. When you make changes in either the main data table or in the data view, the changes are reflected in both. You can make minor changes (such as changing some data or adding a column) in the data view. If you want to make major changes (like adding a formula) you must make those changes in the main data table.

The Data Grid

Figure 3.8 illustrates how to select rows and columns, find rows, and use keyboard arrows. See "Selecting Rows and Columns," p. 88, for details about making selections.

Figure 3.8 The Data Grid

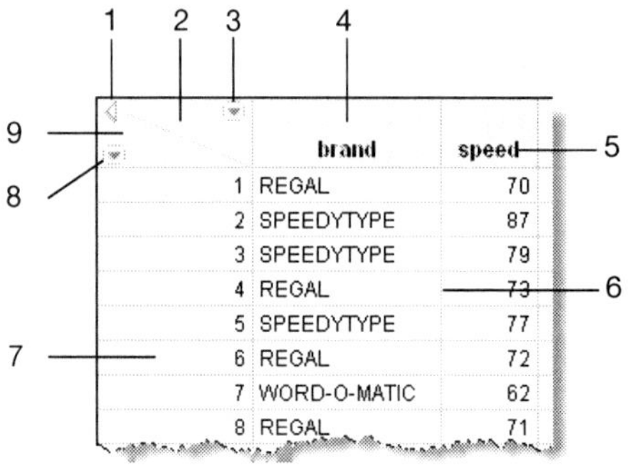

Table 3.2 Explanation of the Data Grid

1	Hides or shows the data table panels.
2	Click to deselect any selected columns. Click and hold down the SHIFT key to quickly select all columns.
3	Shows the columns menu. See "The Cols Menu," p. 425.
4	• Click to select the column. • Double-click to view the Column Info window. See "Assigning Properties to Columns," p. 145. • Right-click for column options. See "Context Menu for Columns," p. 54.
5	Click twice to edit the column name.
6	Click and drag to adjust the width of the column. To simultaneously adjust the widths of all of the selected columns, hold down the ALT key as you click and drag.
7	Selects a row.
8	Shows the rows menu. See "The Rows Menu," p. 423.
9	Click to deselect all selected rows. To select all rows, click and hold down the SHIFT key.

Context Menu for Columns

Table 3.3 describes the options that appear when you right-click in a column heading.

Table 3.3 Description of Context Options for Columns

Column Info	Opens the Column Info window. See "Assigning Properties to Columns," p. 145.

Table 3.3 Description of Context Options for Columns *(Continued)*

Column Properties	Enables you to choose from a list of column properties. This list is also available from the Column Info window. See "Assigning Properties to Columns," p. 145.
Modeling Type	Enables you to change the column's modeling type. See "About Modeling Types," p. 58.
Preselect Role	Assigns an analysis role to the column. See "Giving Columns a Preselected Analysis Role," p. 143.
Formula	Opens the Formula Editor. See the "Formula Editor" chapter.
Color Cells	Enables you to color the selected cells in the column. To revert back to the original color, select black.
Data Filter	Opens the Data Filter. See "The Data Filter," p. 95.
Sort	Sorts the values in the column by ascending order. To choose descending order, select Sort again.
Delete Columns	Deletes the column.
Copy Column Properties	Copies all of the column properties for the selected column.
Paste Column Properties	Pastes all of the copied column properties into the selected column.

Context Menu for Rows

Table 3.4 describes the options that appear when you right-click in a row heading.

Table 3.4 Description of Context Options for Rows

Exclude/Unexclude	Excludes or includes selected rows from analyses. See "Excluding Rows and Columns," p. 135.
Hide/Unhide	Hides or shows selected rows in all plots and graphs. See "Hiding Rows and Columns," p. 136 in the "Properties and Characteristics of Data" chapter.
Label/Unlabel	Labels or unlabels selected rows in all plots. See "Labeling Rows and Columns," p. 136.
Colors	Changes the color for the selected rows, which is then reflected in any plots. See "Giving Rows a Color," p. 137.
Markers	Adds markers (user-selected symbols) to the selected rows, which is then reflected in any plots by replacing points with markers. See "Adding Markers to Rows," p. 137.

Table 3.4 Description of Context Options for Rows *(Continued)*

Color Rows by Row State	Colors the row the same as the current row state color. For more details, see "Assigning Colors or Markers to Rows According to Column Values," p. 138.
Select Matching Cells	Selects rows in the active data table with values that are similar to the highlighted row(s).
Invert Selection	Selects all previously deselected rows.
Clear Row States	Clears all active row states in the data table. All rows become included, visible, unlabeled, and show in plots as black dots. It does not affect row states saved in row state columns.
Delete Rows	Removes all selected rows from the data table. Use the **Undo** command on the **Edit** menu to undo an accidental deletion. See "Adding and Deleting Rows," p. 63.

Cursor Forms

The cursor has different forms, and the actions that it performs depend on its location in the data grid. See "The Tools Menu," p. 433, for details about cursor forms.

Arrow Cursor The cursor appears as the standard arrow cursor when it appears in the following locations:

- In the panels area to the left of the data table
- In the triangular rows and columns area, located in the upper left corner of the data grid
- In the middle or bottom of a column heading

You can perform the following actions with the arrow cursor:

- To select a column using the arrow cursor, click its name in the columns panel.
- To show the Column Info window, double-click on a column name.
- Click the table name on the title bar of the tables panel to edit it.
- Double-click a column name in the column panel to edit it. Or, in the column heading, click twice on the column name to edit it.
- Click the triangular areas in the upper left corner of the data grid to deselect rows and columns.

Selection (Large Plus) Cursor When the cursor is at the top of a column heading, or in a row number area, it becomes a large plus, indicating that it is available to select rows or columns. When you click, that row or column is highlighted. Click and drag to highlight multiple rows or columns, and hold down the CTRL key and click to select discontiguous rows or columns.

- Double-click a column heading area to see the Column Info window for that column.
- Select a column to change the column name. The column highlights. Begin typing (if it is not in a locked column or locked data table).
- Double-click the row number area to edit the rows using the row editor.

Standard I-beam Cursor When you select editable text, the cursor becomes a standard I-beam. To edit text, position the I-beam within highlighted text. Click to mark an insertion point, or drag to select text for replacement. The I-beam deposits a blinking vertical bar to indicate a text insertion point or a highlighted area of text to be replaced.

Double Arrow Cursor ↔ The cursor changes to a double arrow when it is on a column or a panel boundary. Drag this cursor left or right to change the width of a column or panel. Changing the width of a column does not affect the column field width specified in the Column Info window (accessed by double-clicking a column name).

Note: You can adjust widths of all selected columns at once by pressing the ALT key as you drag the double arrow cursor on any of the selected column boundaries.

List Check Cursor The cursor changes form when you move the mouse over values in columns that have data validation in effect. It becomes a small, downward-pointing arrow on a column with list checking. When you click, the value is highlighted and the cursor becomes the standard I-beam; you enter or edit data as usual with any values defined as valid text or valid numbers. See “List Checking,” p. 148, for details.

Popup Pointer Cursor The cursor changes to a pointer over these objects:

- Red triangle menus for options
- Diamond-shaped disclosure buttons that open or close panels
- Data table titles for editing
- Table script titles for opening
- Modeling type icons for changing

Specifying Data Types and Modeling Types

A column in a JMP table can contain different types of information. However, all information in a single column must be of the same data and modeling types.

- When you import data into JMP, it guesses which data and modeling types to use. Therefore, you should check to make sure that JMP has guessed correctly.
- When you manually insert data into JMP, you should assign a data type and a modeling type as you insert the data.

About Data Types

The data type of a column determines how its values are formatted in the data grid, how they are stored internally, and whether they can be used in calculations. The three data types are as follows:

- **Numeric** Columns only contain numbers, with or without a decimal point.
- **Character** Columns contain any characters, including numbers. In character columns, numbers are seen as characters only and are treated as discrete values instead of continuous values. The maximum field width for character values is 32,766 bytes.
- **Row State** Columns contain row state information—information that tells you if the rows are excluded, hidden, labeled, colored, or marked. See "Using Row State Columns," p. 173, for details.

About Modeling Types

The modeling type of a column applies only to columns whose data types are numeric or character. The modeling type tells JMP how to treat its values during analyses. Changing the modeling type lets you look at a variable in different ways in an analysis. Figure 3.9 illustrates the three modeling types.

Figure 3.9 Modeling Types

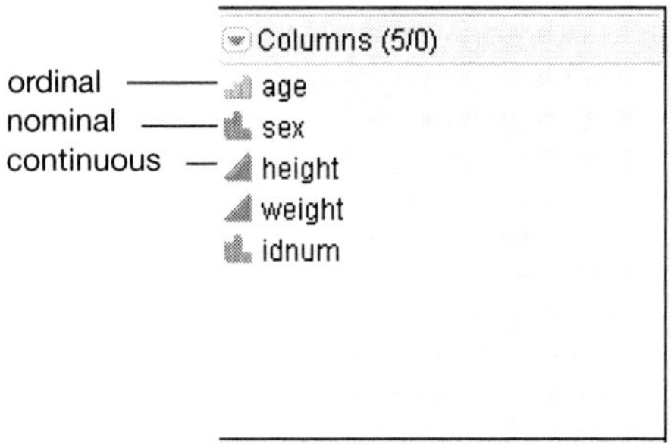

Continuous Columns only contain numeric data types. Continuous values are treated as continuous measurement values. JMP uses the numeric values directly in computations.

Ordinal Columns contain either numeric or character data types. JMP analyses treat ordinal values as discrete categorical values that have an order. If the values are numbers, the order is the numeric magnitude. If the values are character, the order is the sorting sequence.

Nominal Columns contain either numeric or character data types. All values are treated in JMP analyses as if they are discrete values with no implicit order.

How to Assign Data and Modeling Types

There are three methods to assign data and modeling types or to make sure JMP has automatically given your data the correct types. The following sections describe these three methods.

Method 1: Create a New Column

One way to assign data and modeling types is when you create a new column. When you create a new column (by selecting **Cols > New Column**), the New Column window appears, as shown in Figure 3.10. Specify the data type and modeling type in this window.

Figure 3.10 Add a New Column and Specify Data and Modeling Type

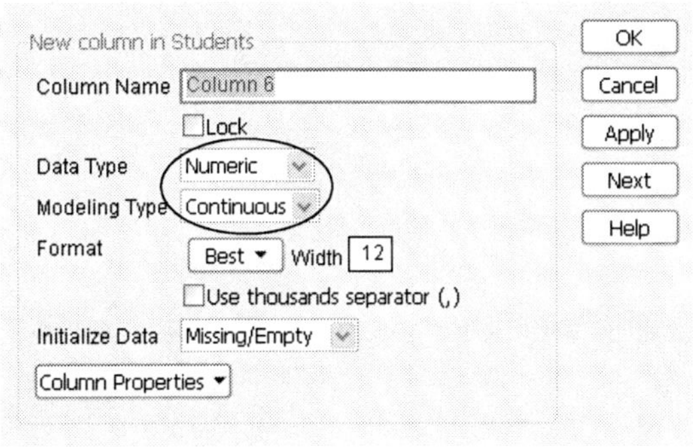

Method 2: Double-click the Area Above an Existing Column Name

If your data is already in JMP and you want to assign data and modeling types, double-click the empty area above the name of the column. The Column Info window appears, as shown in Figure 3.10. Specify the data type and modeling type in this window.

Method 3: Use the Columns Panel Icons

If your data is already in JMP, click the icon beside the column name in the column panel that indicates its modeling type. Select a different type from the drop-down menu.

Figure 3.11 Using the Columns Panel Icons

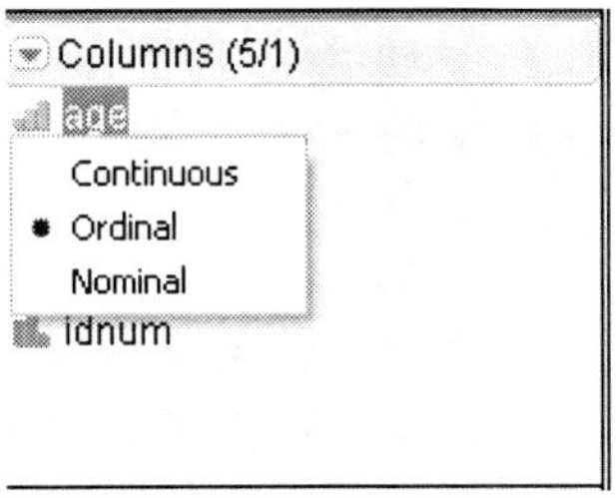

Note: You can select **Continuous** only if your data type is numeric. If the **Continuous** option is dimmed on the menu and you want to make the column continuous, first change its data type in the Column Info window. (You can access this window by highlighting the column and selecting **Cols > Column Info**).

Using the Short-Integer Format

JMP has the ability to store numeric data in as few as 8 bits (one byte). To implement this, you need to tell JMP to first make short-integer formats available to you in the Column Info window's **Data Type** drop-down menu. Then you should specify how many bytes you want the integers to hold. When you use the correct short-integer format for your data, you do not see any difference in how the numbers appear, but the numbers occupy less disk space and use less memory.

To store numeric data in short-integer format:

1. Select **File > Preferences** and click the **Tables** option on the left panel.
2. Click the **Allow short numeric data format** option.
3. Click **OK** to return to the data table.
4. Double-click the column name whose values you want to be short-integer. The Column Info window appears.
5. Click the **Data Type** drop-down menu, and select **1-byte integer**, **2-byte integer**, or **4-byte integer**. JMP now stores values as integers in the range corresponding to your selection. For example, when **1-byte integer** is selected, the range of numbers you can enter is from -126 to 127. When **2-byte integer** is selected, the range of numbers you can enter is from -32,766 to 32,767. When **4-byte integer** is selected, the range of numbers you can enter is from -2,147,483,646 to 2,147,483,647.

Choosing Numeric Formats

When you assign a variable's data type to be numeric, you can also specify the display format for it to use. To select the numeric format:

1. Select **Cols > Column Info** or double-click the area above the name of the column in the data grid. Or, if you are creating a new column, select **Cols > New Column**.
2. Click the drop-down menu beside **Format** and select the option that specifies how you want JMP to display numeric values. See "Numeric Format Options," p. 61, for descriptions of options.
3. If you selected:
   - **Date** or **Time** from the **Format** drop-down menu in Step 2, click the **Input format** menu that appears and select which format dates or times have when they are entered into JMP. See "Numeric Format Options," p. 61, for descriptions of options. See "Changing the Numeric Format of an Axis," p. 205, for a description of how to change the way a date or time appears in a graph without changing the way it appears in a data table.
   - **Fixed Dec** from the **Format** drop-down menu in Step 2, the **Dec** text box appears. Type the number of decimal places that you want JMP to display in the data table. If the value does not have as many numbers after the decimal as the number that you specify, JMP adds zeros to the number to give it that many decimal places. For example, if a value is 1.23 and you type **5** in the **Dec** box, JMP displays the number with five decimal places: 1.23000.
4. In the **Width** text box, type the number of characters that you want the cell in the data table to accommodate. Be sure to type a number that is the largest number of digits or characters that you think a value in the column could have. If your format is **Fixed Dec** and you set the width of the cell to be smaller than the number of decimal places that you want in the values, JMP ignores the number of

decimal places that you specify and sets the column width to the number of characters that you specify in the **Width** box.

5. Click **OK**.

Numeric Format Options

The options in the **Format** and **Input Format** drop-down menus are numeric formats that tell JMP how to display numbers in the data grid.

Available options are:

Best Lets JMP consider the precision of each cell value and select the best way to show it. By default, the physical width of the column is 12 characters. To increase the column width and show large values with many decimal places, in the Column Info window next to **Width**, type the number of characters that you want to appear.

Fixed Dec Displays all column values rounded to the number of decimal places that you specify. You can set the number of decimal places to zero to see only whole numbers. Type the number of decimal places that you want JMP to display into the **Dec** text box in the Column Info window.

Percent Multiplies numeric values by 100 and displays the result followed by a percent sign.

PValue Displays probability values. It has a default width of six and shows four decimal places. If a number is less than 0.0001, the number is displayed as <.0001. The format is mostly used in JSL scripts and rarely needed for a data table column.

Scientific Displays a number in standard scientific notation. If you enter the number 123456, it appears as 1.23456e+5.

Currency Formats values with two decimal positions, separators, and the currency sign appropriate for your computer's locale settings. For example, if you enter the number 123456, it shows in the data table on an American system as $123,456.00. The default width of the **Currency** format is 17. If you have a number that requires a wider field width, the format defaults to the **Best** format. To show large values with many decimal places, increase the column width by typing the number of characters in the **Width** text box found in the Column Info window.

Date Displays all column values as a date. If you assign a date format to a numeric column that already contains data, then the numeric values are treated as the *number of seconds since January 1, 1904*. For example, if you have a numeric column with a cell value of 1,234,567,890 and you change the format to **Date > m/d/y**, the cell value appears as 02/13/1943. You can leave the date values displayed as described above, or select one of these formats: (The examples below show formats for reading or displaying the date, December 31, 2004. Its unformatted (numeric) value is 3,187,296,600, which represents 12:10am.)

- **m/d/y** has the form of *mm/dd/yyyy*, giving 12/31/2004.
- **mmddyyyy** has the form of *mmddyyyy*, giving 12312004.
- **m/y** has the form of *mm/yyyy*, giving 12/2004.
- **yyyyQq** has the form of *yyyyQq*, giving 2004Q4.
- **d/m/y** has the form of *dd/mm/yyyy*, giving 31/12/2004.
- **ddmmyyyy** has the form of *ddmmyyyy*, giving 31122004.
- **ddMonyyyy** has the form of *ddMonyyyy*, giving 31Dec2004.
- **Monddyyyy** has the form *Monddyyyy*, giving Dec312004.
- **y/m/d** has the form *yyyy/mm/dd*, giving 2004/12/31.
- **yyyymmdd** has the form *yyyymmdd*, giving 20041231.
- **yyyy-mm-dd** has the form *yyyy-mm-dd*, giving 2004-12-31.
- **Date Long** shows a date value as *weekday, month day, year*, which gives Friday, December 31, 2004. When using this option, you might need to increase the column width to see the formatted value in its entirety. Do this by double-clicking the column name and entering the needed number of digits or characters into the box labeled **Width**.
- **Date Abbrev** shows the month as a three-character abbreviations, such as Dec 31, 2004. When using this option, you might need to increase the column width to see the formatted value in its entirety. Do this by double-clicking the column name and entering the needed number of digits or characters into the box labeled **Width**.
- **Locale Date** displays a JMP date value according to the settings of your operating system. For example, the local OS setting for date in the United states is the format **mm/dd/yyyy**.

Time displays all column values as an instance in time using one of these formats. An example of an instance is a specific date in time, such as 12/2/03 at 2:23 PM.

- **m/d/y h:m**, **m/d/y h:m:s**, **d/m/y h:m**, **d/m/y h:m:s**, **y/m/d h:m**, **y/m/d h:m:s**, **ddMonyyyy h:m**, **ddMonyyyy h:m:s**, **ddMonyyyy:h:m**, **ddMonyyyy:h:m:s**, **Monddyyyy h:m**, **Monddyyyy h:m:s** show the number of hours, minutes, and seconds after midnight of the prepended date. The date appears in the formats described above. The hours, minutes, and seconds are separated by a colon. The example of December 31, 2004, which has a numeric value of 3,187,296,600 is equivalent to 12/31/2004 12:10 AM.
- **:day:hr:m, :day:hr:m:s** show the number of days, hours, minutes, and seconds since January 1, 1904. The results for December 31, 2004 are :36890:00:10: and :36890:00:10:00.
- **h:m:s, h:m** gives hours, minutes, and seconds portion of the date in the date field. The example used here (December 31, 2004) has a time of 0:10 AM, so the result is 01:10:00 AM.
- **yyyy-mm-ddThh:mm, yyyy-mm-ddThh:mm:ss** show the year, month, day, and time. For example, 2004-12-31T12:10:00. T is a literal, representing itself.
- **Locale Date Time h:m** and **Locale Date Time h:m:s** display a JMP time value according to the settings of your operating system. For example, the local OS formats for time in the United states are the formats **mm/dd/yyyy hh:mm** and **mm/dd/yyyy hh:mm:ss**.

Duration displays all column values as a duration of time. An example of duration is hours, minutes, and seconds.

- **:day:hr:m, :day:hr:m:s** show a duration of time, such as 52:03:01:30 (:day:hr:m:s), which reads fifty-two days, three hours, one minute, and thirty seconds.
- **hr:m, hr:m:s, min:s** shows a duration of time, such as 17:37 (hr:m), which reads seventeen hours and thirty-seven minutes.

Geographic displays latitude and longitude number formatting for geographic maps. Latitude and longitude options include the following:

- DDD (degrees)
- DMM (degrees and minutes)
- DMS (degrees, minutes, and seconds)

In each format, the last field can have a fraction part. You can specify the direction with either a signed degree field or a direction suffix. Uncheck the **Direction Indicator** field to show a signed degree field, such as *-59°00'00"*. To show the direction suffix instead (as in *59°00'00" S*), check **Direction Indicator.**

To use spaces as field separators, uncheck **Field Punctuation**. To use degrees, minutes, and seconds symbols instead, check this option.

Working With International Formats

If you are importing or entering data that contains formatting specific to country standards, you might need to make sure your number formats are interpreted correctly. To do this on Windows, access the Control Panel's region and language option, and select the country for which the number should be formatted. On the Macintosh, select **System Preferences** from the apple menu, and then select **International**, and then select **Numbers**, and select the correct country.

Entering Data

The following sections describe how to enter data into a table, including entering data one cell at a time or adding multiple rows and columns of data at once.

Adding and Deleting Rows

To add new empty rows, take one of the following actions:

- Select **Rows > Add Rows**. Then enter the number and location of rows that you want to add. By default, if you selected rows, the new rows appear after the last currently selected row. If no rows are currently selected, the new rows appear at the end of the table.
- Click in a cell anywhere below the last row in a table and begin typing. Then press the Enter key to automatically generate new rows up to and including the row with the value that you typed.
- Double-click an empty row number area below the last row to add that many empty rows.

- Double-click the lower triangular area in the upper left corner of the data grid, as shown in Figure 3.12. Then enter the number of rows to add and specify where you would like to add them.
- Right-click in an empty row below the last row, and select **Add Rows**. An abbreviated version of the Add Rows window appears. Enter the number of rows you want to add.

Figure 3.12 Double-click the Lower Triangular Area

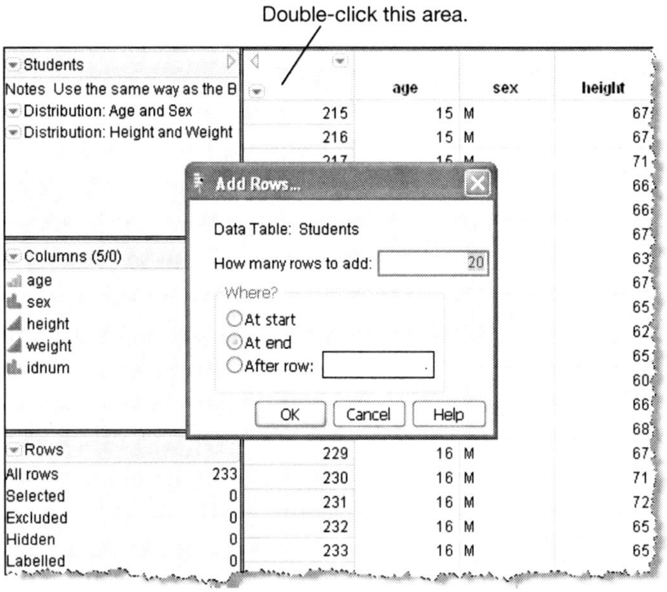

To delete rows from the data grid:

1. Highlight the rows that you want to delete.
2. Select **Rows > Delete Rows,** or right-click on the row numbers and select **Delete Rows**.

Caution: When you try to delete thousands of rows, an alert might appear if your computer has insufficient memory. Click **No** to cancel the deletion, and then try to delete fewer rows.

Adding and Deleting Columns

To add new empty columns, take one of the following actions:

- Double-click the empty space to the right of the last data table column.
- Select **Cols > New Column**. The Column Info window prompts you to name the new column and provide column characteristics, including the type of data with which you initially want to fill the column. Click **Next** to add another column and click **OK** to see the new column(s) appear in the data table. You can also assign properties to columns using the **Column Properties** button. See "Assigning Properties to Columns," p. 145, for details.

- Select **Cols > Add Multiple Cols** (or double-click the upper triangular area in the upper left corner of the data grid). Using the **Add Multiple Columns** command to define multiple columns is different from using the **New Column** command. All of the columns that you add using the Add Multiple Columns window have the same data characteristics. By default, the new column names are Column 1, Column 2, and so on. However, you can specify other text that becomes the prefix of the new column names by typing into the **Column prefix** box. Then, as shown in Figure 3.13, proceed as follows:
  - Enter the number of columns to add
  - Specify if they are to be grouped
  - Select a data type
  - Enter their location
  - Select the initial data values.

Figure 3.13 The Add Multiple Columns Window

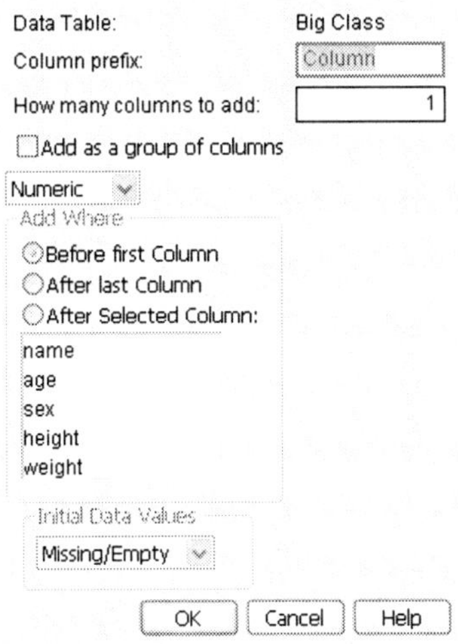

To delete columns from the data table

1. Highlight the columns to delete.
2. Select **Cols > Delete Columns**, or right-click and select **Delete Columns**.

Setting Up Initial Data Values

When you add a new column to a data table (select **Cols > New Column**), you can specify the type of initial data values that you want to appear in the column, as shown in Figure 3.14.

Figure 3.14 Initial Data Values

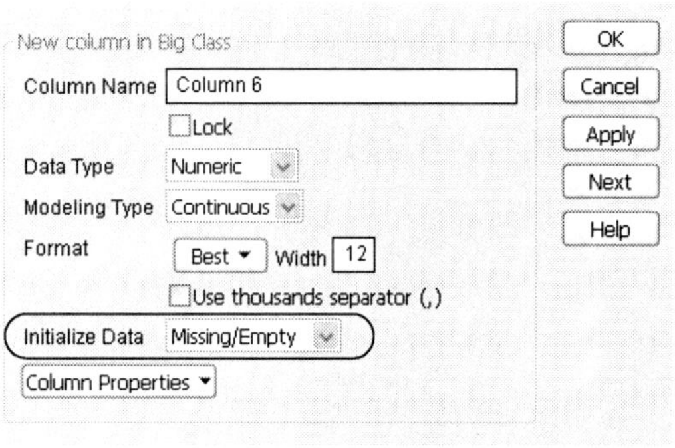

Missing/Empty Places missing values in the column, represented by a black dot (•) for numeric data and a blank for character data.

Constant Places one number in all the column's rows for numeric data. Select **Constant**, and then specify what number you would like to use by typing it into the box that appears. For character data, any number of characters can be entered into this box.

Today Places today's timestamp in the column for each row. Used only with the Date or Time formats.

Sequence Data Inserts sequential data based on the parameters that you specify in the boxes that appear. For numeric data, select **Numeric** in the Data Type box and select **Sequence Data** for the Initial Data Values. In the **From** and **To** boxes, assign a starting and ending point and then assign the sequence in the **Step** box. For example, if you want the column to contain even numbers from 2 to 60, type **2** in the **From** box, **60** in the **To** box, and **2** in the **Step** box. In the **Repeat each value N times** box, type the number of times that you want each numeric value repeated. Click **OK.**

For character data, select **Character** in the Data Type box and select **Sequence Data** for Initialize Data. In the box next to **Add**, type the character data and click **Add**. In the **Repeat each value N times** box, type the number of times that you want each character value (such as red) repeated. Click **OK.**

Random Inserts random data into the column. Click the button beside the type of random number that you want to use, and then enter one of the following:

- A range for random integers
- A range for random uniform numbers
- The mean and standard deviation for random normal numbers
- Values and proportions for random indicators

Filling Columns with Sequential Data

You can fill columns with either a repeating sequence of data or with a continuation of values:

1. Create a sequence of data in a column, as shown in Figure 3.15.

Figure 3.15 Sequence of Data

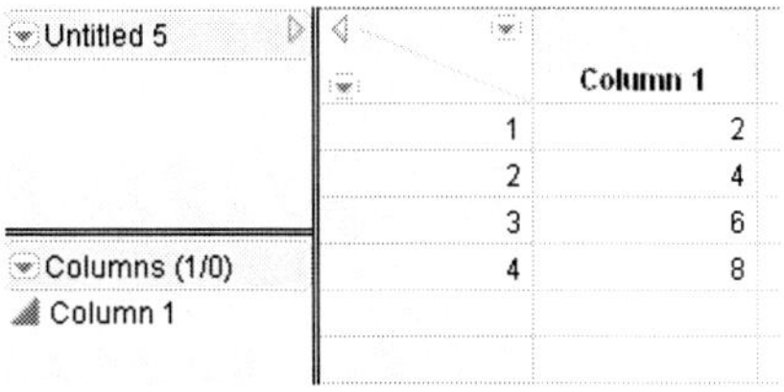

2. Highlight the cells containing the sequenced data. The cells can be in different columns.
3. Right-click the selected cells and select an option under **Fill**, shown in Figure 3.16.

Figure 3.16 Data Filling Options

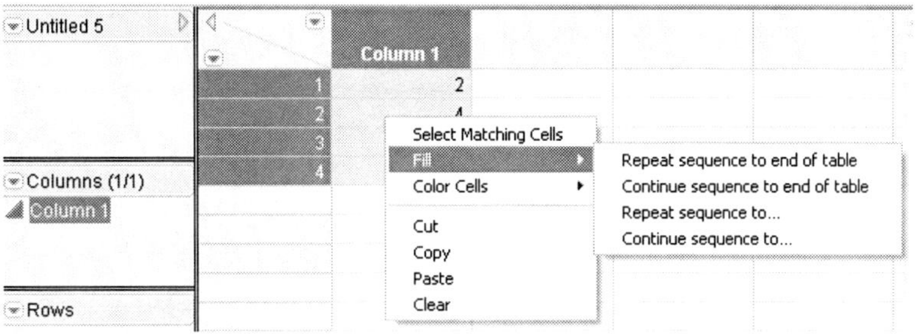

Repeat sequence to end of table Cells below the selection are filled with repeats of the selected cells.

Continue sequence to end of table Cells below the selection are filled with a continuation of the pattern found in the selected cells. For example, if the selected cells contain the numbers 1 and 2, then the remaining cells are filled with 3, 4, 5, 6, and so on. If the selected cells contain the numbers 2 and 4, then the remaining cells are filled with 6, 8, 10, 12, and so on.

Repeat sequence to... JMP repeats the pattern found in the selected cells to the row number that you specify.

Continue sequence to... JMP continues the pattern found in the selected cells to the row number that you specify.

Entering Cell Formulas

In numeric columns, you can enter cell expressions preceded by an equal sign (=). JMP immediately evaluates the expression and stores the new number as the cell's value. Unlike column formulas, a cell expression is not stored.

Cell expressions can contain operators, constants, and global and column variables.

To enter an expression:

1. Click the cell in which you want to enter the expression.
2. Type an equal sign (=) and then type the expression. See Table 3.5.
3. Press the Enter key.

Table 3.5 Examples of Expressions in Table Cells

Example expression	Cell value
`=sqrt(2)`	1.41
=456+890	1346
`=height+weight`	Sums the values of cells in columns height and weight located in the same row as the cell that you entered the expression.
`=height[1]`	Displays the value found in row 1 of the height column

Editing Data and Tables

The following sections describe how to edit data that is in a table, including editing cells themselves and making changes to rows and columns.

Editing Cells

To edit the contents of a cell:

1. Click a cell that already has a value. It is highlighted.
2. Click the cell a second time. The cursor becomes a flashing text insertion bar.

Figure 3.17 Editing the Contents of a Cell

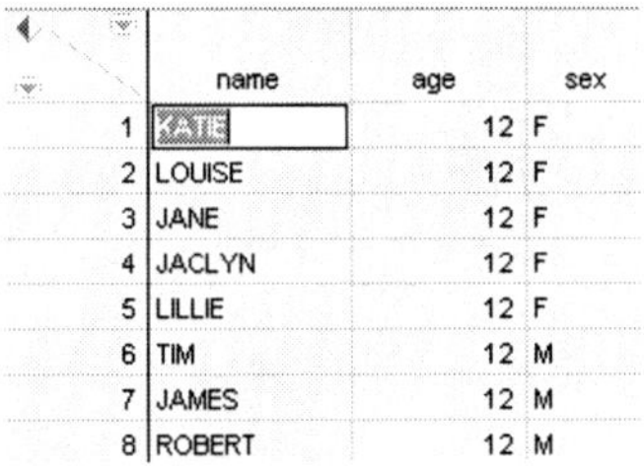

3. Edit the cell's value.

Editing Column Names

There are many ways to edit a column name. Choose a way from the list below:

- Select the column and begin typing.
- Double-click or right-click the empty area around the column name and select **Column Info**. Edit the name in the Column Info window that appears.
- Click twice on the column name in the column panel and begin typing.

Recoding Data

If you have data that is coded incorrectly (for example, contains typos or incorrect wording), you can quickly recode it so it follows a consistent format. You can also use recode to replace missing or empty values.

JMP displays a window that contains selected columns' unique values and enables you to enter new values to replace them:

1. Highlight the column(s) that contain data you would like to recode.
2. Select **Cols > Recode**. The Recode window appears (Figure 3.18). In this window, the Count column lists the number of occurrences each value has in the data table.

Figure 3.18 Enter New Values

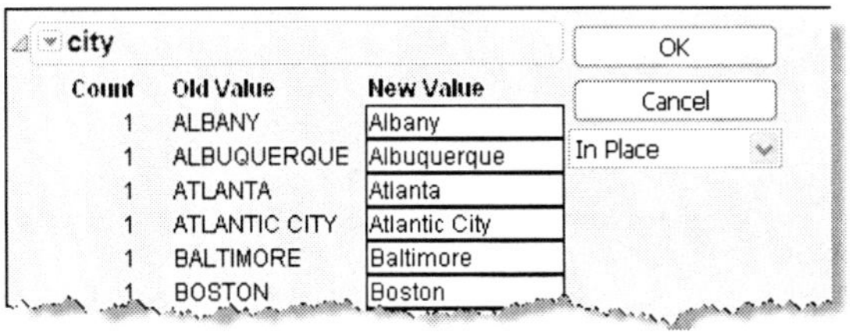

3. Type the new value in the text box corresponding to the old value that you want to change.
4. Select how you would like the text to be recoded:

   In Place applies any change to the original data column.

   New Column creates a new column for the changed data and retains the original column.

   Formula Column creates a new column with the changes as a formula instead of values.

   Script creates a new script called Recode in the data table. You can apply this script to perform recoding in-place. If you add more recodes later, the script updates (assuming that you select Script as the destination). You can apply this script later, to new data, and copy it to other data tables, or run it from your own scripts.
5. Click **OK**.

Command Options

The red triangle menu contains options for the Recode window. The first three commands apply to all items in the data table list.

Convert to Titlecase converts to title case, that is, an initial uppercase character and subsequent lowercase characters.

Convert to Uppercase converts all items to uppercase.

Convert to Lowercase converts items to lowercase.

Often, when importing data tables, tab characters, space characters, and line separators are imported with the data table. These two commands are used to remove these characters.

Trim Whitespace removes leading and trailing whitespace characters. For example, if an extra space was imported before and after the name John, this command would delete the spaces.

Collapse Whitespace trims leading and trailing whitespace characters and removes duplicate interior whitespace. That is, if more than one whitespace character is present, the **Collapse Whitespace** command replaces the two spaces with one space.

Viewing Patterns of Missing Data

If your data table contains missing data, you might want to determine whether there is a pattern that the missing data creates. The pattern might help you make discoveries about your data.

Example of Viewing Patterns of Missing Data

1. Open the Missing Data Pattern.jmp sample data table.
2. Select **Tables > Missing Data Pattern.**

Figure 3.19 The Missing Data Pattern Window

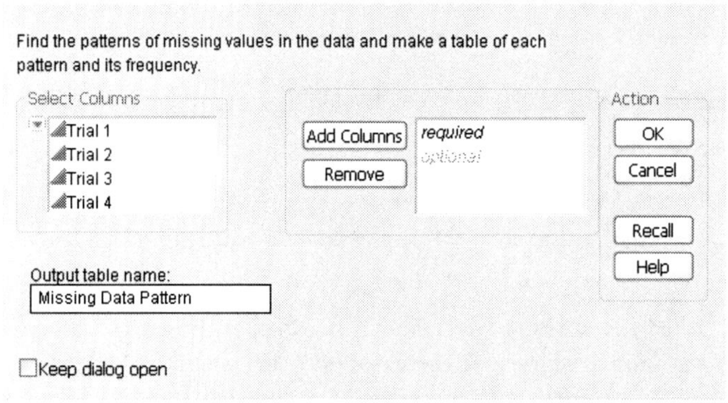

3. Highlight the columns from which you would like to find missing data. For this example, highlight all of the columns.

Note: For details about the options in the red triangle menu, see "Columns Filter Menu," p. 10 in the "Preliminaries" chapter.

---

4. Click **Add Columns**.
5. Click **OK**.

Figure 3.20 A Missing Data Pattern Table

	Count	Number of columns missing	Patterns	Trial 1	Trial 2	Trial 3	Trial 4
1	2	0	0000	0	0	0	0
2	2	1	0001	0	0	0	1
3	2	2	0011	0	0	1	1
4	1	3	0111	0	1	1	1

Figure 3.20 shows the following patterns:

- Row 1 shows that there are two instances where all rows in Trial 1, Trial 2, Trial 3, and Trial 4 have no missing values.
- Row 2 shows that there are two rows in the source table whose one missing value is in the Trial 4 column.
- Row 3 shows that there are two rows in the source table whose missing values are in the Trial 3 and Trial 4 columns.
- Row 4 shows that there is one row in the source table whose three missing values are in the Trial 2, Trial 3, and Trial 4 columns.

In the Missing Data Pattern table, JMP automatically assigns the **Count** column the analysis role of *frequency*. If you now use the Missing Data Pattern data table to run an analysis, JMP automatically uses Count as a frequency, so that you do not have to specify its role each time. See "Giving Columns a Preselected Analysis Role," p. 143, for details.

Finding and Replacing Cell Values

You can find and replace cell values with the **Edit > Search** options. The following rules apply to searching for values:

- To find values in hidden columns or locked tables, unhide the column or unlock the table.
- Values found in locked columns cannot be modified.
- The **Undo** command works only with **Replace**. You cannot undo **Replace All**.
- If your data table contains value labels, using the **Search** commands searches for actual values, but does not search for labels. (See "Using Value Labels," p. 151.)
- If your data table contains formatted values (such as dates, times, or durations) using the Search command searches for the formatted values, not the actual values.

To find or find and replace values:

1. Select **Edit > Search > Find**. The window in Figure 3.21 appears.
2. Enter a value in the **Find what** box.

Note: To find missing character values, leave the **Find** field empty and check **Match whole words only**. To find missing numeric values, insert a period into the **Find** field and check **Match whole words only.**

Figure 3.21 The Search Feature (Windows)

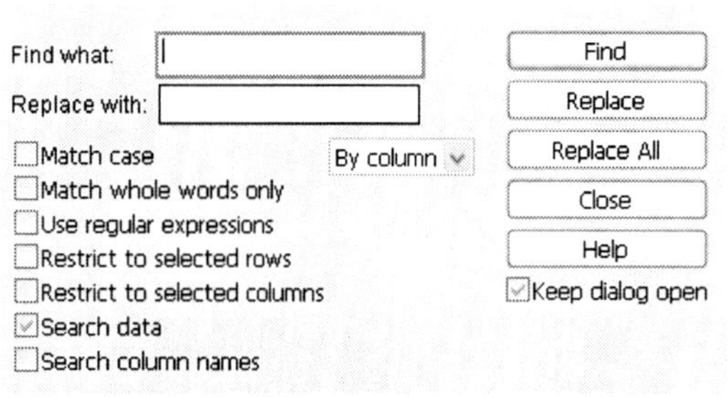

3. Type a value in the **Replace with** field, if you would like to replace the values found with a different value.
4. Refine your search. Select any of the following:

   Match Case Gives a case sensitive search, useful for locating proper nouns or other capitalized words.

   Match whole words only Detects empty spaces, which lets you search for a series of words in a character column, or locate strings with unwanted leading or trailing empty spaces. Using the **Match whole words only** check box also locates words with at least one leading and one trailing empty space.

   Use regular expressions assumes the find string to be a regular expression instead of the literal string that you enter in the Find What box. The regular expressions follow standard semantics.

   Restrict to selected rows restricts the search to selected rows.

   Restrict to selected columns restricts the search to selected columns.

   Search data searches only data cells (omitting column names).

   Search column names searches only column names (omitting data cells).

   By column searches the table column by column, from top to bottom, until it reaches the last cell in the rightmost column, or until you stop the search.

   By row searches the data table row by row from left to right, to the rightmost cell in the last row or until you stop the search.
5. Use the table below to start searching:

Table 3.6 Search Actions

Action	Instructions
Search the active table for the find value.	Click the **Find** button. The search begins with the first cell in the first column and searches every cell until it locates the value or reaches the end of the table.
Replace the currently highlighted cell value with the contents of the **Replace with** field, which contains the replace value.	Click the **Replace** button. Or, if the Search window is closed, select **Edit > Search > Replace.** If the replace value is a missing value, the currently highlighted cell content becomes a missing value.
Replace all occurrences of the specified value with another value.	Click the **Replace All** button. Or, if the Search window is closed, select **Edit > Search > Replace All.**
Automatically replace the value and search for the next value.	Click the **Replace** button. Or, if the Search window is closed, select **Edit > Search > Replace and Find Next.** Or press CTRL-L (Windows) or Command-L (Macintosh).
Use the selected value for **Find what.**	On Windows and Macintosh, go to the data table and click in a cell. Now that you have selected a value, select **Edit > Search > Use Selection for Find.** Next, select **Edit > Search > Find.** The value that you selected in the data table is already entered in the **Find what** field.
Use the selected value for **Replace with.**	On Windows, you can select **Edit > Search > Use Selection for Replace** to populate the **Replace with** field.
Find the next value in the table when the Find window is closed.	Select **Edit > Search > Find Next.** Or, press CTRL-G (Windows) or Command-G (Macintosh).
Find a missing value.	In the Search window, choose from these options: • To find missing character values, leave the **Find** field empty and check **Match whole words only.** • To find missing numeric values, type a period into the **Find what** text box and check **Match whole words only.** **Note:** Typing a period into the text box without clicking the **Match whole words only** box searches for a period.

Reordering Columns

You can rearrange, or sort data table columns by their name, data type, or modeling type, or you can reverse their current order. You can also move selected columns to a particular place in the data table. To reorder columns:

1. Select **Cols > Reorder Columns**.
2. Make a selection from the submenu options, which include the following:

Move Selected Columns Moves the selected columns to a particular place in the data table. When you select **Move Selected Columns**, the Move Selected Columns window appears.

Figure 3.22 Move Selected Columns Window

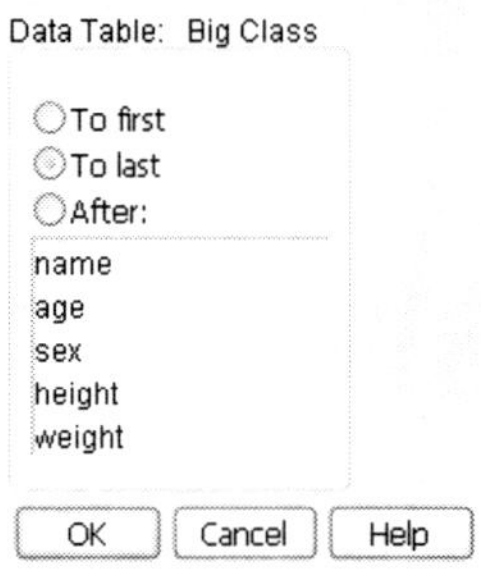

The Move Selected Columns window contains the following choices:

- **To first**: Moves the selected columns so they are in the left-most position in the data table.
- **To last**: Moves the selected columns so they are in the right-most position in the data table.
- **After**: Moves the selected columns so they are after a column that you identify.

Original Order Returns the columns to the order they were in at the time the data table was last saved.

Reorder by Name Arranges the columns from left to right in alphabetical order by column name.

Reorder By Data Type Arranges the columns from left to right in alphabetic order by data type (row state, character, numeric).

Reorder By Modeling Type Arranges the columns from left to right in alphabetic order by modeling type (continuous, ordinal, nominal). Row state columns have no modeling type, and are shown last.

Reverse Order Reverses the order of the data table columns.

Rows and Columns Context Menus

When you right-click in the row number area, or at the top of a column in the column name area, the context menus shown in Figure 3.23 appear. These menus give you quick access to selected **Rows** and

Columns menu commands. For details, see "The Rows Menu," p. 423 and "Context Menu for Columns," p. 54.

Note: The **Sort** command in the context menu for columns gives you a quick way to sort a data table by a single column in ascending values of that column. Sort twice to have the column sorted in descending values. The context menu **Sort** command does not display a window with sort options.

Figure 3.23 Context Menus for Rows and Columns

row context menu column context menu

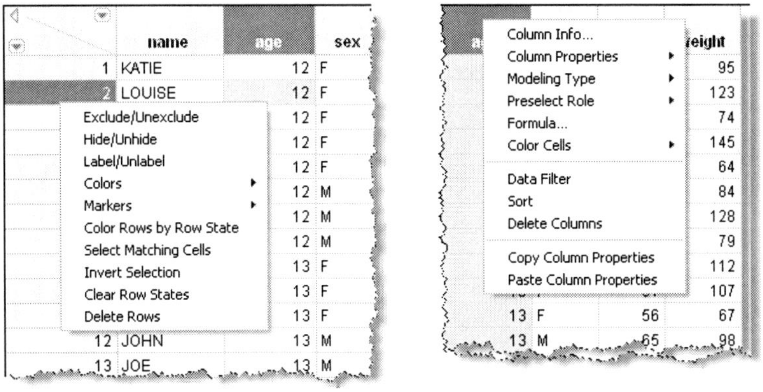

Copying, Cutting, and Pasting

You can copy or cut and paste in JMP data tables by selecting **Edit > Copy** or **Edit > Cut,** and then **Edit > Paste.** Data that you cut or copy to the clipboard can be pasted into data tables or other applications.

If one cell is selected, **Edit** commands apply only to that cell. If no columns are selected, **Edit** commands apply to all selected rows. Likewise, if no rows are selected, **Edit** commands apply to all selected columns. If you select both rows and columns, **Copy** copies the subset of cells defined by their intersection.

The **Copy** and **Paste** commands in the **Edit** menu function as listed in Table 3.7:

Table 3.7 Using Copy and Paste

Action	Instructions
Cut highlighted rows, columns, a single cell, or combination of cells from the active data table	Select **Edit > Cut.**

Table 3.7 Using Copy and Paste *(Continued)*

Action	Instructions
Copy highlighted rows, columns, a single cell, or combination of cells from the active data table to the clipboard	Select **Edit > Copy**. On the clipboard, fields are delimited by tab characters and a return character indicates an end of row.
Copy using the first line of data as column headings	Select **Edit > Copy With Column Names**.
Paste information from the clipboard to the highlighted area in a JMP data table	Select **Edit > Paste.** This command can be used with the **Copy** command to duplicate rows, columns, or any subset of cells defined by highlighted rows and columns.
Duplicate an entire row or column	Highlight the row or column to be duplicated and select **Edit > Copy**. Then highlight an existing row or column to receive the values and select **Edit > Paste**.
Duplicate a subset of values	Highlight the cells and select **Edit > Copy**. Then highlight an existing row or column to receive the values and select **Edit > Paste.** You must highlight the same arrangement of rows and columns to receive the copied values as originally contained them. • If you paste data with fewer rows into a destination with more rows, the source values recycle until all receiving rows are filled. • If you paste more rows or columns than you have highlighted, the excess values are lost unless you are pasting to the end of a data table. • If you highlight no rows or columns before you paste, rows and columns are added as needed.
Transfer data from another application into a JMP data table by copying and pasting	First copy the data to the clipboard from within the other application. Then select **Edit > Paste** to paste the values into JMP. Pasting automatically creates rows and columns as needed.
Paste using the first line of data as column headings	Select **Edit > Paste With Column Names**.

Moving and Duplicating Values

Dragging and dropping is a convenient way to move or duplicate values in a data table.

Figure 3.24 Moving Columns

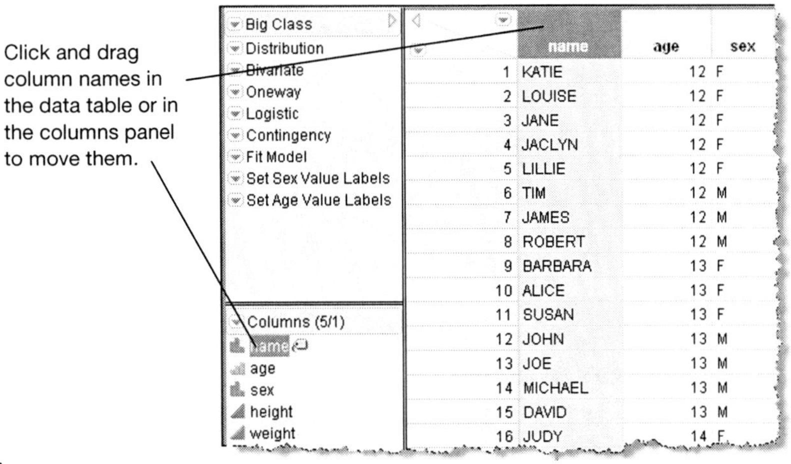

To move columns, cells, or rows:

Note: Follow this procedure to move columns or cells inside a data table or to another data table.

1. In the data table, select the columns, cells, or rows that you want to move. You can also select columns in the Columns panel.

   To select adjacent items, hold down the SHIFT key and click each item.

   To select non-adjacent items, hold down the CTRL key (or the COMMAND key on Macintosh) and click each item.
2. Release the mouse button.
3. Click the highlighted columns, cells, or rows, hold down the mouse button for a second, and then drag the items to the new position. If you do this with discontinuous columns, cells, or rows, they become contiguous when dropped. Note that when dragging and dropping *cells*:
   - They retain all of their characteristics and column properties.
   - After you move cells, missing values appear in the cells that you initially selected.
   - The selected cells and the destination cells must have the same data type.
   - If you drag a set of cells to an empty area of the table, new columns are automatically created.

Also note that another way to move *rows* is to:

1. Highlight the rows that you want to move.
2. Select **Rows > Move Rows**.
3. Specify where you would like to move the rows in the Move Rows window. Click the appropriate button to move highlighted rows to the beginning of the table (**At start**), to the end of the table (**At end**), or after a specific row number (**After row:**).

To duplicate columns, cells, or rows:

Note: Follow this procedure to copy columns or cells inside a data table or to another data table.

---

1. In the data table, select the columns, cells, or rows that you want to duplicate. You can also select columns in the Columns panel.

   To select adjacent items, hold down the SHIFT key and click each item.

   To select non-adjacent items, hold down the CTRL key (or the COMMAND key on Macintosh) and click each item.

2. Release the mouse button, and hold down the CTRL key (or COMMAND key on Macintosh).
3. Click the highlighted columns, cells, or rows, hold down the mouse button for a second, and then drag the items to the new position. If you do this with discontinuous columns, cells, or rows, they become contiguous when dropped. Note the following:
   - The new columns have the original columns' display format and their modeling types.
   - If you drag a set of cells to an empty area of the table, new columns are automatically created. Your values are copied to the destination, leaving the original values intact.

Using the Row Editor

The row editor lets you browse or edit cells one row at a time. You can access the row editor from an output report or from the data table. (See "Pasting Reports into Another Program," p. 185.) To use the row editor from the data table:

1. Highlight a row in the data table.
2. Select **Rows > Row Editor** or double-click in a row number area.

Note: If you are inside of a plot instead of a data table, right-click inside the plot and select **Row Editor** to access the row editor.

Figure 3.25 Row Editor

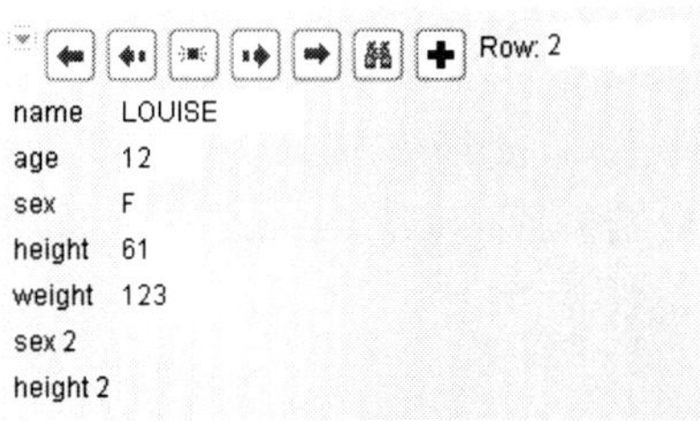

---

3. Edit the data. Note that if your data table contains value labels, the row editor displays the label, and when the cell is highlighted for editing, it shows the actual value. (See "Using Value Labels," p. 151.)

4. Click the arrow buttons to browse through selected rows or the entire data set if no rows are selected. Table 3.8 describes the function of each button.

Table 3.8 Row Editor Buttons

	Shows the previous row.
	Shows the previously selected row.
	Makes the row blink in graphs.
	Shows the next selected row.
	Shows the next row.
	Enables you to search for a row. See "Selecting Cells That Contain Specific Values," p. 91.
	Creates a new row at the end of the data table.

5. Click the red triangle icon in the row editor to select one of the following:

Next Selected Displays information for the selected row that is located after the current one.

Prev Selected Displays information for the selected row that is located before the current one.

Next Displays information for the row that is located after the current one, regardless of whether the row is selected.

Prev Displays information for the row that is located before the current one, regardless of whether the row is selected.

Save Saves the data table and any changes that you have made to it via the row editor.

New Row Creates a new row in the data table.

Find Displays the same window as if you had selected **Rows > Row Selection > Select Where**. Select one of the options on the **Action on currently selected rows menu**, and then highlight the column whose rows you want to select. Type in the value for which you want JMP to search. See "Selecting Cells with Specific Values," p. 90, for details.

Blink Causes the current row's highlight to flash at a rapid rate.

Note: Text in a locked column or a locked data table cannot be edited. See "Locking Columns," p. 147, and "Locking Tables," p. 80, for details.

Changing Table Names

A data table's name appears at the top of its window, in the table panel, and on all analysis reports. It is the name by which it has been or will be saved.

To change a table's name:

1. Click twice on the table name in the table panel.

2. Type the name that you want.

On Windows, you can also change a data table's name by selecting **Window > Set Title**.

Locking Tables

There are two ways you can lock a JMP data table:

- You can lock a table so that its values cannot be edited. However, you can still run analyses, assign characteristics, and so on. To do this, select **Lock Data Table** from the table panel menu (Figure 3.26).

Figure 3.26 Lock the Table so Values Cannot be Edited

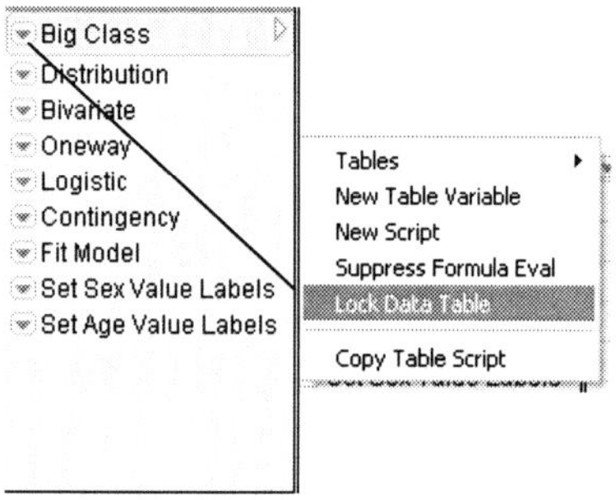

A lock icon appears next to the data table name. To unlock the file, select **Lock Data Table** again.

- You can lock a table so it can be edited but not saved. To do this, change the file's properties according to your computer's operating system. For example, change the file to be read-only. Then, when you open it in JMP, you see the words *Locked File* in the table panel underneath the table's name, as shown in Figure 3.27.

  To unlock the file, remove the read-only property according to your computer's operating system.

Figure 3.27 A Locked File

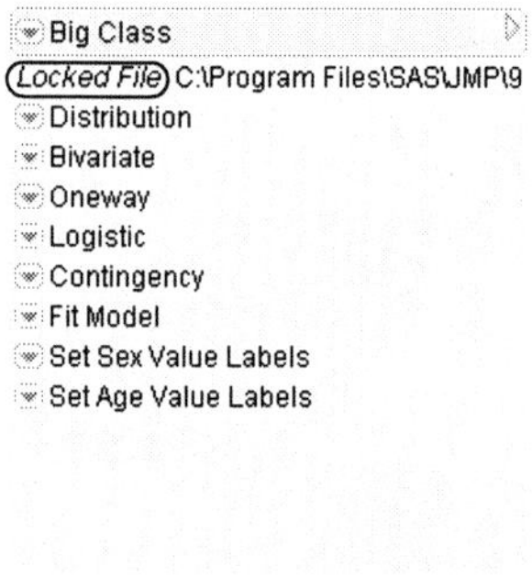

Note: You can lock a column in place so when you scroll horizontally, the column remains visible. See "Locking Columns in Place," p. 143, for details.

Adding Table Variables

Note: Starting in JMP 9, the Table Variable window is non-modal, meaning that you can access other windows and perform other operations while you are in the Table Variable window.

Table variables are character strings that are available to the entire table. They are any constant value that is always available in the data table. Table variable names are displayed in the table panel at the left of the data grid, as shown in Figure 3.28.

Figure 3.28 Table Variables are Located to the Left of the Data Grid

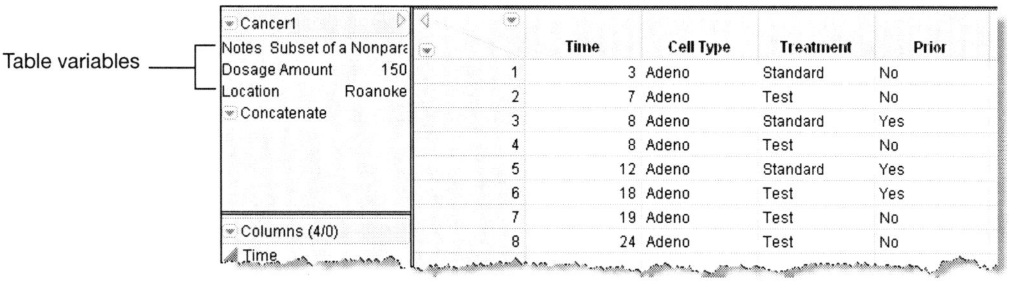

Table variables are mostly used to document tables. For example, when you installed JMP, a folder named **Sample Data** was also installed. Many files in the **Sample Data** folder contain a table variable called **Notes**, which contain descriptive information. The example in Figure 3.28 shows a data table that contains **Notes** as one of its table variables. JMP also automatically creates table variables when you create a design table with the Design of Experiments commands in JMP. The design table has a table variable named **Design** with the name of the design type as its value.

Table variables can also be incorporated in formulas that you build using the formula editor. These formulas calculate values for a column by referring to a table variable. See "Referencing Columns and Table Variables," p. 304, for details about constructing a formula that uses table variables.

Another use for table variables is to incorporate them into JSL scripts. See the *Scripting Guide* for details.

Adding New Table Variables

To add new table variables, proceed as follows:

1. In the tables panel, click the red triangle icon to the left of the data table name.
2. Select **New Table Variable**. The Table Variable window appears.
3. Give the variable a name and value in the boxes labeled **Name** and **Value**, as shown in Figure 3.29.

Figure 3.29 Naming and Defining the Table Variable

4. Click **OK**. The table variable appears in the tables panel.

Viewing or Editing Table Variables

1. Click twice on the content of an existing variable in the tables panel.
2. Edit its value.

Editing Table Variable Names

1. Double-click the table variable name. Or, right-click the table variable name and select **Edit**. The window in Figure 3.29 appears.
2. Edit the name.

Deleting Table Variables

To delete a table variable, right-click its name or value and select **Delete**.

Example of Using a Table Variable

Table variables can be especially useful when you need to combine two or more data tables. For example, suppose two cancer trials were conducted at two different hospitals. One of the trials' data is in a table called Cancer1.jmp, and the other is in Cancer2.jmp.

Consolidate the data and the variables into one table, as follows:

1. Open the Cancer1.jmp and Cancer2.jmp sample data tables.

   Notice that there are two distinct table variables: Dosage Amount and Location.
2. From the Cancer1.jmp data table, select **Tables > Concatenate**.
3. Select Cancer2 and click **Add**.
4. Click **OK**.

Figure 3.30 Data and Variables Concatenated

Cancer1.jmp

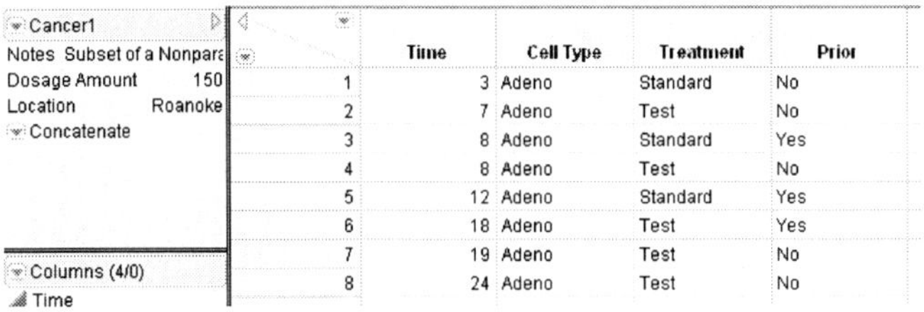
Cancer1
Notes Subset of a Nonpara
Dosage Amount 150
Location Roanoke
Concatenate
Columns (4/0)
Time

	Time	Cell Type	Treatment	Prior
1	3	Adeno	Standard	No
2	7	Adeno	Test	No
3	8	Adeno	Standard	Yes
4	8	Adeno	Test	No
5	12	Adeno	Standard	Yes
6	18	Adeno	Test	Yes
7	19	Adeno	Test	No
8	24	Adeno	Test	No

Cancer2.jmp

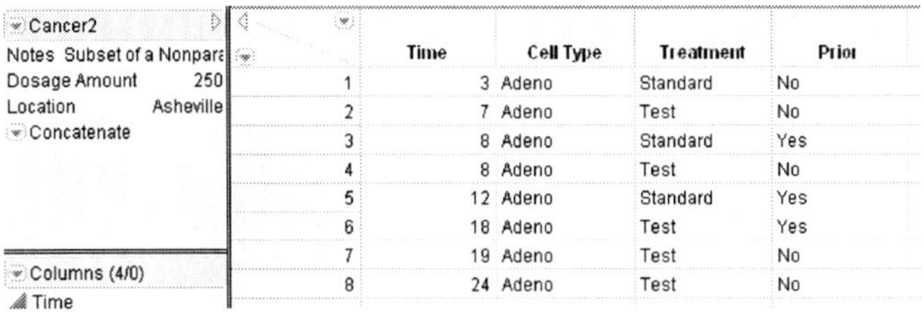
Cancer2
Notes Subset of a Nonpara
Dosage Amount 250
Location Asheville
Concatenate
Columns (4/0)
Time

	Time	Cell Type	Treatment	Prior
1	3	Adeno	Standard	No
2	7	Adeno	Test	No
3	8	Adeno	Standard	Yes
4	8	Adeno	Test	No
5	12	Adeno	Standard	Yes
6	18	Adeno	Test	Yes
7	19	Adeno	Test	No
8	24	Adeno	Test	No

Concatenated data table

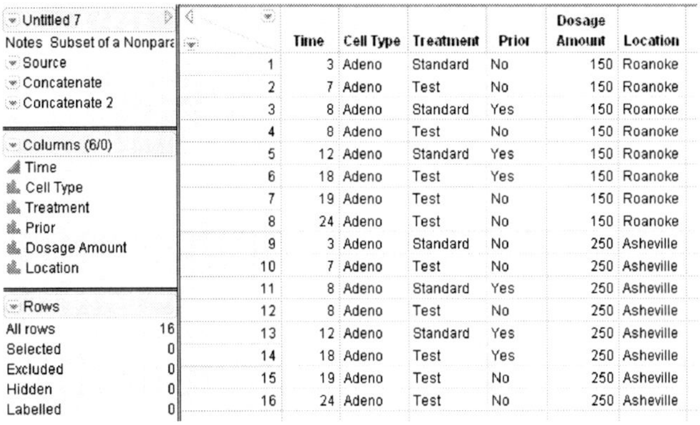
Untitled 7
Notes Subset of a Nonpara
Source
Concatenate
Concatenate 2
Columns (6/0)
Time
Cell Type
Treatment
Prior
Dosage Amount
Location
Rows
All rows 16
Selected 0
Excluded 0
Hidden 0
Labelled 0

	Time	Cell Type	Treatment	Prior	Dosage Amount	Location
1	3	Adeno	Standard	No	150	Roanoke
2	7	Adeno	Test	No	150	Roanoke
3	8	Adeno	Standard	Yes	150	Roanoke
4	8	Adeno	Test	No	150	Roanoke
5	12	Adeno	Standard	Yes	150	Roanoke
6	18	Adeno	Test	Yes	150	Roanoke
7	19	Adeno	Test	No	150	Roanoke
8	24	Adeno	Test	No	150	Roanoke
9	3	Adeno	Standard	No	250	Asheville
10	7	Adeno	Test	No	250	Asheville
11	8	Adeno	Standard	Yes	250	Asheville
12	8	Adeno	Test	No	250	Asheville
13	12	Adeno	Standard	Yes	250	Asheville
14	18	Adeno	Test	Yes	250	Asheville
15	19	Adeno	Test	No	250	Asheville
16	24	Adeno	Test	No	250	Asheville

The data and the variables are concatenated. The variables appear as columns in the concatenated table.

Note: Variables that are named as Notes are not concatenated. In this case, the notes attached to the main table, or the table from which you invoke the **Tables > Concatenate** command, are the notes that are maintained.

Creating Scripts

Note: Starting in JMP 9, the Script window is non-modal, meaning that you can access other windows and perform other operations while you are in the Script window.

You can create a JSL script to save with the data table (Figure 3.31). These scripts are used to automatically complete various tasks and analyses. For full explanations of scripts, see the *Scripting Guide*.

Figure 3.31 Scripts Saved With the Data Table

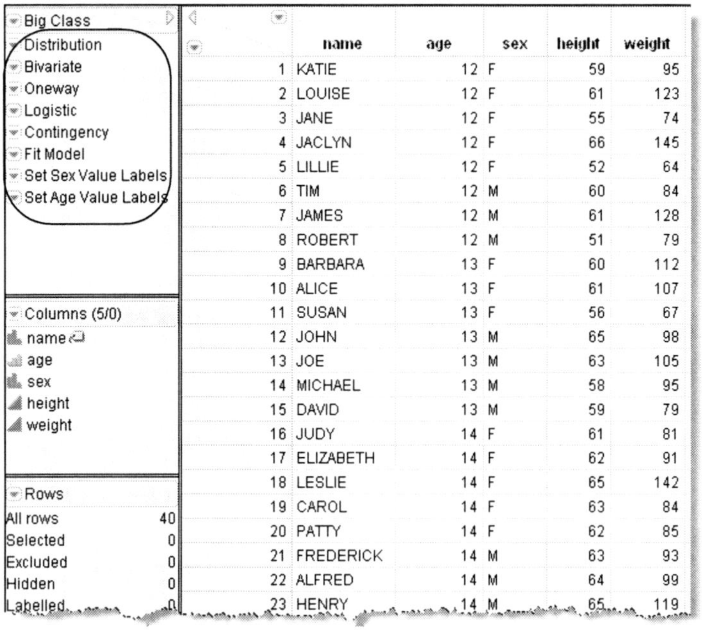

Adding Scripts to Data Tables

1. Click the red triangle icon to the left of the data table name in the tables panel. (See Figure 3.32.)
2. Select **New Script**. The Script window appears.

Figure 3.32 Creating a Script

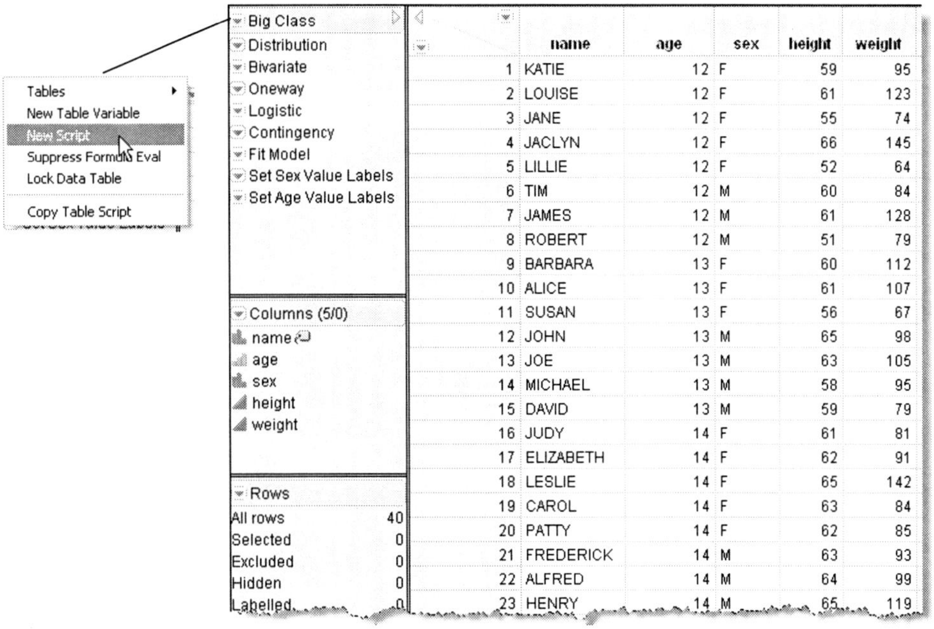

3. Give the script a name by typing it into the box beside **Name**.
4. Add the script by entering JSL code into the box beside **Script**.

Tip: If you want a particular script to run automatically every time the data table is opened, name the script **OnOpen**. Only one script saved in the data table can be set to run automatically.

5. Perform one of the following actions:
   - If you are finished editing the script, click **OK**. The script appears in the tables panel and the window closes.
   - If you want to run the script at any time, click **Run**.
   - If you are not finished editing the script and want to save it, click **Save**. The script appears in the tables panel and the window remains open for further editing.

Adding Scripts without Entering JSL Code

1. Run an analysis and create an output report using the steps that you would like the script to duplicate.
2. Click the red triangle icon beside the report title name, as shown in Figure 3.33.

Figure 3.33 Click the Red Triangle

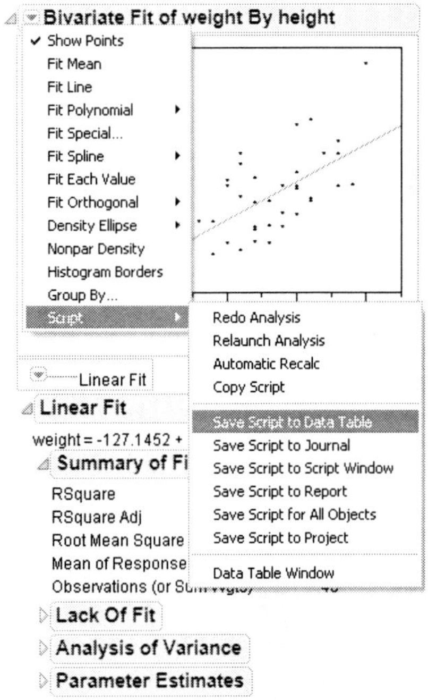

3. Select **Script > Save Script to Data Table.** The script is created and added to the tables panel.

Running Scripts

1. In the tables panel of the data table, click the red triangle icon beside the script's name, as shown in Figure 3.34. Alternatively, you can right-click on the script name.

Figure 3.34 Running a Script

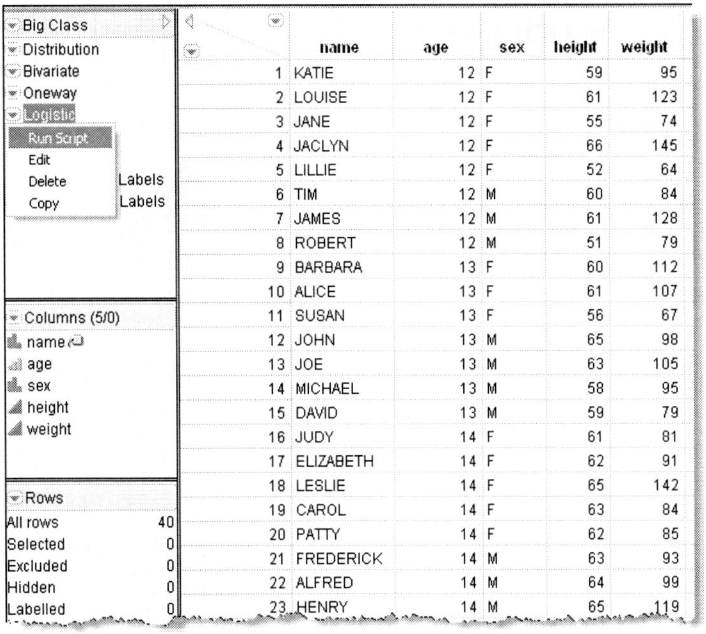

2. Select **Run Script**.

Viewing or Editing Scripts

1. In the tables panel of the data table, click the red triangle icon beside the script's name, as shown in Figure 3.34. Alternatively, you can right-click on the script name.
2. Select **Edit**.

Deleting Scripts

1. In the tables panel of the data table, click the red triangle icon beside the script's name, as shown in Figure 3.34. Alternatively, you can right-click on the script name.
2. Select **Delete**.

Copying Scripts

1. In the tables panel of the data table, click the red triangle icon beside the script's name, as shown in Figure 3.34. Alternatively, you can right-click on the script name.
2. Select **Copy**.

You can then paste the script into a script window or into the tables panel of another data table.

Selecting Rows and Columns

You can select rows, columns, or both rows and columns:

- To select rows, see the following:
  - Table 3.9.
  - "Selecting Excluded, Hidden, or Labeled Rows," p. 90
  - "Selecting Cells with Specific Values," p. 90
  - "Selecting a Particular Row or Column," p. 93
  - "Randomly Selecting Rows," p. 93
  - "Inversely Selecting and Selecting All Rows," p. 93
  - "Locating Next and Previously Selected Rows," p. 94
- To select columns, see the following:
  - Table 3.10
  - "Selecting a Particular Row or Column," p. 93
- You can also use the keyboard to select columns and rows, and to edit a value. See "Using the Keyboard to Select Rows and Columns," p. 95.

Table 3.9 Selecting Rows

Action	Instructions
Select an entire row	Click the empty space that contains the row number.
Select multiple rows	• To select, click and drag the cursor over the row number in the data grid. • Hold down the SHIFT key and click the first and last rows of the desired range. • To make a discontiguous selection, hold down the CTRL key and click on each row.
Select or deselect all rows	• Select **Rows > Row Selection > Select All Rows.** or • Select **Rows > Clear Row States.** or • Hold down the SHIFT key and click the lower triangular area in the upper left corner of the data grid to select. Click again to deselect all rows. To clear all highlights in the data table, press the ESC key on your keyboard.
Select a block of cells formed by the intersection of rows and columns	Drag the arrow cursor diagonally across the cells.

Table 3.9 Selecting Rows *(Continued)*

Action	Instructions
Select a certain row number	Select **Rows > Row Selection > Go to Row** and type in the desired row number.
Invert the row selection	Select **Rows > Row Selection > Invert Row Selection**. See "Inversely Selecting and Selecting All Rows," p. 93 for details.
Select random rows	Select **Rows > Row Selection > Select Randomly**. See "Randomly Selecting Rows," p. 93, for details.
Select the Pareto Frontier	Select **Rows > Row Selection > Select Dominant**. A window appears that lets you choose one or more columns. Then a second window appears that lets you select high or low values for the Pareto frontier selection for each column separately. In the column selection window, hold the CONTROL key to select multiple columns. In the Select dominant high values window, select the check box for any column to search for high values, and clear the check box for any column to search for low values.
Save the current selection	Select **Rows > Row Selection > Name Selection in Column.** The current selection is saved in a new column.

Table 3.10 Selecting Columns

Action	Instructions
Select a column	• Click the empty space around the column name. • Click the column name in the column panel to the left of the data grid.
Select multiple columns	• To select, click and drag the cursor over the column name in the data grid. • Hold down the SHIFT key and click the first and last columns of the desired range. • To make a discontiguous selection, hold down the CTRL key and click the desired columns.
Extend a selection of columns	• Hold down the SHIFT key and click the first and last column of the desired range. • To make a discontiguous selection, hold down the CTRL key and click the desired columns.

Table 3.10 Selecting Columns *(Continued)*

Action	Instructions
To select or deselect all columns at once	Hold down the SHIFT key and click the upper triangular area in the upper left corner of the data grid to select. Click in the upper left corner to deselect all columns. To clear all highlights in the data table, press the ESC key on your keyboard.
Make a discontiguous selection	Hold down the CTRL key and click on each column.
Select a block of cells formed by the intersection of rows and columns	Drag the arrow cursor diagonally across the cells.

Selecting Excluded, Hidden, or Labeled Rows

Sometimes you need to automatically highlight or select certain types of rows so that you can see or manipulate them among the many rows of a data table. To select rows that have been marked as excluded, hidden, or labeled:

1. Select **Rows > Row Selection.**
2. Select one of the following:

   Select Excluded Selects all excluded rows regardless of their current selection status and deselects any other previously selected rows.

   Select Hidden Selects all hidden rows regardless of their current selection status and deselects any other previously selected rows.

   Select Labeled Selects all labeled rows regardless of their current selection status and deselects any other previously selected rows.

Selecting Cells with Specific Values

If you are looking for a specific value in a data table, there are several ways to quickly select it, as explained in the sections below.

Selecting Cells That Match the Currently Highlighted Cell

JMP can find all cells whose values are the same as the one(s) you currently have highlighted. You can do this within one data table or throughout all open data tables. To select cells that contain the same values as highlighted ones:

1. Highlight the cells that contain the value(s) that you want to locate.

2. To find all matching cells within the active data table, select **Rows > Row Selection > Select Matching Cells.** Or, right-click (hold the CTRL key and click on Macintosh) one of the highlighted row numbers and select **Select Matching Cells.**

   To find all matching cells across all open data tables, select **Select All Matching Cells.**
3. The rows that contain the same values as the selected ones are highlighted.

Example of Selecting Cells That Match the Currently Highlighted Cell

1. Open the Big Class.jmp and New Heights.jmp sample data tables.

   Suppose that you want to select all of the cells in the height column that contain the value 61.
2. In the Big Class.jmp data table, select the first instance of 61 (Louise).

   Find the cells that have the value 61 in both data tables.
3. Select **Rows > Row Selection > Select All Matching Cells.**

Figure 3.35 Select Rows Using the **Select Matching Cells** Window

Big Class.jmp

	name	age	sex	height	weight
1	KATIE	12	F	59	95
2	LOUISE	12	F	61	123
3	JANE	12	F	55	74
4	JACLYN	12	F	66	145
5	LILLIE	12	F	52	64
6	TIM	12	M	60	84
7	JAMES	12	M	61	128
8	ROBERT	12	M	51	79
9	BARBARA	13	F	60	112
10	ALICE	13	F	61	107
11	SUSAN	13	F	56	67
12	JOHN	13	M	65	98
13	JOE	13	M	63	105
14	MICHAEL	13	M	58	95
15	DAVID	13	M	59	79
16	JUDY	14	F	61	81
17	ELIZABETH	14	F	62	91

New Heights.jmp

	name	height
1	KATIE	62
2	LOUISE	61
3	JANE	58
4	JACLYN	68
5	LILLIE	52
6	TIM	64
7	JAMES	63
8	ROBERT	56
9	BARBARA	62
10	ALICE	61
11	SUSAN	58
12	JOHN	68
13	JOE	64
14	MICHAEL	60
15	DAVID	61
16	JUDY	62

Selecting Cells That Contain Specific Values

JMP can search for a specific value (or text string) and highlight all of the cells in the data table that contain the specific value.

To select cells that contain specific values:

1. Select **Rows > Row Selection > Select Where.** The window in Figure 3.36 appears.

Figure 3.36 Select Rows Using the **Select Where** Window

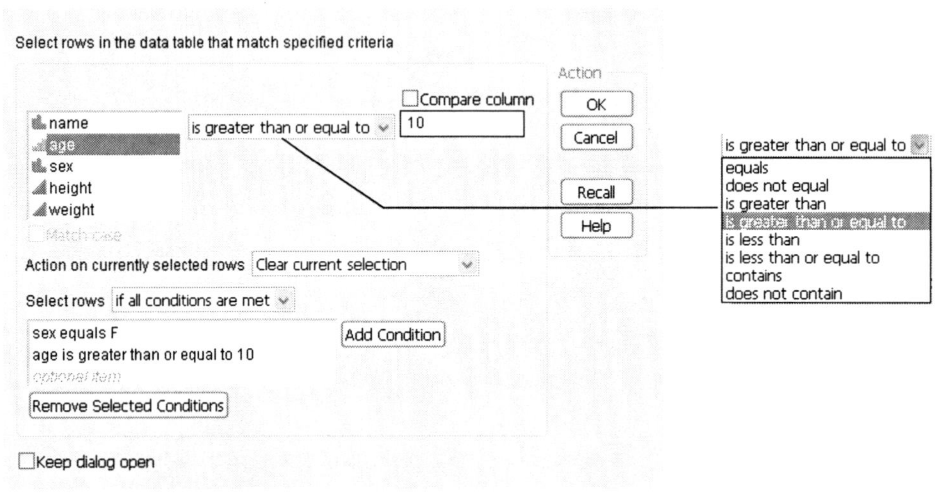

2. From the list, highlight the name of the column whose rows you want to select.
3. Use the drop-down menu to select a condition from the list (equals, does not equal, and so on), as shown in Figure 3.36.
4. Type the search value. To search for missing values, leave the box empty.
5. (Optional) If you would like to compare the values of two columns, click the **Compare column** check box. Select from the list of columns for comparison.
6. (Optional) If you would like the search to be case-sensitive, click the box beside **Match Case.**
7. If you currently have rows selected in the data table, click an option under **Action on currently selected rows** to tell JMP how to handle that current selection:

   Clear Current Selection Removes the highlight from currently selected rows and selects all rows that contain the specified value.

   Extend Current Selection Keeps the currently selected rows selected and also selects the rows in which the specified value has been found.

   Select From Current Selection Selects the rows in the currently selected array that contain the specified values.
8. Click **Add Condition**. The condition is added to the list.
9. (Optional) To add more conditions to the search, repeat the previous steps. Click the appropriate item in the **Select Rows** area to specify if you would like JMP to select rows only if all conditions are met, or if any of the conditions are met.
10. If you want the window to stay open after the selection, select **Keep dialog open**.
11. Click **OK**.

Using the Search and Find Commands

You can also select cells by selecting **Edit > Search.** To find (or find and replace) values:

1. Select **Edit > Search > Find.**
2. Enter a value in the **Find what** box. To find missing character values, leave the **Find** field empty and check **Match Whole Words Only.** To find missing numeric values, insert a period into the **Find** field and check **Match whole words only.** See "Finding and Replacing Cell Values," p. 71, for details.

Selecting a Particular Row or Column

If you know the number of the row that you want to highlight, instead of scrolling through many rows to find it, you can quickly find a row number using the **Go to Row** command.

To find a row:

1. Select **Rows > Row Selection > Go to Row.**
2. Enter the row number and click **OK.**

To find a column:

1. Select **Cols > Go to.**
2. Enter the column number or name and click **OK.**

Randomly Selecting Rows

You can have JMP randomly select a specific number of rows or a percentage of the total rows:

1. Select **Rows > Row Selection > Select Randomly.**
2. To randomly select a specific number of rows, enter the absolute sample size in the text edit box. Or, to randomly select a percentage of the total rows, enter the proportional sample size.

   Note the following:

   - If you enter a number that is less than one, JMP assumes that the number is a proportion and that you want to randomly select that percentage of rows.
   - If you enter a number that is higher than one, JMP assumes that the number that you entered is the number of rows that you want randomly selected.

   For example, entering 10 randomly selects 10 rows. Entering 0.1 randomly selects 10% of the rows.

Inversely Selecting and Selecting All Rows

To select all of the rows in a JMP data table, select **Rows > Row Selection > Select All Rows.**

To deselect all selected rows and select all previously deselected rows, select **Rows > Row Selection > Invert Row Selection.**

Locating Next and Previously Selected Rows

You can locate the next selected row after the current row and cause it to flash by selecting **Rows > Next Selected**. Similarly, you can locate the previously selected row before the current row and cause it to flash by selecting **Rows > Previous Selected.**

Each time you select **Rows > Next Selected** or **Rows > Previous Selected**, the next or previously selected row is found and flashes. A beep signals when the last selected row is located.

Example of Locating Next and Previously Selected Rows

For example, you might want to use this feature when you are dealing with a large amount of data, and you have selected rows intermittently, and want to look through the selected rows in the data table.

1. Open the Diamonds Data.jmp sample data table.
2. Select **Analyze > Fit Y by X.**
3. Select Carat Weight and click **Y, Response**.
4. Select Price and click **X, Factor**.
5. Click **OK.**
6. Select **Tools > Lasso.** If you cannot see the menu bar, hover over the blue bar below the title bar to reveal it.
7. Lasso some of the points near the 10,000 dollar price at the bottom and the top of the plot. See Figure 3.37.

Figure 3.37 Points Selected

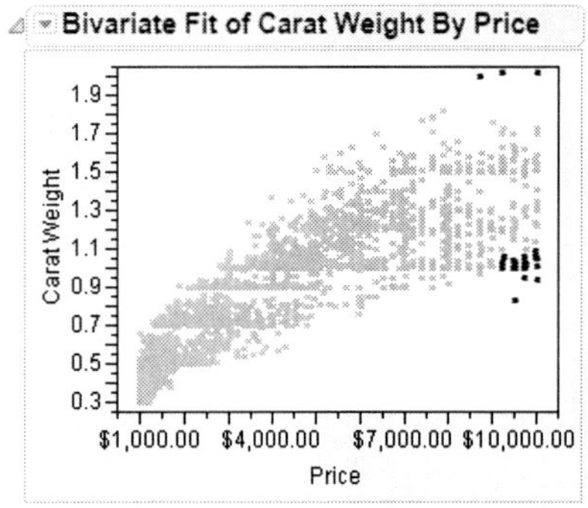

8. In the data table, select **Rows > Next Selected.** Or you can press the F7 key.

You can easily navigate through the selected rows to see the data for each.

Using the Keyboard to Select Rows and Columns

Table 3.11 describes the keyboard selections that you use to navigate, select rows and columns, and edit a value.

Table 3.11 Keyboard Selections and Actions

Action	Keyboard Selection
Enter the data grid from a panel	If you are currently in a panel, and you want to move into the data grid, press the TAB key until the data grid is selected. A dark line then appears above the first row of the grid.
Select a row	Once you are in the data grid, use the up and down arrows to move down and up between rows.
Select a column	Once you have a column selected, use the left and right arrows to move back and forth between columns.
Select multiple rows	Hold down the SHIFT key and press the up or down arrows.
Select multiple columns	Hold down the SHIFT key and press the left or right arrows.
Edit a value	Select a single column and a single row, highlighting the value that you want to edit. Press the Enter key. or Select a single column. Use the up and down arrows to move between values.
Edit a column name	Select a single column. Press the Enter key.
Deselect all selections	Press the ESC key.

The Data Filter

The Data Filter gives you a variety of ways to identify subsets of data. Using **Data Filter** commands and options, you interactively select complex subsets of data, hide these subsets in plots, or exclude them from analyses.

When you select **Rows** > **Data Filter**, the initial Data Filter window appears. As you work with the data table, this window is attached to the table. You can change this preference by selecting **File > Preferences > Tables** and deselecting **Use a Floating Window for Data Filters**.

Figure 3.38 Initial Data Filter Window

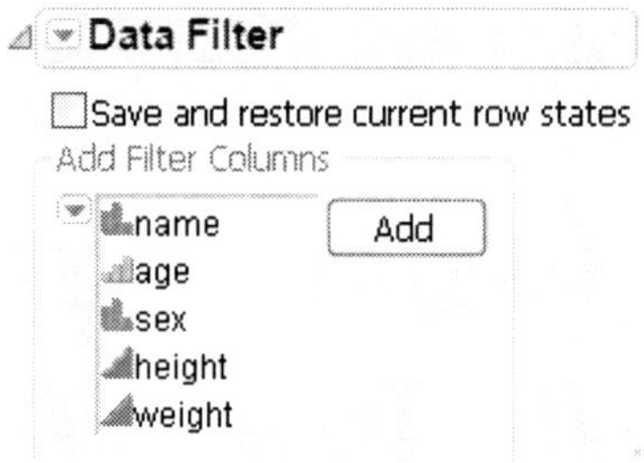

Table 3.12 Description of the Initial Data Filter Window

Data Filter red triangle menu	The red triangle menu next to Data Filter contains options such as **Animation** and **Make Subset**. For a description of all options, see "Data Filter Options," p. 102.
Add Filter Columns red triangle menu	For details about the options in the red triangle menu, see "Columns Filter Menu," p. 10 in the "Preliminaries" chapter.
Save and restore current row states	This option restores your current row states when the Data Filter window is closed.
Add Filter Columns panel	Lists all included variables in the current data table. You select columns in this list whose values you want to use as filters that identify subsets of data. You can re-open the Add Filter Columns panel and add more filter variables to the process at any time.
Add	Adds the selected columns to the Data Filter.

The Data Filter Control Panel

Once you have selected columns in the initial window and clicked **Add**, the Data Filter control panel appears.

Figure 3.39 Data Filter Control Panel

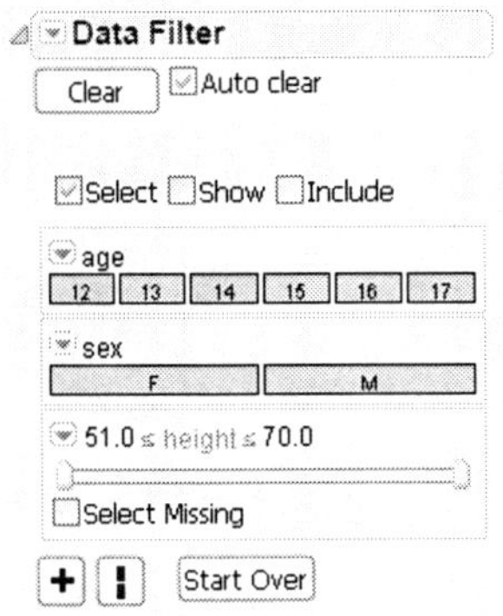

Table 3.13 Description of the Data Filter Control Panel

Red triangle menu for the Data Filter	The red triangle menu next to Data Filter contains options such as **Animation** and **Make Subset**. See "Data Filter Options," p. 102.
Clear	Clears all selections that you have made on variables in the Data Filter window.
Auto clear	Enables you to select values from one variable at a time. If you want to select values from multiple variables at the same time, deselect the **Auto clear** check box.
Select	Highlights the rows (or observations) in the data table that correspond to the selections that you make in the Data Filter. See "Changing the Row State in the Data Table After Making Data Filter Selections," p. 100.
Show	Shows the rows that you identified using the Data Filter in plots and charts. All other rows have the hidden row state icon assigned to them. The hidden row state icon ( ) appears for those rows in the data table. For details about row states, see the "Properties and Characteristics of Data" chapter.
Include	Includes rows that you identified using the Data Filter in statistical analyses. All other rows have the excluded row state assigned to them. The excluded row state icon ( )appears next to the row number in the data table for these rows. For details about row states, see the "Properties and Characteristics of Data" chapter.

Table 3.13 Description of the Data Filter Control Panel *(Continued)*

Red triangle menu for a nominal or ordinal variable	Contains the following options: **Delete** removes the variable from the Data Filter control panel. **Display Options** which include the following: **Blocks Display** shows each level as a block. **List Display** shows each level as a member of a list, followed by its frequency. **Single Category Display** shows each level, followed by its frequency, in a menu. **Order by Count** orders the variable values by frequency of occurrence whether the variable is shown as a bar of values or in list form. **Clear Selection** clears any selection in effect for that variable only. **Invert Selection** deselects any selected values, and selects all values previously not selected, for that variable only.
Red triangle menu for a continuous variable	Contains the following options: **Delete** removes the variable from the Data Filter control panel. **Clear Selection** clears any selection in effect for that variable only. **Invert Selection** deselects any selected values, and selects all values previously not selected, for that variable only.
Nominal or ordinal variables	Values appear in blocks, in a list, or in a menu. If the variable contains only a small number of categories, the values appear in blocks. If the variable contains a large number of categories, the values appear in a list or in a menu.
Continuous variables	Values appear in a range with a slider that you can adjust in one of the following ways: • Click and drag the slider bar. You can drag from either end of the slider bar. The selected values appear above the slider bar. • Click anywhere in the empty (not selected) part of the slider to set the filter range at that point. • Click on the number to type in the value that you want. By default, the range of values includes an equal sign, which includes the endpoints. You can remove the equal sign by holding down the SHIFT key and clicking on the ≤ or the ≥. The **Select Missing** option highlights any missing values in the data table.
Plus button	The plus button (+) at the bottom of the panel opens the Add Filter Columns list. This option uses an *and* operator, which restricts the selection. You can add variables to the filter process at any time.

Table 3.13 Description of the Data Filter Control Panel *(Continued)*

Dashed bar	The dashed bar ( ) at the bottom of the panel opens the OR Add Filter Columns list. This option uses an *or* operator, which extends the selection. You can add variables to the filter process at any time.
Start Over	Removes all of the filter columns.

Data Filter Context Menus

The Data Filter contains three different right-click menus. Figure 3.40 shows the right-click areas.

Figure 3.40 Right-Click Areas

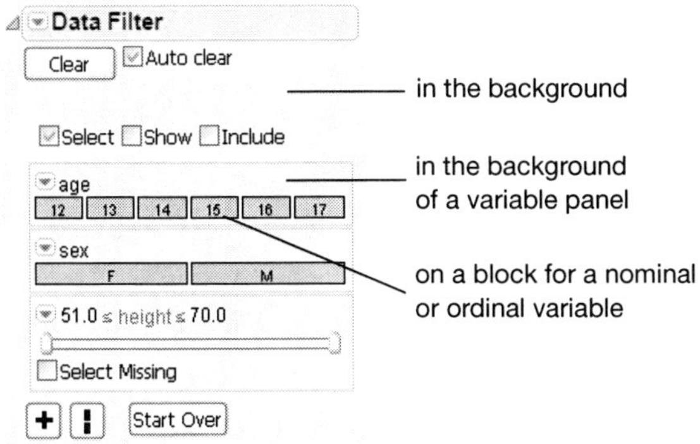

Context Menu for the Background

The options that appear when you right-click in the background are the same as the options in the red triangle menu next to Data Filter. See "Data Filter Options," p. 102.

Context Menu for the Background of a Variable Panel

The options that appear when you right-click in the background of a variable panel are the same as the options in the red triangle menu for a nominal/ordinal or continuous variable. See "The Data Filter Control Panel," p. 96.

Context Menu for a Block

Table 3.14 describes the available options when you right-click in a block for a nominal or ordinal variable. This option is not available if the values are in a list.

Table 3.14 Description of Block Options

Color	Changes the background color of the block.
Cell Labeling	**Unlabeled** removes the label from the block. **Show Counts** shows the count of each group within the variable. **Show Percents** shows the percentage of each group within the variable. **Show Labels** shows the value of each group within the variable.

Example of Changing Cell Labeling

In the following example, change the values for the age variable to show counts (of age groups) instead of labels (of ages).

1. Open the Big Class.jmp sample data table.
2. Select **Rows > Data Filter**.
3. Select age, sex, and height and click **Add**.
4. Right-click on any of the values in the age variable and select **Cell Labeling > Show Counts**.

Figure 3.41 Changing Categories from Labels to Counts

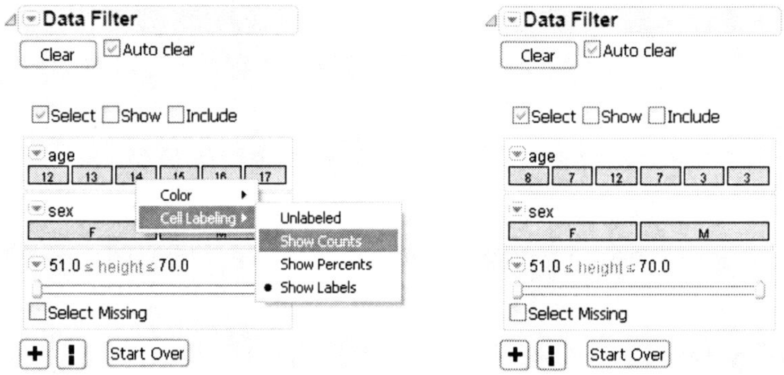

Counts of each age group appear instead of labels. For example, there are eight twelve-year olds, seven thirteen-year olds, twelve fourteen-year olds, and so on.

Changing the Row State in the Data Table After Making Data Filter Selections

If you modify a row state that you set in the Data Filter and subsequently alter row states in the data table, or select points in a graph or a plot, the selections in the Data Filter might not match the selections in the data table. The Data Filter shows a warning message that says: Your selection was changed in another window. The **Reset Selection** button appears. Clicking the **Reset Selection** button changes the data table selections back to reflect the selections in the Data Filter.

Example of Modifying Selections

1. Open the Big Class.jmp sample data table.
2. Select **Analyze > Distribution.**
3. Select sex and age and click **Y, Columns.**
4. Click **OK.**
5. Go back to the Big Class.jmp data table.
6. Select **Rows > Data Filter.**
7. Select sex and click **Add.**

In the Data Filter control panel, select all of the males.

8. Click on the M block.

Figure 3.42 Rows Containing Males Highlighted in Data Table and Histograms

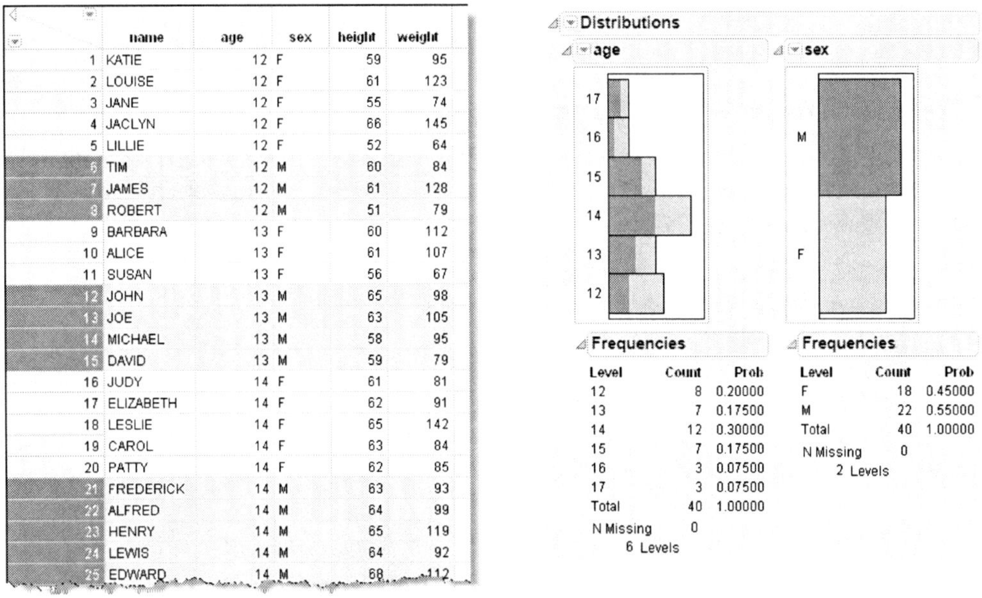

	name	age	sex	height	weight
1	KATIE	12	F	59	95
2	LOUISE	12	F	61	123
3	JANE	12	F	55	74
4	JACLYN	12	F	66	145
5	LILLIE	12	F	52	64
6	TIM	12	M	60	84
7	JAMES	12	M	61	128
8	ROBERT	12	M	51	79
9	BARBARA	13	F	60	112
10	ALICE	13	F	61	107
11	SUSAN	13	F	56	67
12	JOHN	13	M	65	98
13	JOE	13	M	63	105
14	MICHAEL	13	M	58	95
15	DAVID	13	M	59	79
16	JUDY	14	F	61	81
17	ELIZABETH	14	F	62	91
18	LESLIE	14	F	65	142
19	CAROL	14	F	63	84
20	PATTY	14	F	62	85
21	FREDERICK	14	M	63	93
22	ALFRED	14	M	64	99
23	HENRY	14	M	65	119
24	LEWIS	14	M	64	92
25	EDWARD	14	M	68	112

Frequencies (age)

Level	Count	Prob
12	8	0.20000
13	7	0.17500
14	12	0.30000
15	7	0.17500
16	3	0.07500
17	3	0.07500
Total	40	1.00000
N Missing	0	
6 Levels		

Frequencies (sex)

Level	Count	Prob
F	18	0.45000
M	22	0.55000
Total	40	1.00000
N Missing	0	
2 Levels		

You can see that all of the rows containing males are highlighted in the data table and in the histograms. Now, you decide that you only want to see the students who are age 12.

5) In the age histogram, select the bar representing age 12.

Now the selection does not match the Data Filter selection. A warning message and a **Reset selection** button appear in the Data Filter window.

Figure 3.43 Data Filter Warning Message and Reset Button

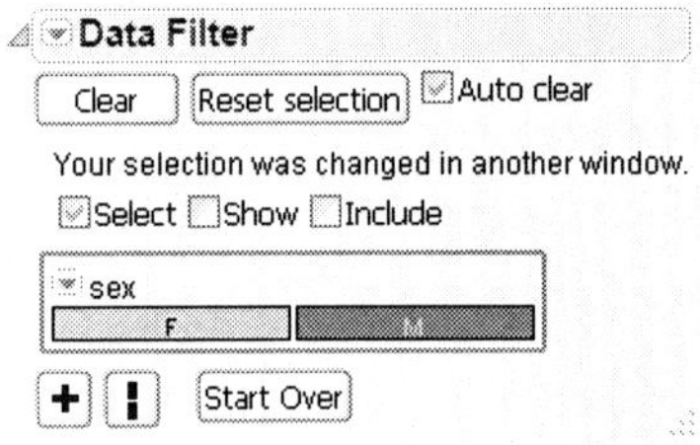

Data Filter Options

The red triangle menu next to Data Filter contains options such as **Animation**, **Make Subset**, and so on.

Animation

The animation feature sequentially highlights the values of a single variable. The variable's values highlight in the data table. However, patterns are more interesting if you first create a plot and then animate a variable using the Data Filter to see how it behaves on the plot.

To use the animation feature, from the red triangle menu next to Data Filter, select **Animation**. Then select the variable that you want to animate. The highlighted frame around the variable indicates which variable is selected for animation.

Figure 3.44 Animation Control Panel in the Data Filter

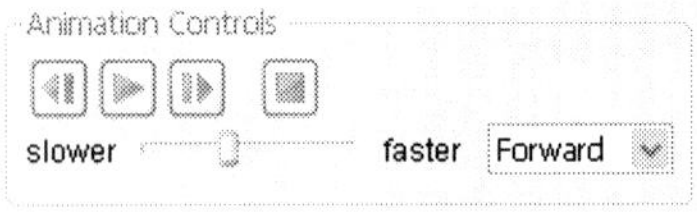

The Animation Control panel (Figure 3.44) has these controls:

- The middle button ( ) starts and stops the animation. After you start the animation cycles, the button changes to a stop button ( ). By default the animation begins with the first value of the topmost variable.
- The backward arrow ( ) moves the animation backward one cycle. Click more than once to go backward more than one cycle.
- The forward arrow ( ) moves the animation forward one cycle. Click more than once to go forward more than one cycle.
- The square button ( ) hides the Animation Control section on the Data Filter Window. Select **Animation** from the menu on the Data Filter title bar again to see the Animation Control.

The Animate mode drop-down menu has these options.

Forward highlight values forward from first to last.

Backward highlight values from last to first.

Bounce highlight forward and then backward repeatedly.

Use the slider to adjust the speed of the animation.

Make Subset

The **Make Subset** option is similar to the **Tables** > **Subset** command, only without subsetting options. When you choose **Make Subset**, JMP creates a new data table that contains only the following:

- The rows identified by the Data Filter.
- The columns selected in the active data table. If no columns are selected, then all columns are included.

Save WHERE Clause

Once you have filtered variable values in the Data Filter, that information can be expressed as a JMP WHERE clause. The WHERE clause is used in JSL (JMP Scripting Language) programs to identify specific rows of data for processing or analysis. The Data Filter builds a WHERE clause based on the value selections that you make.

The options in the **Save Where Clause** menu include the following:

to Clipboard creates a WHERE clause from the filter criteria and puts it on the clipboard.

to Row State Column creates a row state column in the data table that has a formula equivalent to the filter criteria.

to Data Table creates a WHERE clause from the filter criteria and saves it as a JSL command with the current data table in a table property called Filter.

to Script Window creates a WHERE clause from the filter criteria and appends it to the current script text window, or creates a new script if one does not exist already.

to Journal creates the WHERE clause from the filter criteria and appends it to the current journal, or creates a new journal if one does not already exist.

Example of Saving a WHERE Clause

1. Open the Big Class.jmp sample data table.
2. Select **Rows > Data Filter**.
3. Select age, sex, and height and click **Add**.
4. Select all females who are twelve and fourteen years old and whose height is between 56 and 60 inches.
5. Hold down the CTRL key and click on the 12 and 14 blocks and the F block.
6. Click on 51 and type 56.
7. Click on 70 and type 60.
8. From the red triangle menu next to Data Filter, select **Save Where Clause > to Script Window**.

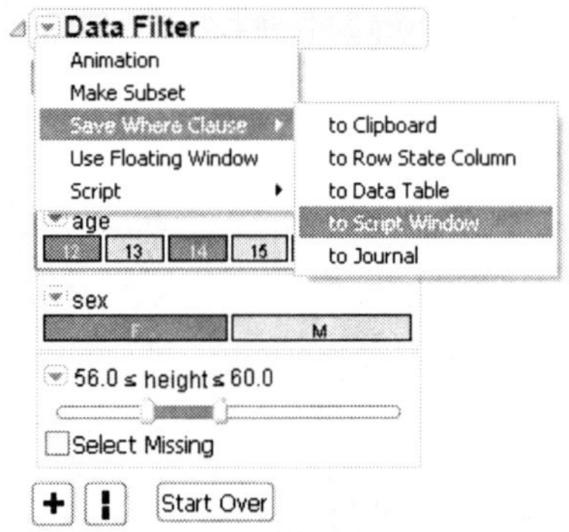

Figure 3.45 Save WHERE Clause

The WHERE clause that is created from this example appears in a script window, as follows:

```
Select Where(
(:age == 12 | :age == 14) & :sex == "F" & (:height >= 56 & :height <= 60)
);
```

Use Floating Window

If you want the Data Filter window to remain on top of its associated data table, select the **Use Floating Window** option. If you do not want the Data Filter window to remain on top, deselect this option.

Script

The **Script** menu provides options for saving scripts. For details, see "Saving Your Steps as a Script," p. 187.

Chapter 4

Saving Tables, Reports, and Sessions

Different Saving Methods

There are several methods for saving your analyses in JMP. The method that you choose depends on how you want to interact with your analyses in the future.

This chapter describes how to save data tables and reports in JMP.

Contents

Saving Data Tables

JMP saves data tables in the formats listed in Table 4.1.

Here are the basic steps for saving a data table:

- On Windows, select **File > Save As** to save in multiple formats.
- On Macintosh, select **File > Save As** to save as a JMP file (.jmp). Select **File > Export** to save in Text (.dat), Microsoft Excel (.xls), and SAS Transport (.xpt) formats.

In addition to these file types, JMP can save files for types that have a corresponding ODBC driver. See "Saving Data Tables to a Database," p. 111, for details. Note that the Excel ODBC driver is required for exporting to Excel.

Table 4.1 File Types and Extensions

File Type(s)	Extension	OS Availability
JMP Data Table	.jmp	All
SAS Transport File	.xpt	All
JMP Report	.jrp	All
JMP Journal	.jrn	All
Excel File **Note**: Due to a limitation of the Open Database Connectivity (ODBC) driver, each cell must contain no more than 255 characters. Additional characters do not show up in the data table. To prevent this issue, divide the 255 characters into two or more cells.	.xls	All
Text File	.txt .dat	All
SAS Version 6 and 7 Data Sets	.sd2 and .sas7bdat	Windows
MS Access Database, CBA_EH_DB, CBA_TL_DB	.mdb	Windows
dBASE File	.dbf, .ndx, .mdx	Windows

The maximum length of the data table's name depends on your computer's operating system. See your operation system documentation for details.

Note: JMP tries to save any modified files when a crash is detected.

Save a Data Table to Open in JMP 5.1.2 or Earlier

JMP 9 uses the Unicode character set, which supports special characters such as *é* and ½. You can save data tables from JMP so that they can be opened in older applications that do not support Unicode (such as JMP 5.1.2 and earlier). This works if all of the data table's character data is all convertible to a legacy character set that the application can read.

There are two ways to maintain backward compatibility:

- For data tables, select **File > Save As**. On Windows, select **Use JMP V5 Format (No Unicode).** On Macintosh, the check box is **JMP 5 Compatible (without Unicode text)**. Then click **Save.**
- For text files, disable the Unicode feature in the preferences as described below:

1. Select **File > Preferences** (Windows) or **JMP > Preferences** (Macintosh).
2. Click the **General** category.
3. Deselect **Save Text Files as Unicode.**
4. Click **OK.**

   Your text files are saved as plain text.

Save as a Text File

JMP can convert data from a JMP data table to standard text format with rows and columns. To save a data table as a text file, follow the steps in Table 4.2 based on your operating system and how you want to save the file.

Table 4.2 Saving a Data Table as a Text File

Windows	1. Select **File > Save As.** 2. Select **Text Export File** from the **Save as type** list. 3. Click the **Options** button and specify end-of-line and end-of-field characters, and choose whether to export column headings as text. 4. Click **OK.** 5. Click **Save.** An alert appears regarding loss of formulas, formatting information, and other metadata. 6. Click **Yes.**
Macintosh	1. Select **File > Export.** 2. Select **Text** and click **Next.** 3. Type a name for the file in the **Save As** box and change the extension to .txt. 4. Specify end-of-line and end-of-field characters, and choose whether to export column headings as text. 5. Click **Export.**

Text formatting options are:

- **Export Column Names to Text File** To save column headings in the first line of text, and to save labels or header information with the data, select **Export Table Headers.**
- **End of Field** Select the radio button next to the character that marks the end of a field (or cell). Select **Other** and enter a character if the appropriate character is not listed.
- **End of Line** Select the radio button next to the character that marks the end of a line (or row). Select **Other** and enter a character if the appropriate character is not listed.

Note: If double quotation marks are found when importing text data, the delimiter rules change, and JMP looks for an end double quotation marks. Other text delimiters, including spaces embedded within the quotation marks, are ignored and treated as part of the text string.

Save as a SAS Transport File

You can save a JMP data table in SAS transport file format. JMP replaces spaces in filenames and column names with underscores, converts column headings to uppercase SAS variables, and makes other changes to follow the transport file specifications.

When you save the data table as a transport file, you can opt to append it to an existing transport file. When you reopen the transport file in JMP, the two data tables open in separate windows.

Note: The maximum length for the file name is 32 characters. JMP warns that characters beyond that limit are omitted.

Windows

1. Select **File > Save As.**
2. Select **SAS Transport File** from the **Save as type** list.
3. (Optional) Append the data table to an existing SAS transport file.
   - (Optional on Windows Vista and 7) Select the file to which you want to append the data table, and then click the **Save** arrow. Select **Append To**, and then click **Yes** to overwrite the selected data table.
   - (Windows XP) Select the **Append To** check box, select the file to which you want to append the data table, and then click **Save.** Click **Yes** to overwrite the selected data table.

   An alert appears regarding loss of formulas, formatting information, and other metadata.
4. Click **Yes.**

   Alerts appear regarding other SAS format modifications.
5. Click **OK** on each alert.

Macintosh

1. Select **File > Export.**

2. Select **SAS Transport**.
3. (Optional) To append the data table to an existing SAS transport file, select the **Append** check box.
4. Click **Next**.
5. Do one of the following:
   – If you chose to append the data table, select the file to which you want to append the data table and click **Append**.
   – Enter a name for the file in the **Save As** box, and then click **Export**.

   Alerts appear regarding SAS format modifications.
6. Click **OK** on each alert.

Save as a SAS Data Set (Windows Only)

On Windows, you can save data tables as SAS version 6 and higher data sets. JMP writes columns as SAS variables, writes rows as SAS observations, and makes other changes to follow the SAS data set specifications.

When you export data to a SAS data set, JMP date columns become SAS date values with the appropriate SAS format.

Save a Data Table as a Data Set

1. Select **File > Save As**.
2. Type a name for your file in the **File Name** box. The maximum length for the file name is 32 characters. JMP warns that characters beyond that limit are omitted.
3. Select **SAS Data Set** from the **Save as type** list.
4. (Optional) To save the file using SAS version 6 format (.sd2), select the **Use SAS V6 Format (.sd2)** check box. If you are using SAS version 7 or higher, do not select this option.
5. (Optional) To save SAS variable names or SAS formats, select the **Preserve SAS column names** (**Preserve SAS variable names** in Windows XP) or **Preserve SAS formats** check boxes.
6. Click **Save**.

   An alert appears regarding loss of formulas, formatting information, and other metadata.
7. Click **Yes**.
8. If other alerts appear, click **OK** on each alert.

Save as a Microsoft Excel File

You can save a JMP data table in Microsoft Excel format if you have an Excel ODBC driver installed. The document is saved in the format that is specified in the ODBC driver properties. 64-bit JMP requires a 64-bit ODBC driver.

Note: The maximum length of the Excel file name is determined by your computer's operating system.

Windows

1. Select **File > Save As.**
2. Enter a name for your file in the **File Name** box.
3. Select one of the **Excel** file types from the **Save as type** list.
4. Click **Save**.

Macintosh

1. Select **File > Export.**
2. Select **Excel**.
3. Click **Next**.
4. Enter a name for the file in the **Save As** box.
5. Click **Export**.

Saving Data Tables to a Database

You can save a data table to any database on your system that has a compliant ODBC (Open DataBase Connectivity) driver:

1. Select **File > Database > Save Table**. Figure 4.1 shows the window that is initially displayed.

Figure 4.1 The Database Save Table Window

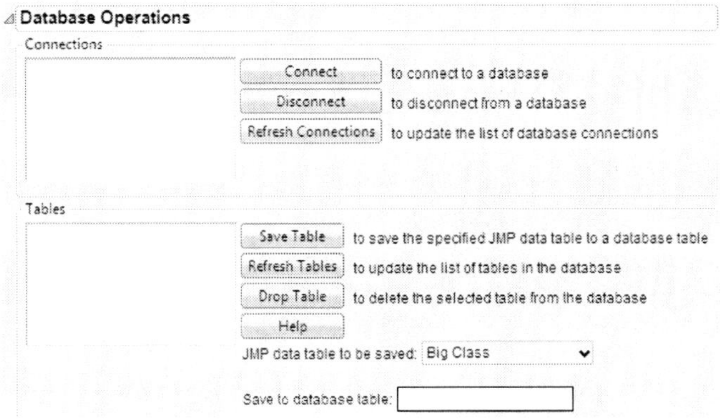

2. In the Connections box, highlight the name of the database to which you want to save the file. The Connections box contains a list of databases to which your system is connected (Figure 4.2).

   If you are *not* connected to the needed database, create it by using another application, or see “Adding a new database connection,” p. 112.
3. From the **JMP data table to be saved menu**, select the open JMP data table that you want to save to the database.

Figure 4.2 Specify Which Table to Save

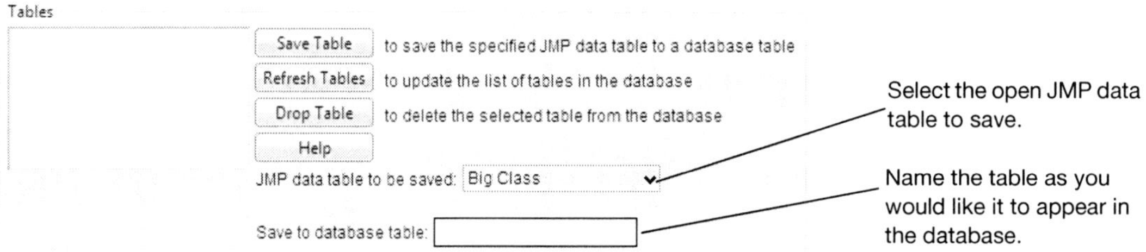

4. In the **Save to database table** box, type the name that you want the table to have when you save it in the database.
5. Click **Save Table**.

Adding a new database connection

1. Click **Connect**.
2. Select the data source that you want and click **OK**. Or, to create a new source, click the **New** button (Windows) or **Add** button (Macintosh).

   Depending on which data source you select (and which database drivers you have installed on your computer), you might be presented with a variety of windows. Use them to create the database source.
3. Select the database to which you want to save the file.

Saving Reports

JMP saves reports in the formats listed in Table 4.3.

Here are the basic steps for saving a report:

- On Windows, select **File > Save As** to save in multiple formats.
- On the Macintosh, select **File > Save As** to save as a JMP report (.jrp). Select **File > Export** to save in Text (.txt), HTML (.html), and Rich Text (.rtf) formats.

Table 4.3 Report File Types and Extensions

File Type	Extension	Description	OS Availability
CompuServe Graphics Interchange Format	.gif	Compressed bitmap pictures.	Windows
Encapsulated PostScript File	.eps	Line drawing (or *vector* image) that can show a bitmap preview of the image.	All

Table 4.3 Report File Types and Extensions *(Continued)*

File Type	Extension	Description	OS Availability
Enhanced Metafile	.emf	Can contain both vector and bitmap components.	Windows
Hypertext Markup Language	.htm, .html	Browser format; marked up text and references to separate picture files.	All
JMP Journal	.jrn	Analysis report duplicated in a separate window titled Journal. You can edit it or append other reports to it. Note that on Macintosh, select **Edit > Journal**, and then **File > Save**. On Windows, choose **File > Save**, and then choose the journal format.	All
JMP Report	.jrp	Analysis report originally created in JMP. It can be reopened for continued analysis.	All
Joint Photographics Expert Group	.jpg	Compressed bitmap pictures; standard for photographs.	Windows
Microsoft Word	.doc	Word processing format; mixture of pictures, text, and tables.	Windows (You must have Microsoft Word 2000 or later installed.)
Portable Document Format	.pdf	Format for sharing documents regardless of the operating system or application in which they were created.	Windows
Portable Network Graphics	.png	Compressed bitmap pictures; successor to GIF.	All
Rich Text Format	.rtf	Word processing format; mixture of pictures, text, and tables.	All
Scalable Vector Graphic	.svg	Pictures stored as text; best used for two-dimensional graphics.	Windows
Text Format	.txt	Plain text format; no pictures.	All

Note: When saving as .html or .rtf in Windows, there are options at the bottom of the window (Windows) that let you select a graphics file format. Select the format that you want JMP to save graphics in within the .html or .rtf file.

Saving Using the Layout Command

You can edit or manipulate the report before you save, enabling you to combine several reports into one or rearrange the report elements. You do this by selecting **Edit > Layout**. Using this command is different from using the **Edit > Journal** command; the **Layout** command provides additional options that let you ungroup parts of a report and restructure it to best fit your needs.

To create a layout:

1. Select **Edit > Layout**. (Or hold down the CTRL key and press L.)

   The report window is duplicated in a separate window titled Layout.

Figure 4.3 Report and Layout Windows

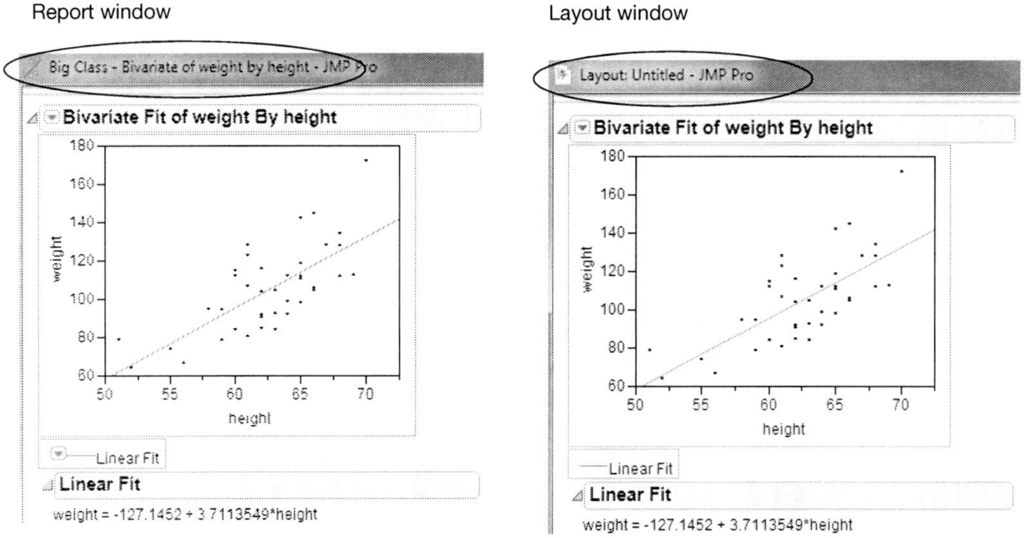

In the Layout window, the **Layout** menu appears between the **Edit** menu and the **Tables** menu. Items in the **Layout** menu are available only when you select an item with the arrow tool. See "The Layout Menu," p. 441, for details.

2. Click inside the layout window. The entire report becomes highlighted. To show the **Layout** menu, hover your cursor over the bar at the top.
3. Select **Layout > Ungroup**, or right-click (hold the CTRL key and click on Macintosh) inside the report and select **Ungroup**.

   This performs the first stage of ungrouping report elements, which ungroups (or disconnects) the topmost title bar from its reports and subreports. Each time you ungroup a report outline level title from its reports, the disclosure button for that level disappears and you can no longer close it. However, you can do many of the surface operations available, use the context menu in plots, tailor axes, rerun the analysis in a new window, edit scripts, and so on.

4. Select **Layout > Ungroup** again to ungroup the next level of the report outline. Note that only selected items are ungrouped.
5. Continue to select specific report elements (or all elements) and ungroup them until each title and each piece of a report or plot is an object.

Note: You can ungroup a report from its title bar and ungroup all of its major pieces, but you cannot ungroup a column in a report table from its column heading.

---

6. Click an object to select it and move it anywhere in the layout window. The layout window has as many pages as you want, outlined with gray boundary lines.
7. Select **File > Save As.** JMP saves the file as a journal file (.jrn).

Table 4.4 describes other layout options.

Table 4.4 Working in a Layout Window

Action	Instructions
Edit a title bar	Double-click the title bar or report table column heading.
Quickly ungroup a layout to its smallest objects	Repeatedly press Ctrl-U (Command-U on the Macintosh).
Rerun the report in a new window	Click the red triangle icon on the title bar and select **Rerun in new window** from the menu.
Edit a script	Click the red triangle icon on the title bar and select **Edit Script** from the menu.
Access **Layout** menu items (on the main menu bar)	Select the arrow cursor.
Insert a page break	Right-click a disclosure button (on Windows and on the Macintosh) on the title bar and select **Edit > Page Break.**
Select and deselect items	Right-click a disclosure button (on Windows and on the Macintosh) on the title bar and select **Edit > Select** or **Edit > Deselect.**

The example in Figure 4.4 shows results combined and organized from a bivariate analysis, distribution analysis, and one-way analysis (means and *t*-test). The result gives a neat-looking summary of oxygen uptake as a function of run time in an exercise experiment. This was done by forming a layout window for one analysis, dragging the other desired analysis elements into the layout window, arranging the desired pieces, and deleting unwanted elements.

Figure 4.4 Multiple Analysis Results in a Layout Window

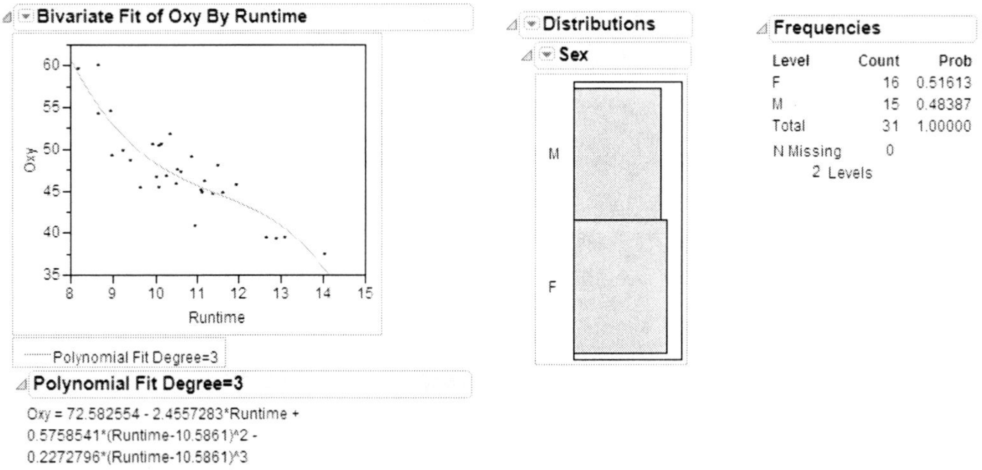

Saving Parts of a Report in a Graphic Format

You can save part of an analysis report window as a graphic. On Windows, you can save the selection in .png, .jpg, .gif, .eps, .emf, .pdf, and .svg formats. On the Macintosh, you can save the selection in .png format.

Save a Selection

1. Click the selection tool (✥).
2. Highlight the area that you want to save.
3. On Windows, select **Edit > Save Selection As**. On Macintosh, hold down the CTRL key and select **File > Save Selection As**.

Select the graphics file format to which you want to save the selection.

Note: Right-click the selection and choose **Copy**, and then paste the selection into another part of JMP or another software program. (See "Pasting Reports into Another Program," p. 123.)

Create Journals

Journals consist of JMP graphs and reports, graphics, text, and links to items such as Web pages and files. Using journals has several advantages:

- Journals let you save relevant graphs and reports and then arrange the content as you want.

- Journals let you store information for presentations. You can then edit and interact with the content as you would in JMP (such as scrolling an axis or changing marker sizes). This interactivity is not available in a slide presentation. See "Example: Making a Journal for a Presentation," p. 121 for details.
- Journals are an intermediate format to export to Microsoft Word, HTML, and other formats.
- Journals also let you store data tables and reports from a session, close and reopen JMP, and then open the journaled files instantly.

You can create a journal that groups items under headings (or outline levels). To close the groups, click the gray disclosure icon next to the heading.

Figure 4.5 shows some of the items that you can add to a journal.

Figure 4.5 Example of a Journal

Note: Links to directories of files, windows, and all open files look like other links, so they are not included in the preceding figure.

Create a New Journal

1. Close all open journals, and do one of the following:
   - To create an empty journal, select **File > New > Journal.** Or, from the JMP Starter window, select **New Journal.**
   - To create a journal from an entire report or data table, select **Edit > Journal.**
   - To create a journal from a specific report or text in a report, click and drag the Selection ( ) tool to select adjacent items in a report or data table. To select discontinuous items, press SHIFT and click the items with the Selection ( ) tool. Then select **Edit > Journal.**
   - To create a journal from specific text in a data table, select the text, and then select **Edit > Journal.**
   - To create a journal from most graphs, right-click on the graphic, and then select **Edit > Journal.** (Not available for surface plots and 3D scatterplots.)
   - Write a JSL script to create a new journal. See the *Scripting Guide* for more information.

Tip: If you prefer keyboard shortcuts, hold down CTRL (Windows) or COMMAND (Macintosh) and press J to create the new journal.

2. Select **File > Save** and save the journal in .jrn format.

Prevent Modifications

When a journal is opened, and then journal another report or data table, the second report or data table is added to the end of the first journal.

To prevent modifications to a journal, right-click the blank area at the bottom of the journal and select **Lock**. Additional reports are not added to the journal (as described in the following section). JMP instead finds an open unlocked journal or creates a new journal when you select the **Edit > Journal** command again.

To unlock the journal, right-click the blank area at the bottom of the journal and deselect **Lock**.

Alternatively, place separate reports in separate layout windows by using the **Layout** command, as described in "Saving Using the Layout Command," p. 114.

Appending Reports to a Journal

To append other reports to a report that you already journaled, select **Edit > Journal** again. If an area of an analysis window is selected, **Edit > Journal** saves only the selected area instead of the entire window.

When manipulating the report, remember that:

- The journal window has the same functions as the report window: you can click icons, click and drag, and right-click to access menus.
- When a report is journaled, the journaled copy is no longer connected to the data table.

Add and Edit Outline Levels

You can group text and links in an outline level and then click the gray disclosure icon to expand and collapse the group.

Note: Links to files have absolute paths. If you move the journal to another location on your computer, the links continue to work. When adding links to files in the JMP Sample Data folder, use the pathname $SAMPLE_DATA/xx. *xx* is the absolute path from the Sample Data directory. (Right-click the link, select **Edit > Set Script**, and then edit the path.)

- To add an outline level, right-click in the journal, and then select **Append Item > Add Outline Item**. You can also add nested outline levels to outline levels that you have created. (When you click the red triangle icon of a graph that you have journaled, the **Add Outline Item** option is unavailable.) Click the outline level's red triangle icon and select **Add Outline Item**.
- To edit the outline level heading, double-click the heading, enter the new name, and press **Enter**.
- To add text or links within an outline level, click the outline level's red triangle icon and select an option described in Table 4.5. Note that these options are not available when you click a scripted item's red triangle icon.
- To add text or links outside an outline level, right-click on the journal and then select one of the options described in Table 4.5.

- To edit a link, right-click the link, select **Edit > Set Script**, edit the path, and then click **OK**.
- To edit the name of the link displayed in the journal, right-click the link, select **Set Button Name**, enter the new name, and then click **OK**.

Tip: In Windows, show the preceding options by holding down the ALT key and right-clicking the outline level's gray disclosure icon.

Table 4.5 Journal Item Options

Add Outline Item	Groups text and links into collapsible and expandable lists.
Add Text Item	Enter up to six paragraphs with the option to add a bullet or hide the text. To unhide, right-click the hidden text and deselect **Hide**. **Note:** To add text outside an outline level, you can also double-click at the end of the journal and enter text in the box that appears. Click outside the box to exit the editing mode.
Add Window Reference	Creates a link to a window already opened in JMP.
Add File Reference	Creates a link to a file on your computer.
Add Directory of Files	Creates links to all files (including non-JMP files) in the selected directory.
Add All Open Files	Creates outline levels and links to open files in JMP (except for the log). For example, a link to the open Big Class.jmp data table and chart is displayed below a collapsible heading named Big Class. Files such as journals, scripts, and tutorials are grouped below a collapsible heading called *Other Files*.
Add URL Reference	Creates a link to files that are delivered through an Internet protocol such as HTTP, FTP, or FILE://
Add Script Button	Creates a link to a JSL script.

Control the Display of Outline Levels

You can instantly control which outline levels are displayed (for example, closing all sublevels below the selected level). Click the red triangle icon for that level, select **OutlineBox**, and then select one of the **Open** or **Close** commands. For details on the **Open** and **Close** commands, see "Access Report Display Options," p. 180 in the "Output Reports" chapter.

Tip: On Windows, show the OutlineBox options by holding down the ALT key and right-clicking the outline level's gray disclosure icon.

Add a Graph or Graphic

Do one of the following:

- Copy a graph from within JMP or a graphic from another application and then select **Edit > Paste**. JMP places the graphic at the end of the journal.
- Drag and drop the graph or graphic from another window to the journal. (See "Adding Graphics," p. 223 in the "Output Reports" chapter.) A blue line indicates where you can drop the graphic.

Customize Journal Items

- To click and drag journal items to different positions, select the selection tool ( ), select the item, and then drag the item to the new position. A line indicates where you can drag the item.
- To adjust text wrapping, right-click the text and select **Set Wrap**. Enter the number of desired pixels per line and click **OK**. To apply the wrap to all text items, select **Set Wrap,** select **Extend this to other text boxes**, and then click **OK**.
- To modify a plot axis, double-click or drag the axis. See "Customizing Axes and Axis Labels," p. 202 in the "Output Reports" chapter for details.
- To add text or shapes anywhere in the report, use the drawing tools (Annotate, Line, Polygon, and Simple Shape). For details about the drawing tools, see "Adding Elements to a Report," p. 217 in the "Output Reports" chapter.
- To resize plots and graphs, do one of the following:
  - Click and drag the edge.
  - Right-click the plot or graph, and then select **Size/Scale > Frame Size**. See "Resizing Plots and Graphs," p. 198 in the "Output Reports" chapter for details.

Save the Journal in Another Format

You can save a journal in another format (such as HTML or PDF). When you save the journal as HTML, buttons and links are active when the file is viewed in a web browser.

Windows

1. Select **File > Save As.**
2. Select the format.
3. Click **OK.**

Macintosh

1. Select **File > Export.**
2. Select the format.
3. Click **Next.**
4. Enter a name for the file in the **Save As** box.
5. Click **Export.**

Print a Journal

Select **File > Print.**

Delete Items from a Journal

1. Click the selection tool ( ).
2. Select the item that you want to delete and press DELETE.

Example: Making a Journal for a Presentation

Most people use a presentation application like PowerPoint to give presentations. With JMP journals, you can avoid using a presentation application: all your bullet points can be combined with live links and buttons to help automate the analyses that you want to show. Interactivity is also maintained in reports.

Follow this example to create a presentation using a journal.

1. Select **File > New > Journal**.
2. Right-click in the journal. Start your outline by selecting **Add Outline Item**.
3. Enter the title of the presentation "My Bivariate Demo."

Figure 4.6 Add an Outline Item

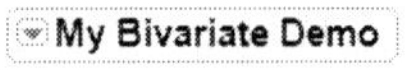

A presentation should have outline nodes, which are nested, opened, and closed in sequence, as you give the presentation.

4. Add bullet points into the outline by clicking the red triangle in the My Bivariate Demo title bar and selecting **Add Text Item**.
5. Type the text shown in the following figure.
6. Select the **Bullet Point** check box and click **OK**.

Figure 4.7 Add Text Item

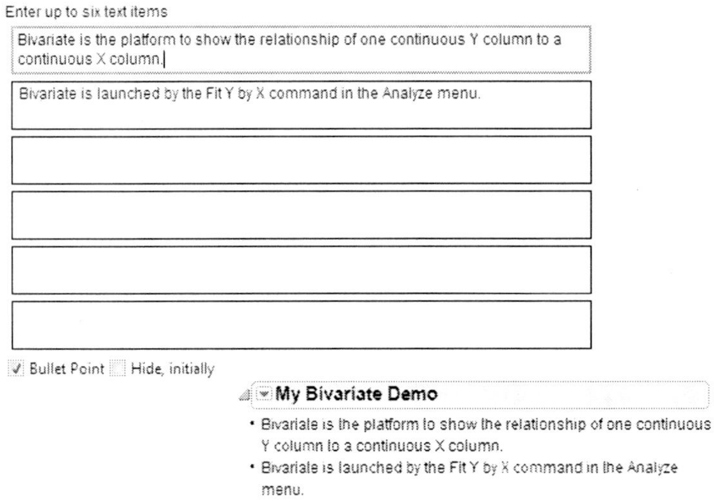

Now add a window reference as a link. These links lets you open a file with one click during the presentation.

7. Open the Big Class.jmp sample data table.
8. Click the red triangle in the My Bivariate Demo title bar and select **Add Window Reference.**
9. Select Big Class and click **OK.**

Figure 4.8 Add Window Reference

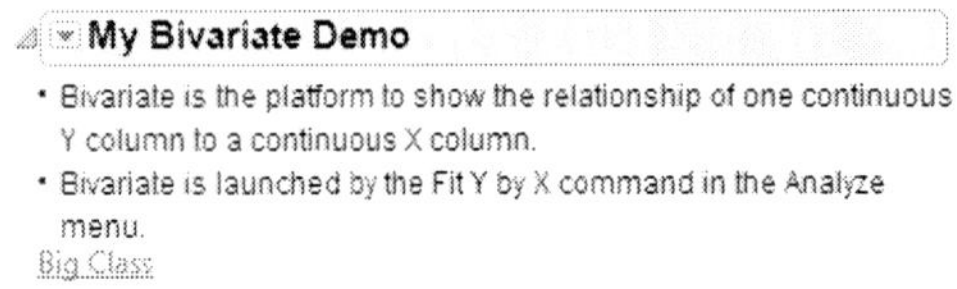

10. Click the window reference to verify the link.
11. Display the window reference as a button by right-clicking the link and deselecting **Underline Style**.

Figure 4.9 Change the Link to a Button

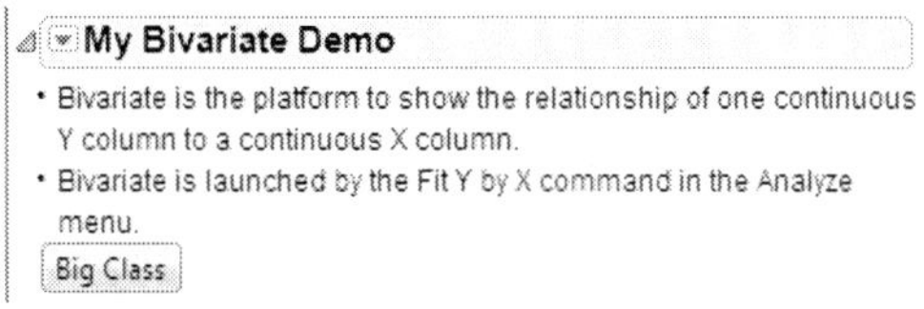

Pasting Reports into Another Program

When you need to use JMP reports or data tables in another program, you copy and paste parts of it into the document, such as Microsoft Word or PowerPoint file. Then, save the document in that application. Dragging and dropping the item is also an option.

1. Click the selection tool ( ).
2. Click and drag (or hold down SHIFT and click) to select items in a report window or data table. Clicking near the edge of the report window selects the entire report.
3. Click the selected items and drag them from JMP to the other program. Or, copy the selected items in JMP and paste them into the other program. When you paste an item into another application, the format used depends on the application into which you paste. If you use an application's **Paste Special** command, you can select a format to use, such as text (.rtf), unformatted text (.txt), or picture bitmap (.bmp), or Enhanced Metafile (.emf).

Note: To copy all text (no graphs) from the active report window as unformatted text, select **Edit > Copy As Text**. To copy only the graph (no text), right-click the graph and select **Edit > Copy Picture.**

Saving JMP Sessions

Each time you use JMP is called a *session*. A *saved session* is a JSL script that re-opens documents and re-runs analyses to restore JMP's state when the session script was saved. A saved session can help get you back to a previous state without having to manually re-open files and re-run analyses.

If you are an advanced user, it is important to understand what session information is preserved in a saved session. Any documents (such as data tables, scripts, and journals) that have been saved are re-opened. JMP windows that support script saving are re-run (equivalent to clicking the red triangle in a report and selecting **Script > Redo Analysis**). Side effects of running scripts, such as global variable values or custom windows, are not saved. The state of highly interactive analyses is also not saved.

Saving Sessions Upon Exiting

The most common use of saved sessions is to save the state when JMP exits so it can be restored when JMP restarts. By default, JMP asks whether you would like to save the state of your session each time you exit the program (Figure 4.10). This enables you to quit JMP, and then return to it later without having to open the files with which you were previously working.

Figure 4.10 Saving Session Upon Exiting

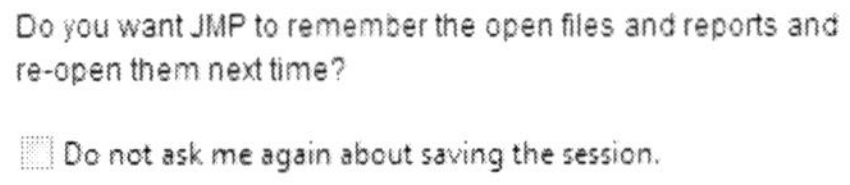

To always save the session upon exiting, select **Do not ask me again about saving the session** and click **Yes** or **No.** This option also prevents JMP from saving the session upon exiting,

You can change this option later in Preferences:

1. Choose **File > Preferences**.
2. Select the **Reports** icon.
3. Next to the **Save the session when exiting** option, choose when you want JMP to save the session. Always, never, and asking whether you want to save are the options.

Saving Sessions Manually

You can also save a session to a location of your choosing and continue working, so you can restore the saved state whenever you like. Manually saving sessions is useful when you want more control of session saving and restoring. This option is especially helpful when you want to maintain multiple independent session states, each with a different set of files and analyses.

Create a Script of a JMP Session

1. Select **File > Save Session Script**.
2. Enter the name of your script in the window and click **Save**.

Another way of manually saving a session is by creating a journal of each session. A journal can be a notebook-style or project-style file. With it, you can collect references to files in a project, develop presentation launch pads, document projects, and store many scripts in one place. See "Create Journals," p. 116 for details.

Create a Journal of a JMP Session

1. Open the files that you would like to include in the journal.
2. Select **File > New > Journal**. Or, to append your open files to an existing journal, open that journal.
3. Right-click in the empty journal and select **Add All Open Files**.

Figure 4.11 Creating Journals of JMP Sessions

Working with JMP Projects (Windows Only)

A JMP Project is useful when you want to save multiple JMP file types (data tables, reports, scripts, or other JMP-supported formats) into a single file. The single JMP project file contains everything needed to re-open all the included files. This is its major difference from saved session scripts. The scripts save the state of the software, while the projects save the state along with the files.

Projects can also include non-JMP documents (such as Microsoft Word or Adobe PDF files). This option lets you group and quickly open files.

Note: Though you can script and open projects on Macintosh, this section only covers working with projects on Windows.

Creating a JMP Project

To create a new project, select **File > New > Project**. The Projects window appears and shows the untitled project along with any other open projects. The Projects window also appears in the JMP Home Window's Window List.

Figure 4.12 JMP Home Window and Projects Window

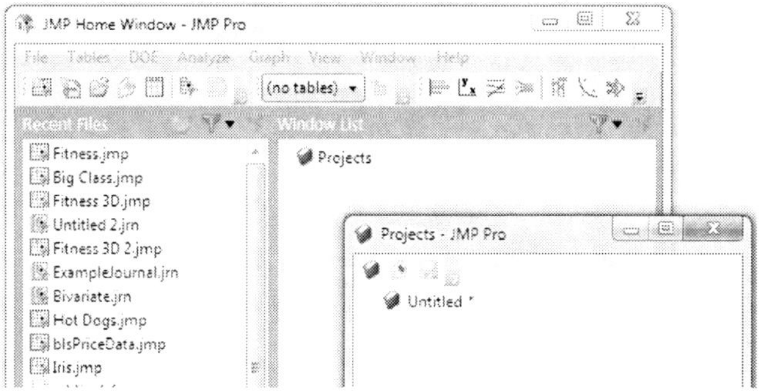

To name the project, right-click the project name in the Projects window, select **Rename**, enter the name, and then press ENTER. You can also press F2, enter the name, and press ENTER.

Saving a JMP Project

After you create and name a project, save it using one of the following methods:

- Right-click the project name in the Projects window and select **Save** "*project name*".
- Select the project name and click the **Save Selected Project** button on the Projects window toolbar.
- In another open JMP window (such as the JMP Home Window), select **File > Save Projects**.

When you save a project, JMP creates a compressed file with the .jmpprj extension and saves a *copy* of each item in the project. You actually modify the archived items rather than the originals. Other JMP users on Windows and Mac can then open the project on their computers, and the links remain intact.

JMP maintains the links by creating a folder structure that mirrors the location of the original files. Say that you save a project in the following folder:

C:\MyProjects

The project has a link to the MyDataTables folder.

When you open this project, JMP creates the following file structure in the project's subdirectory:

C:\MyProjects*Project Name* Dependencies\C\MyDataTables

At times, you might want to continue to modify the original files rather than the archived copies. Here are some examples:

- When you add a file to multiple projects, you are actually modifying three different copies of that file. If sales figures in one of the files need to be updated, you must update three different copies of that file.
- When you add files to a project from a network drive, you and other JMP users can edit those files without having to open the JMP project file.

In these instances, turn off archiving by right-clicking the project in the Projects window and deselecting **Archive all files and folders when project is saved** (shown in Figure 4.13).

Figure 4.13 Project Menu

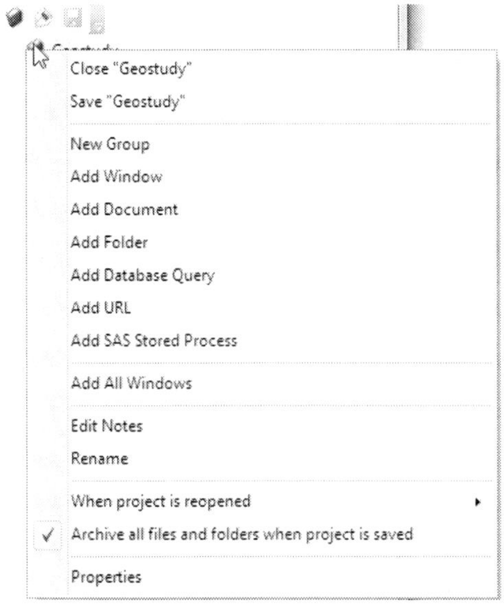

When you save a project, JMP notifies you when files in the project archive are not found. See "Fix Broken Links," p. 130 for details about fixing the links.

Closing a JMP Project

A project remains open even if you close the Projects window or exit JMP. To close the project, right-click the project name in the Projects window and select **Close**.

Opening a JMP Project

On Windows, open a project by selecting **File > Open** and selecting **JMP Projects** from the list next to **File name** (or the **Files of type** list on Windows XP). The file also shows up in the JMP Home Window Recent Files list if you previously opened it with the **File > Open** command or the **Open** toolbar button.

On Macintosh, open a project by selecting **File > Open** and selecting the file. Though there is no project window on Macintosh, the files that are set to open or run automatically appear.

Adding Items to a JMP Project

You can add either saved or open files to a JMP project. If a file has not already been saved, you are prompted to save it before it can be added to the project.

Using the Project Menu

To add items using the **Project** menu, right-click on the project name in the Projects window and select the desired command.

New Group Adds a collapsible and expandable container that organizes related items.

Add Window Opens a window that lets you select which open JMP window to add to the project. If the contents of the window has not been saved, you are prompted to do so before the window is added to the project.

Add Document Adds a copy of a file (if the archiving option is turned on) or a link to the file (when the archiving option is turned off). After selecting this option, the standard File Open window appears.

Add Folder Adds a Windows folder to the project. You browse to select the folder from your computer.

Add Database Query Opens the JMP database query Window, where you can create or open a query to a database. This query is then added to the project. See "Connecting to Databases," p. 380 in the "External Data and Analytical Sources" chapter for detailed information about creating queries.

Add URL Opens the Select URL window, where you specify a URL to add to the project.

Add SAS Stored Process Lets you add a SAS Stored Process. If you are not connected to a SAS Metadata Server, you are either connected to the server using your saved profile, or the SAS Server Connection window is opened for you. Once you are connected, you can select the SAS Stored Process to add to the project. See the "External Data and Analytical Sources" chapter for details about SAS Stored Processes.

Add All Windows Adds links to all open windows.

Dragging and Dropping Files

You can drag and drop files into a project from the following programs:

- Windows Explorer
- Microsoft Outlook (such as e-mails, attachments, notes, and calendar entries)
- JMP windows (Hold down the CTRL key, click the title bar, and drag the window to the project.)

Figure 4.14 shows an example of two files being dragged from Windows Explorer into the Projects window. Note that you drop the files onto the project name inside the window, not into the empty space.

Figure 4.14 Dragging Files from Windows Explorer into the Projects Window

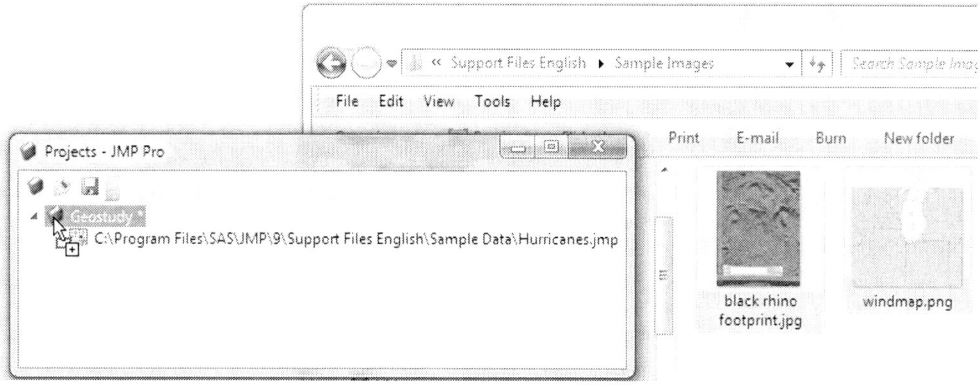

Customizing the Project

JMP has several commands that let you organize and customize your project.

Opening Files

When you open a project, JMP can automatically open most files or run scripts in that project.

To open all files or run scripts or database queries automatically, right-click the project name and select **When project is reopened > Restore all items in the project**.

To open or run specific items automatically, do the following:

1. Right-click each file that you want to open or run and select **Restore this *item* when the project is opened**. The name of the options depends on the file type. For documents, the option is named **Restore this document when the project is opened**. For windows, the option is named **Restore this window when the project is opened**.
2. Right-click the project name and select **When project is reopened > Restore only marked items**. (This is the default project setting.)

Though you cannot create a project on Macintosh, you can open files in a project if you set the files to open automatically.

Note: Be careful setting *all* items to open automatically. When you add a folder full of files, opening all of those files might cause a delay.

When you double-click a file in a project, most files open in their default programs. For example, data tables, journals, and scripts open in JMP. Graphics open in your default graphics program. Files that can be imported into JMP (such as Excel files or SAS data sets) are opened based on your text import preferences.

You can also choose how to open a file by right-clicking it and selecting an option. The options differ based on the file type. For example, files that are imported as data have text import options, or you can open them outside of JMP.

Groups

Adding a group to a project lets you organize files into collapsible and expandable containers. Add files through the **Project** menu, or drag and drop existing project files into a group. See Figure 4.15.

Figure 4.15 Project Groups

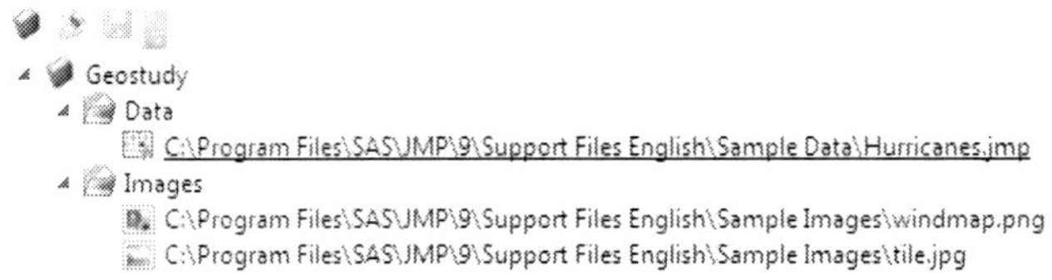

To add a new group, select **New Group** from the **Project** menu (shown in Figure 4.13).

Notes

JMP enables you to add descriptive notes about a project. Select the **Edit Notes** command from the **Project** menu to add or edit a project's notes. The notes are shown in the project properties. Right-click the project name and select **Properties** to view the notes.

Fix Broken Links

When you save a project, JMP warns when the link to a file is broken (Figure 4.16).

Figure 4.16 Missing Project Files

Fix a Broken Link

1. Click **No** on the broken link warning window.
2. In the Projects window, right-click the affected file, and select **Fix Broken Link**.
3. Browse to find the file.
4. Click **OK**.

Saving a Log Window

Selecting **View > Log** displays a window that monitors JMP activities, such as JSL statements as they execute and script errors. You might also encounter instances where running a formula, matrix, or another operation writes information of interest to the log window.

Save a Log

1. Open the Log window.
2. Select **File > Save As.**

Note: (Windows only) To open the log automatically when text is added to it, select **File > Preferences > Windows Specific**. In the **Open the JMP Log window** list, select **whenever text is added**.

Specifying Where to Save Files (Windows Only)

JMP includes a preference that lets you specify default directories for files, such as the location of JMP data tables and Help files. You can also specify the default folder for opening and saving files.

Specify Default Directories:

1. Select **File > Preferences** and select **File Locations** (Figure 4.17).

Figure 4.17 Specifying Default Directories

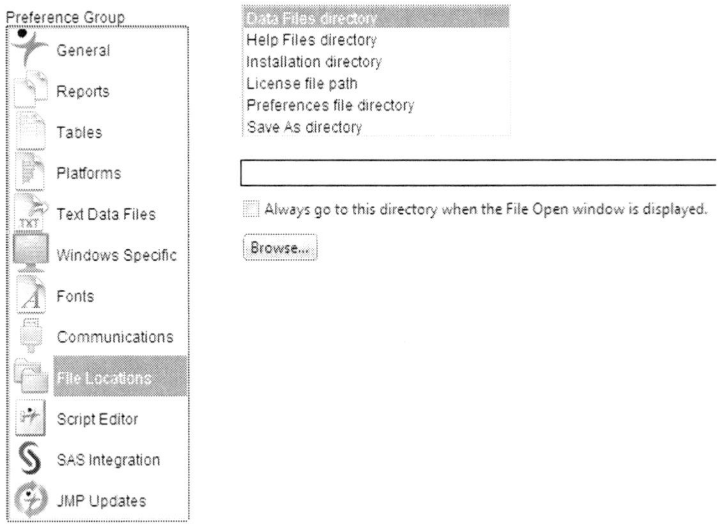

2. Select the directory that you want to specify.
3. Click **Browse**.
4. Select the directory and click **OK**.
5. (Optional) To open the specified data files directory when you try to open a file, select **Data Files directory**. Then select **Always go to this directory when the File Open window is displayed**.
6. (Optional) To always save JMP files in the specified Save As directory, select **Save As directory**. Then select **Always go to this directory when the Data/Journal Save As window is displayed**.

   JMP shows the Save As directory when you save a JMP data table, journal, project, script, report, or layout for the first time. The Save As directory also shows when you select **File > Save As** to save a file.
7. Click **OK**.

Chapter 5

Properties and Characteristics of Data

Customizing Columns and Rows

Beyond the basics of formatting data is customizing your rows and columns to suit your analysis. You can exclude, hide, label, color, mark, or lock rows or columns. Doing so customizes the appearance of points in scatterplots and graphs.

Contents

Assigning Characteristics to Rows and Columns

You can exclude, hide, label, color, or mark rows and columns. Doing so customizes the appearance of points in scatterplots and graphs. Many markers and colors are available. You can also lock columns so they stay in place when you scroll.

The menu for row actions can be accessed from three places:

- the **Rows** menu in the main menu
- the red triangle in the rows panel
- the left red triangle in the upper left corner of the data grid

Similarly, the menu for columns actions can be accessed from three places:

- the **Cols** menu in the main menu
- the red triangle in the columns panel
- the right red triangle in the upper left corner of the data grid

The following sections show you how to assign properties and characteristics to rows and columns.

Excluding Rows and Columns

You can mark rows and columns so that they are excluded from analyses.

- Excluded data are excluded from calculations in text reports and graphs. For most platforms, excluded data are not hidden in plots.
- For certain platforms, to hide excluded observations in plots, you must the **Hide/Unhide** characteristic. (See "Hiding Rows and Columns," p. 136.)
- A circle with a strikethrough ( ) appears beside either the row number or the column name in the column panel. The circle indicates that the row or column is excluded and not analyzed.
- For most platforms, data remain excluded until you select **Exclude/Unexclude** again. This menu option is a toggle. Selecting **Exclude/Unexclude** once sets the highlighted rows or columns to be excluded. Highlighting excluded rows or columns and selecting **Exclude/Unexclude** again turns the option off so that the rows or columns are included in analyses.

To exclude one or more rows from analyses

1. Highlight one or more rows that you want to exclude.
2. Select **Exclude/Unexclude** from the **Rows** menu.

To exclude one or more columns from analyses

1. Highlight one or more columns that you want to exclude.
2. Select **Exclude/Unexclude** from the **Cols** menu.

To unexclude rows or columns

1. Highlight the excluded rows or columns that you want to include in your analyses.
2. Select **Exclude/Unexclude** from the **Rows** menu or **Cols** menu.

Hiding Rows and Columns

You can mark rows and columns so that they are hidden in plots and graphs.

- You suppress (hide) rows and columns so they are included in analyses but do not appear in plots and graphs.
- To exclude hidden observations from analyses, you must also assign them the **Exclude/Unexclude** characteristic.("Excluding Rows and Columns," p. 135.)
- A mask icon ( ) appears beside the hidden row number or the column name, indicating that the row or column is hidden.
- Data remain hidden until you select **Hide/Unhide** again. This menu option is a toggle. Selecting **Hide/Unhide** once sets the highlighted rows or columns to be hidden. Highlighting hidden rows or columns and selecting **Hide/Unhide** again turns the option off so that the rows or columns are shown.

To hide one or more rows

1. Highlight one or more rows that you want to hide.
2. Select **Hide/Unhide** from the **Rows** menu.

To hide one or more columns

1. Highlight one or more columns that you want to hide.
2. Select **Hide/Unhide** from the **Cols** menu.

To unhide rows or columns

1. Highlight the hidden rows or columns that you want to show in your plots and graphs.
2. Select **Hide/Unhide** from the **Rows** menu or **Cols** menu.

Labeling Rows and Columns

When you position the arrow cursor over a point in a plot, the point's label appears. By default, the label that appears is the row number. Customize the label as follows

- You can change the label so that it displays values found in one or more columns instead of the row number.
- You can enable the label to appear always, not just when you position the cursor over points.
- A label, or yellow tag, icon ( ) beside the column name in the columns panel indicates that points on plots are identified by the column value. If there are multiple labeled columns, their values appear on plots separated by a comma.

- Data remain labeled until you highlight the column and select **Label/Unlabel** again. This menu option is a toggle. Selecting **Label/Unlabel** once sets the highlighted rows or columns to be labeled. Highlighting labeled rows or columns and selecting **Label/Unlabel** again turns the option off so that the rows or columns are no longer labeled.

To change the label so that it displays values found in one or more columns instead of the row number

1. Highlight one or more columns whose values you want to appear as the label in plots.
2. Select **Label/Unlabel** from the **Cols** menu.

To enable the label to always appear (**not just when you position the cursor over points**)

1. Highlight one or more rows whose label you want to always appear in plots.
2. Select **Label/Unlabel** from the **Rows** menu.

To turn off labeling for rows or columns

1. Highlight the labeled rows or columns that you no longer want labeled.
2. Select **Label/Unlabel** from the **Rows** menu or **Cols** menu.

Giving Rows a Color

You can assign any colors to highlighted rows so the points in plots appear in the color that you select. The active color assigned to a row appears next to the row number in the data grid.

1. Highlight one or more rows that you want to assign a color to.
2. Select **Colors** from the **Rows** menu.
3. Select one of the available colors.

Note: To clear an assigned color from the selected rows, assign the color black.

Adding Markers to Rows

You can assign a character from the JMP markers palette to replace the standard points in plots. These markers also appear next to row numbers in the data table.

1. Highlight one or more rows that you want to give a new marker shape.
2. Select **Markers** from the **Rows** menu.
3. Select one of the available marker shapes.

Custom markers enable you to use alphabetic characters, numerals, and other symbols as markers. Click the **Other** button and enter the desired marker into the window that appears.

Assigning Colors or Markers to Rows According to Column Values

You can assign colors or markers to your data table rows based on the values found in a particular column. For example, JMP can give all rows whose value in the column **Sex** is **F** a red circle marker and **M** a green plus marker. These colors and markers replace the default black dot in plots and appear next to its row number in the data table.

After a color is assigned to each row based on the column value, you can shade each row in the data table accordingly. For example, if the color *orange* is assigned to a row, the row itself is shaded in orange. If the color *aqua* is assigned to another row, that row is shaded in aqua.

1. Select **Rows > Color or Mark by Column.**
2. Highlight the column to which you would like to assign the color or marker. See Figure 5.1.

Figure 5.1 Color or Mark by Column

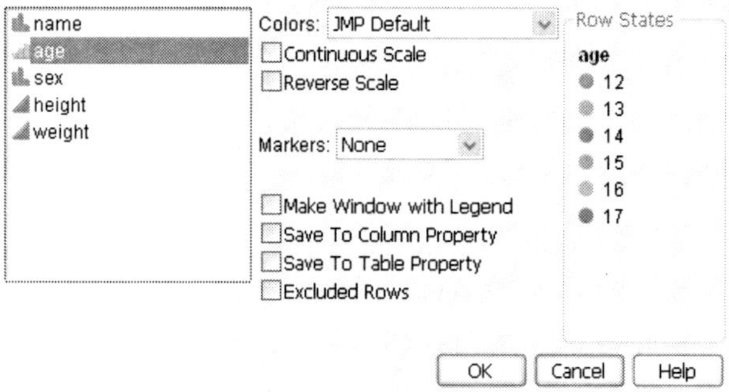

3. Select any of the following options, and then click **OK.**

Action	Instructions
Assign a different color to each row in your data table based on the values found in the column that you highlighted.	Select a color scheme in the **Colors** list.
Assign a different marker to each row in your data table based on the values found in the column that you highlighted.	Select one of the marker schemes from the **Markers** list.
Include a legend with your new characteristics so you can easily identify which colors and markers correspond with which row.	Select the **Make Window with Legend** check box.
Assign colors in a chromatic sequential fashion based on the values in the highlighted column.	Select the **Continuous Scale** check box. (Only available for continuous data.)

Action	Instructions
Assign colors in a reversed chromatic sequential fashion based on the values in the highlighted column.	Select the **Reverse Scale** check box.
Save the color information as a column property.	Select the **Save to Column Property** check box.
Save the color and marker information as a table property.	Select the **Save to Table Property** check box.
Assign colors or markers to rows that are excluded.	Select the **Excluded Rows** check box.

Note that a preview of your selection appears on the right side of the window, under Row States.

4. (Optional) To shade each row according to its row state, right-click the row numbers area in the data table and select **Color Rows by Row State**.

   From then on, the rows are shaded with the color that you assign to the rows.

Create Color Themes

JMP includes several color themes that can distinguish a range of values in a graph. You can also create your own color themes based on an existing color theme or create completely new themes.

The following figure shows examples of the two types of color themes in JMP, continuous and categorical. When a color theme is selected for continuous data, the colors are graduated (as shown on the left). When the same color theme is selected for categorical data, the color consists of distinct blocks of color.

Figure 5.2 Examples of Continuous and Categorical Color Themes

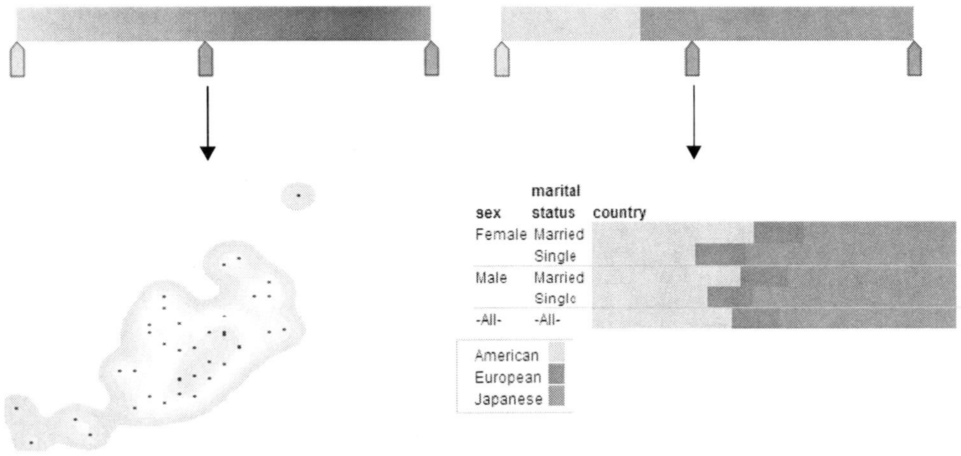

Custom color themes can be applied in the same way as built-in color themes:

- You can select custom color themes as defaults from the **Continuous Color Theme** and **Categorical Color Theme** drop-down menus in the Reports preferences. Only continuous color themes are available for continuous data. All color themes are available for categorical data.
- You can apply the custom color themes to components such as markers and data table rows. See "Assigning Colors or Markers to Rows According to Column Values," p. 138 for details.
- In certain reports, such as tree maps and surface plots, you can select specific custom color themes. See the *Basic Analysis and Graphing* book for details.

Note: When you select a default color theme, the colors are not applied to open reports. You need to rerun the existing reports to format them with the default color theme.

To Create a Color Theme

1. Select **File > Preferences > Reports.**
2. Click the **Customize Themes** button.

   The Customize Color Themes window appears.

Figure 5.3 Customize Color Themes Window

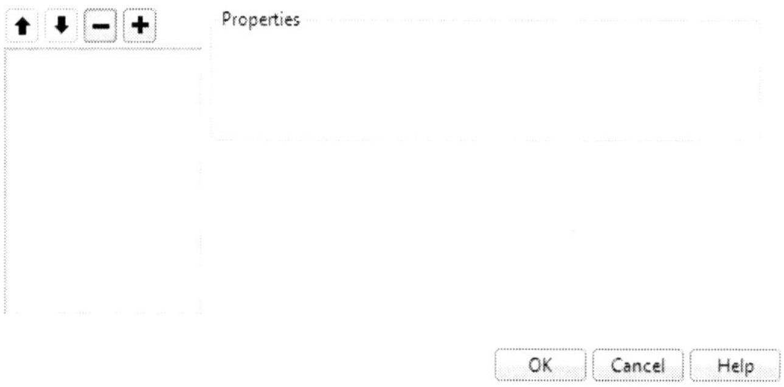

3. Click the  button.

   A new color theme is created based on the default continuous color theme. A temporary name is assigned to the theme.
4. Type a new name in place of the temporary label. (In Windows, do not press ENTER; the window closes if you do so.)
5. (Optional) To base the theme on an existing theme, right-click the color bar, select **Copy From**, then select the color theme.

6. To modify the color theme, do any of the following:
   - To modify the gradient of continuous color, move the sliders left or right.
   - To add more colors to the gradient, double-click the color bar, select a color from the Color window and click **OK**. A new slider is displayed under the color bar.
   - To change the color of a slider, click on the slider to display the Color window and choose another color.
   - To distribute the colors evenly on the gradient, right-click the color bar and select **Space Evenly**.
   - To reverse the order of the colors on the gradient, right-click the color bar and select **Reverse**.
   - To convert the gradient to distinct blocks of color (rather than graduated color), right-click and deselect **Continuous**.
   - To remove a color from the color theme, click the color's slider and drag the slider above or below the color bar.
   - To discard your changes, right-click the color bar and select **Revert**.
7. Click **OK**.

To Reorder Custom Color Themes in the Color Theme Drop-Down Menu

1. Select **File > Preferences > Reports**.
2. Click the **Customize Themes** button.
3. Click the color theme that you want to move.
4. Click the ⬆ or ⬇ button until the color is where you want it.
5. Click **OK**.

To Delete a Color Theme

1. Select **File > Preferences > Reports**.
2. Click the **Customize Themes** button.
3. Click the color theme that you want to delete.
4. Click the ➖ button, and then click **OK**.

Use Custom Color Themes on Multiple Computers (Windows only)

The color themes that you create are defined in the JMP preferences file called JMP.PFS. If you use JMP on more than one computer (for example, at home and at work), you can copy the color theme definitions from one JMP preferences file to another. Custom colors are then available on both computers.

In the preferences file, the code for a custom color theme looks like this:

```
Add Color Theme(
   {"Pink to Blue", {{255, 168, 255}, {255, 0, 255}, {0, 128, 255}}}
   ),
```

In the preceding example, the name of the color theme is "Pink to Blue." The Red/Green/Blue (RGB) values for each color slider are located in brackets. The first slider defines the RGB values 255, 168, and 255. The second and third groups of brackets define colors for the second and third sliders.

In a text editor, such as Notepad, you add this color theme to the preferences file on your other computer. The preferences file is located in the following directories:

- Windows XP: C:\Documents and Settings\\<*username*>\Local Settings\Application Data\SAS\JMP\9
- Windows Vista and 7: C:\Users\\<*username*>\AppData\Local\SAS\JMP\9

Note: To see the preceding directories, you must configure Windows Explorer to show hidden files and folders. For details, refer to the Windows help.

To Transfer Color Themes to Another Windows Computer

1. On the computer that contains the customized JMP preferences, select **File > New > Script.**

   The Script window appears.

2. Type the following JSL function:

   ```
   Show Preferences()
   ```

3. Click the `Run Script` button .

   Your customized preferences are written to the log.

4. Select **View > Log** (or display the open log).

   The custom color theme that you created is displayed as shown in the following example:

   ```
   Add Color Theme(
   {"Pink to Blue", {{255, 168, 255}, {255, 0, 255}, {0, 128, 255}}}
   ),
   ```

   This definition might be in the middle of other customized preferences that appear in the log.

5. Save the log as Log.jsl and open the file on the computer whose preferences you are updating.
6. On the computer whose preferences you are updating, close JMP.
7. Make a backup of JMP.PFS, and then open the original JMP.PFS in a text editor.
8. Copy and paste the custom color definition from Log.jsl to JMP.PFS. The definition goes after `Preferences(` as shown in the following example:

   ```
   Preferences(
   Add Color Theme(
   {"Pink to Blue", {{255, 168, 255}, {255, 0, 255}, {0, 128, 255}}}
   ),
   );
   ```

Note: Be sure to include the closing parenthesis and comma. The code does not need to be indented. You can put the code in any valid location. Pasting it after `Preferences(` helps ensure that you do not delete any necessary parentheses or commas.

---

9. Save the file.

   If you open JMP and the new color definition is not displayed in the preferences, delete the updated preferences file and add the definition to the original preferences file. Make sure that you copy and paste the definition in the correct location.

For details about JSL, see the *Scripting Guide*.

Deleting All Row Characteristics

To clear all row states in the data table, select **Clear Row States** from the **Rows** menu.

All rows become included, visible, unlabeled, and show in plots as black dots. The **Clear Row States** command does not affect row states saved in row state columns, as described in "Using Row State Columns," p. 173.

Locking Columns in Place

You can lock a column in place so that when you scroll horizontally, the column remains visible. The name of a locked column is shown in italics in the columns panel to the left of the data grid.

- Scroll locked columns are moved to the left in the data table. Once you unlock them, they are not moved back to their original locations in the data table, but instead remain on the left.
- Columns remain scroll locked until you highlight the columns and select **Scroll Lock/Unlock** again. This menu option is a toggle. Selecting **Scroll Lock/Unlock** once locks the highlighted columns. Highlighting locked columns and selecting **Scroll Lock/Unlock** again turns the option off so that the columns are no longer locked.

To apply a scroll lock to a column

1. Highlight one or more columns that you want to scroll lock. Note that hidden columns cannot be scroll locked.
2. Select **Scroll Lock/Unlock** from the **Cols** menu.

To unlock a column

1. Highlight the scroll locked columns that you no longer want scroll locked.
2. Select **Scroll Lock/Unlock** from the **Cols** menu.

Giving Columns a Preselected Analysis Role

You can assign an analysis role, such as *x*, *y*, weight, or frequency, to a selected column and save the role with the data table. When you do this and then run an analysis, JMP uses the preselected role to automatically fill in the role boxes in windows. Then you do not have to specify these roles each time you run an analysis. For example, you might want a column named height to take the *x* role in every analysis of that data table. To enforce the *x* role, you assign the preselected role of *x* to the column.

Note that when you select **Freq**, the values in that column are what JMP uses as the frequency of the observation. If *n* is the value of the **Freq** variable for a given row, then that row is used in computations *n* times. If it is less than 1 or is missing, then JMP does not use it to calculate any analyses.

When you select **Weight**, the values in that column provide weights for each observation in the data table. The variable does not have to be an integer, but it is included only in analyses when its value is greater than zero.

To assign a preselected role to a column

1. Highlight the column.
2. Select **Preselect Role** from the **Cols** menu.
3. Select a role: **No Role**, **X**, **Y**, **Weight**, or **Freq** (Frequency).

After selecting the appropriate roles, icons in the columns panel, as shown in Figure 5.4, signify what roles the columns have been assigned. Click the icon to access a list of roles and select a different one.

Figure 5.4 Preselected Roles in the Column Panel

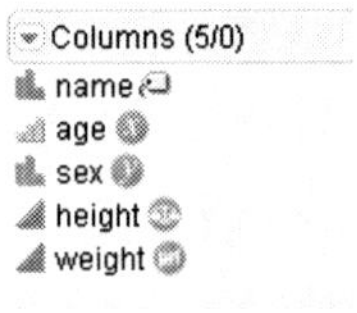

Icon Indicators

After assigning characteristics, icons appear on the data grid to indicate some of the rows' or columns' assigned characteristics. The icons appear to the right of each column name in the column pane and to the left of the row number in the data table. (See Figure 5.5.) You can click some icons on the right of each column name to reveal their contents. See Table 5.1 for a description of all the icons,

Figure 5.5 Column Characteristic Icons (Left) and Row Characteristic Icons (Right)

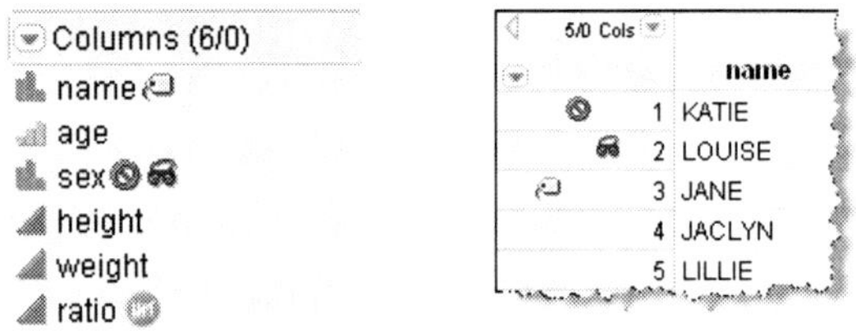

Table 5.1 Characteristic Icons

Icon	Description
	The column or row is excluded.
	The column or row is hidden.
	The column or row is used as a label.
Column Name	If the column name is shown in italics, the column is scroll locked.
	The column has a preselected role: *x*.
	The column has a preselected role: *y*.
	The column has a preselected role: *weight*.
	The column has a preselected role: *freq*.

Assigning Properties to Columns

In addition to the column characteristics described in "Assigning Characteristics to Rows and Columns," p. 135, columns can contain special column properties, such as notes for documentation, formulas that calculate the column's values, and restrictions on values.

After adding a property to a column, the properties icon (*) appears in the data table's columns panel, indicating that the column contains a property. However, when formulas, range checks, and list checks are applied to a column, their own individual icons appear. If the column contains only the notes property or if there are no properties, the properties icon (*) does not appear (Figure 5.6). Table 5.2 describes the column property icons.

Figure 5.6 Elements Indicating Properties

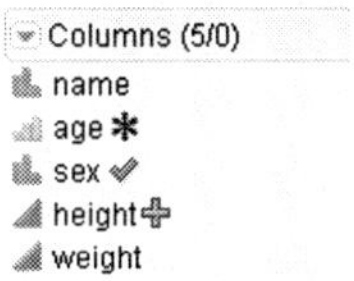

Table 5.2 Column Property Icons

Icon	Description
*	The column contains a property. Click to view a list of properties that the column contains.
	The column contains a list check or a range check property. Click to view and edit the list or range.
	The column contains a formula. Click to view and edit the formula.

The following sections describe properties that you can add to columns.

Assigning Currency Formats

Using the Column Info window, you can assign a specific currency format to a column. Once assigned, the currency symbol is displayed in the column and in graphs that contain the column.

To assign a format

1. Select **Cols > Column Info** for any continuous numeric data.
2. Select **Currency** from the Format menu.
3. Choose the currency unit from the menu that appears.

Giving Columns a Formula to Compute Values

You can insert a formula into a column to compute the values for that column. Once you do this, the column becomes locked so its data values cannot be manually edited (this prevents invalidation of the formula).

To create a formula that calculates the values for your column

1. Double-click the column name in the data grid. The Column Info window appears.
2. Select **Formula** from the **Column Properties** drop-down menu.
3. Click the **Edit Formula** button.
4. Complete the formula. (See "Creating a Formula," p. 303, for details.) Notice that box beside **Lock** becomes checked, indicating that you cannot edit the column's values from the data table. (To edit values, you must first delete the formula.)

Figure 5.7 Formula Editor

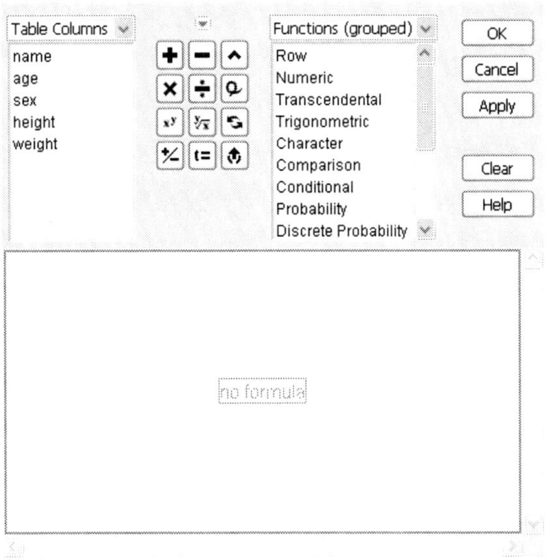

5. If you do not want JMP to evaluate the formula, click the box beside **Suppress Eval**. When formula evaluation is suppressed, part of the yellow plus icon (black on the Macintosh) beside the column name in the column panel becomes gray. See "Assigning Properties to Columns," p. 145 for an example.
6. If you do not want JMP to alert you of any errors in your formula, click the box beside **Ignore Errors**.
7. Click **OK**. The formula icon ( ) now appears next to the column name in the data table's column panel, indicating that the column contains a formula. Click the icon to edit the formula.

Note: In the Column Info window, when the Formula column property is selected, a visual of the formula appears at right. However, if your formula is long, a red X might appear instead. You can click the **Edit Formula** button to view and edit any formula.

Locking Columns

You can lock a column so none of its values can be edited. After you lock a column, the lock icon ( ) appears next to the column name in the data table's column panel.

Note that a column automatically becomes locked when you add a formula. However, the lock icon ( ) does not appear beside that column's name in the column panel. Instead, the formula icon ( ) appears, indicating that the column contains a formula.

To lock a column

1. Double-click the column name in the data grid. The Column Info window appears.
2. Click the box beside **Lock**, as shown in Figure 5.8.

Figure 5.8 Locking a Column

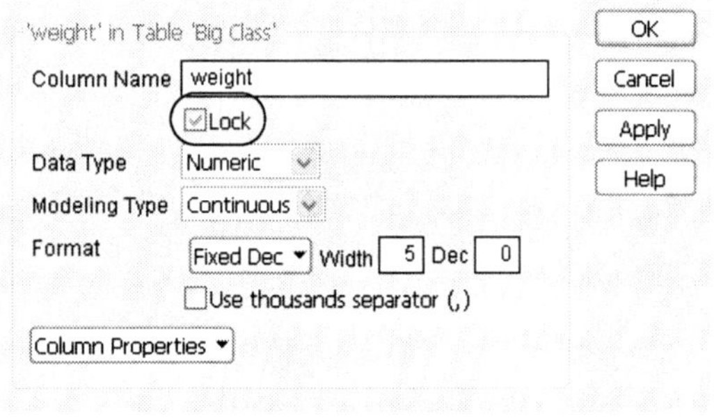

3. Click **OK**.

Adding Notes to Columns

Just as you can add notes to a data table as table variables, you can add notes to individual columns in the form of a column property. For example, you might want to document or keep notes for one or more columns in a data table, such as a description of the source of the data.

To add notes to a column

1. Double-click the column name in the data grid. The Column Info window appears.
2. Select **Notes** from the **Column Properties** drop-down menu.
3. Type into the text box.
4. Click **OK**.

Validating Column Data

You can set up a column so that it accepts only certain values. These values can be individual numbers (list checking), or they can be all numbers that fall within a range (range checking). The following sections describe the two types of column validation.

List Checking

You can set up a column so it can contain only certain individual values that you specify. For example, you can set up a column so it can contain only the numbers 2, 4, or 6. When you do this, the cursor changes to  when positioned over the column. If you try to enter a value not included on the validation list, a warning message appears. If you right-click a cell and select **List Check Values**, a menu of acceptable values appears. You can then select the cell value from the menu instead of typing it into the cell.

List checking can save memory and disk storage when the data have only a few categories. Each distinct category is saved only once, and a one-byte code that maps to a table of categories is saved for each cell.

List checking is also useful when you would like your graphs or plots to order data in a certain way. For example, histogram bars are ordered alphabetically by default. To present them in a different order, create a list check where the values that you enter are in the new order. They are then displayed in an output report in that order instead of the default order. (Note that list checking does not apply to bar charts.)

After you add the List Check column property, the ✔ icon appears next to the column name in the data table's column panel.

To set up list checking

1. Highlight the column to which you want to apply the validation.
2. Access the list check information by taking one of the following actions:
   - Double-click the column name in the data grid. The Column Info window appears. Select **List Check** from the **Column Properties** drop-down menu.
   - Right-click the column name in the Columns panel and select **Validation > List Check**.
   - Select **Cols > Validation > List Check**. See Figure 5.9.
   - Right-click on the column heading and select **Column Properties > List Check**.

Figure 5.9 List Check Window

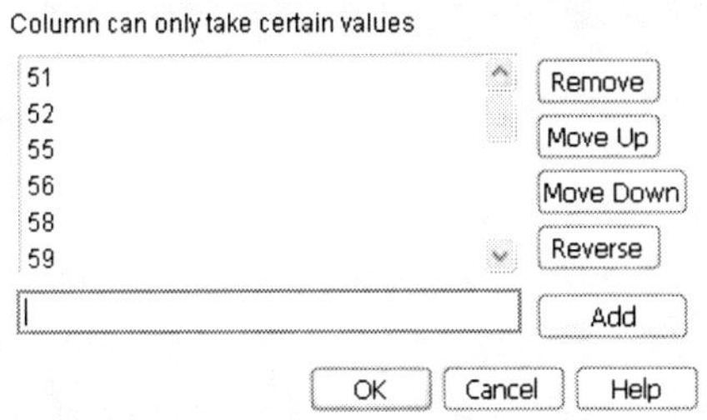

3. The list contains values that JMP accepts. Use the table below to change the values:

Table 5.3 List Check Actions

Action	Instructions
Remove value(s) from the list	Highlight them and click the **Remove** button.
Add values to the list	Type the values into the empty text box and click **Add**.
Move the highlighted number to another position in the list	Click the **Move Up** and **Move Down** buttons, or use drag and drop. Remember that the order in which values appear in this list is the order in which they appear in analysis reports.
Reverse the order of the list.	Click the **Reverse** button.

4. Click **OK**. The icon now appears next to the column name in the data table's column panel, indicating that the column contains a list check. Click the icon to make changes to the list check.

Range Checking

You can set up a column so that it can contain values only *within a range of numbers*. If you try to enter a value outside the validation range, a window prompts you to change the cell value.

To set up range checking

1. Highlight the column to which you want to apply the validation.
2. Access the range check information by taking one of the following actions:
   - Double-click the column name in the data grid. The Column Info window appears. Select **Range Check** from the **Column Properties** drop-down menu.
   - Right-click the column name and select **Validation > Range Check.**
   - Select **Cols > Validation > Range Check**. See Figure 5.10.
   - Right-click on the column heading and select **Column Properties > Range Check.**

Figure 5.10 Range Check Window

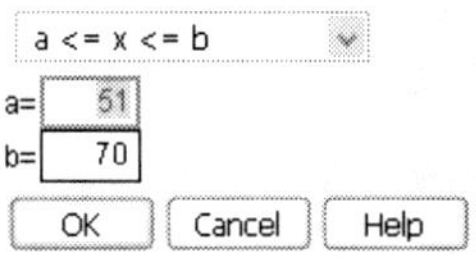

3. Type the lowest value that you want the column to accept into the box beside **a**.
4. Type the highest value that you want the column to accept into the box beside **b**.
5. Select which formula to use to set up the range. Remember that *x* is the value entered into the column, *a* is the beginning of the range, and *b* is the end of the range. For a single-sided range check, leave either **a** or **b** empty.
6. Click **OK**. A checkmark ( ) now appears next to the column name in the data table's column panel, indicating that the column contains a range check. Click the icon to make changes to the range check.

Turning List or Range Checking Off

To turn off list or range checking

1. Right-click the column name.
2. Select **Validation > No Checking.**

Specifying Missing Value Codes

Using the Missing Value Codes column property, you can specify column values that should be treated as missing. For example, sometimes the value 99 is used as a placeholder to represent missing values, or perhaps several values are used to represent different types of missing values.

To specify missing value codes:

1. Double-click the column name in the data grid.

   The Column Info window appears.
2. Select **Missing Value Codes** from the Column Properties list.
3. Next to the **Add** button, type in the column values that should be treated as missing.
4. Click **Add.**
5. Click **OK.**

Using Value Labels

Value labels enable you to display a label instead of a value in each instance that the value appears. When you assign value labels, the labels appear in the data table instead of the original values, but the original values are not lost. They can be displayed by double-clicking the label.

Also note the following points:

- When your data table contains value labels, using the **Search** commands searches for actual values, but does not search for labels.
- When your data table contains value labels, the row editor displays the label, and when the cell is highlighted for editing, it shows the actual value.
- If you copy and paste a cell with a value label, the actual value is pasted.
- Missing values can be labeled using Value Labels.

To set up value labels

1. In the data table, locate the column to which you want to apply the value label and double-click the area above the column name. The Column Info window appears.
2. Select **Value Labels** from the **Column Properties** drop-down menu. The Value Labels panel appears.

Figure 5.11 Value Labels Panel

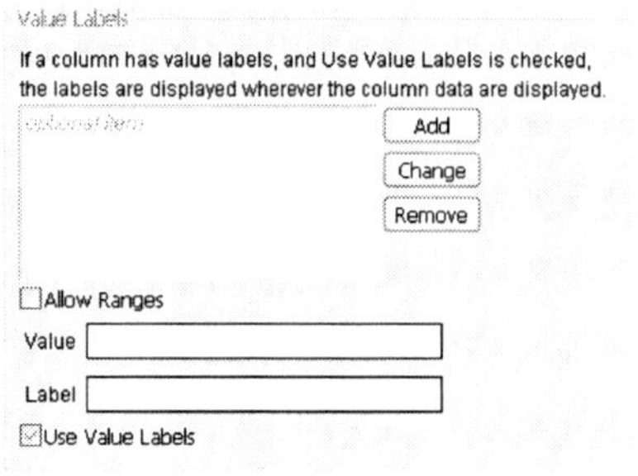

3. In the **Value** text box, type the value that you want to give a label.
4. In the **Label** text box, type what you want to appear as the label.
5. (Optional) To turn value labels off without deleting what you have set up, clear the **Use Value Labels** choice.
6. Click **Add**.
7. To create more value labels, repeat the above steps.
8. Click **OK**. The properties icon (✱) now appears next to the column name in the data table's column panel, indicating that the column contains a property.

To edit a value label that already exists

1. In the data table, locate the column to which you want to apply the value label and double-click the area above the column name. The Column Info window appears.
2. Highlight **Value Labels** in the **Current Properties** box.
3. Highlight the value label that you want to edit.
4. To edit its value, enter a new value into the **Value** box and click **Change**.

   To edit its label, enter a new label into the **Label** box and click **Change**. The new value or label appears in the place of the highlighted one.
5. When finished, click **OK**.

Note: In a formula, when you reference a column using value labels, hover your mouse over the value label to see the actual data value.

To use ranges in a value label

1. In the data table, locate the column to which you want to apply the value label and double-click the area above the column name. The Column Info window appears.
2. Highlight or add **Value Labels** in the **Current Properties** box.
3. Select the **Allow Ranges** check box.
4. You can now enter lower and upper bounds for each label.

Figure 5.12 Value Labels

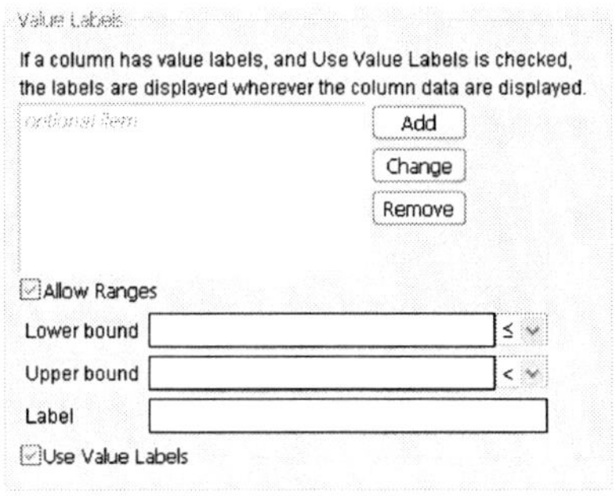

Note: To assign a label to missing values, enter a period (.) for the lower bound and leave the upper bound empty. To assign a label to all other values, enter three periods (...) for the lower bound and leave the upper bound empty.

Ordering Values in Columns

Categorical values in a JMP report might not appear in the order that you prefer. For example, Figure 5.13 shows ages listed from high to low on the left. You might want to list ages from low to high (shown on the right). To do this, add the Value Ordering property to the data table column.

Figure 5.13 Report with the Default Order (Left) and Report with Value Ordering Property (Right)

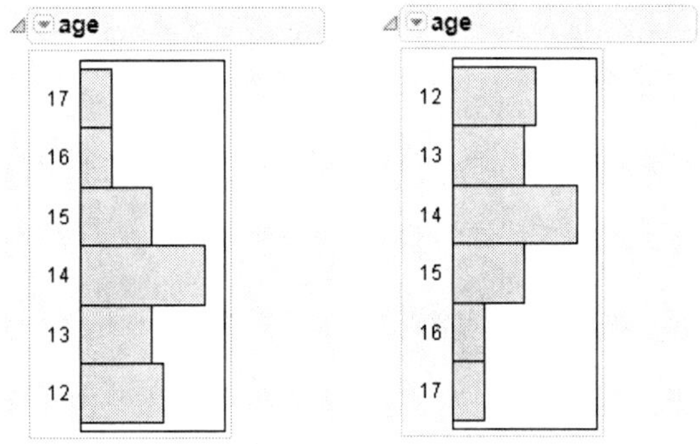

To assign the column the Value Ordering property

1. Double-click the column name in the data grid. The Column Info window appears.
2. Select **Value Ordering** from the **Column Properties** drop-down menu. The Value Ordering panel appears, as shown in Figure 5.14.

Figure 5.14 The Value Ordering Panel

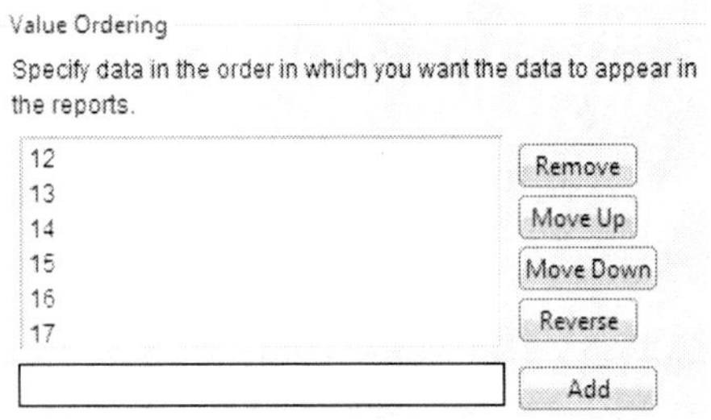

3. The list contains values in the order in which JMP displays them in reports. Use the table below to change the order of the values:

Table 5.4 Value Ordering Actions

Action	Instructions
Remove value(s) from the list	Highlight them and click the **Remove** button.
Add values to the list	Type the values into the empty text box and click **Add**.
Move the highlighted number to another position in the list	Click the **Move Up** and **Move Down** buttons. Remember that the order in which values appear is the order in which they appear in analysis reports.
Reverse the sort order	Click the **Reverse** button.

4. Click **OK**. The properties icon (*) now appears next to the column name in the data table's column panel, indicating that the column contains a property.

Note: If you use both the Value Ordering and Row Order Levels properties, the Value Ordering property overrides the Row Order Levels.

If your values include any of the following, they automatically appear in the appropriate order in reports. You do not need to assign their columns the Value Ordering property.

- January, February, March, April, May, June, July, August, September, October, November, December
- Jan, Feb, Mar, Apr, May, Jun, Jul, Aug, Sep, Oct, Nov, Dec
- Sunday, Monday, Tuesday, Wednesday, Thursday, Friday, Saturday
- Very Low, Low, Medium Low, Medium, Medium High, High, Very High
- Strongly Disagree, Disagree, Neutral, Indifferent, Agree, Strongly Agree
- Failing, Unacceptable, Very Poor, Poor, Bad, Acceptable, Average, Good, Better, Very Good, Excellent, Best

Assigning Value Color Ranges

In mosaic plots, nominal and ordinal values appear color-coded. In other plots with a legend displayed, the values are color-coded. You can assign the values of a nominal or ordinal column a certain color or range of color themes. Once you do that, the column's values appear with an assigned color in all applicable graphs. You can also color the values in the data table column.

To assign a column color values

1. Highlight a nominal or ordinal column in the data table.
2. Select **Cols > Column Info**. The Column Info window appears.
3. Select **Value Colors** from the **Column Properties** drop-down menu. The Value Colors panel appears, as shown in Figure 5.15.

Figure 5.15 The Value Colors Panel

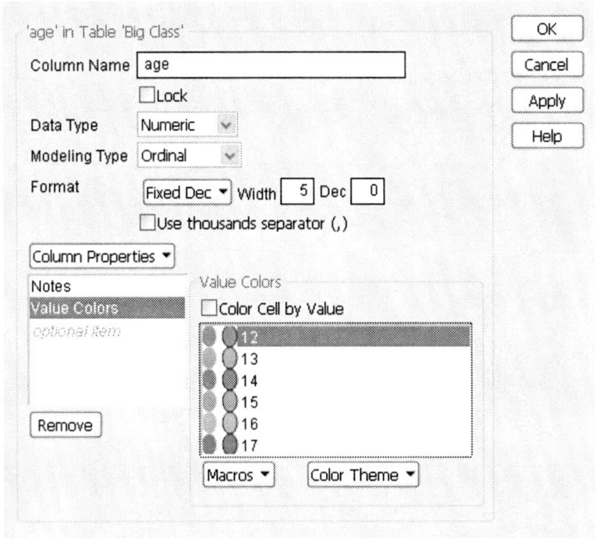

4. To change a color, do one of the following:
   - To change the color of a specific value, right-click a color circle and select a color.
   - To use a color theme, select it from the **Color Theme** menu.
   - To create a custom color theme, select **Custom** from the **Color Theme** menu and then create the color. Click **Save**, click **OK**, and then click **OK** to close the Column Info window. The color is saved in your Preferences. See "Create Color Themes," p. 139 for details.
5. If you want to color the column in the data table, select the **Color Cell by Value** check box. If you do not check this box, the colors apply only to the values in graphs.
6. Use the **Macros** button to do the following:

Gradient between ends Sets the colors of the top and bottom values so JMP can apply a color gradient across the entire range of values. Use this command to make all the colors in between for the other levels.

Gradient between selected points Sets the colors of the top and bottom values so JMP can apply a color gradient to a range of values that you have highlighted in the Value Colors list.

Revert to old colors Sets the colors back to their original color values.

7. Click **OK**. The properties icon (✻) now appears next to the column name in the data table's column panel, indicating that the column contains a property.

Assigning a Color Gradient

If you want to color a continuous column in a plot or in the data table, use the Color Gradient property.

To assign a column a color gradient

1. Double-click the column name in the data grid.

   The Column Info window appears.

2. Select **Color Gradient** from the **Column Properties** list. The Color Gradient panel appears, as shown in Figure 5.16.

Figure 5.16 The Color Gradient Panel

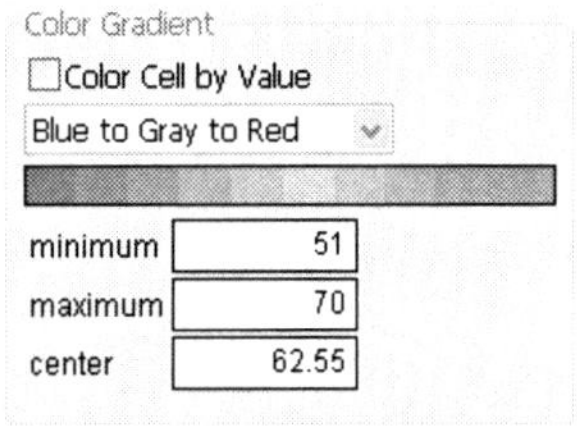

3. Select a color gradient from the menu.
4. Enter the minimum, maximum, and center values:
   - Minimum values reflect the color at the left of the gradient.
   - Maximum values reflect the color at the right of the gradient.
   - (May be optional) Center values reflect the color in the middle of the gradient.

   For example, in Figure 5.16, if you choose Blue to Gray to Red, the minimum value of 51 is green; the maximum value of 70 is red; and the center value of 62.55 is gray.
5. If you want to color the column in the data table, select the **Color Cell by Value** check box. If you do not check this box, the colors apply only to the values in graphs.
6. Click **OK**. The properties icon (✻) now appears next to the column name in the data table's column panel, indicating that the column contains a property.

Changing Columns' Default Axis Settings

You can specify default axis settings for a column so that whenever the column is used in analysis, JMP automatically uses the specified settings. These settings include the minimum and maximum values, number of increments, number of tick marks, and so on.

To specify default axis settings

1. Double-click the column name in the data grid. The Column Info window appears.
2. Select **Axis** from the **Column Properties** drop-down menu. The Axis panel appears, as shown in Figure 5.17.

Figure 5.17 The Axis Panel

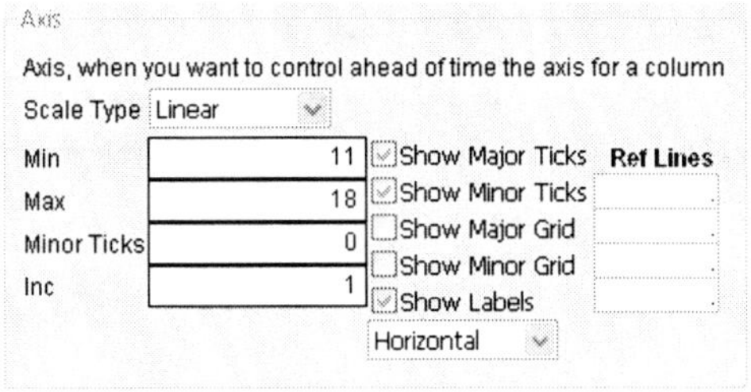

3. Specify axis properties as described in the following table:

Table 5.5 Axis Actions

Action	Instructions
Change the scale type	Click the drop-down menu beside **Scale Type** and select one of the following options: • **Linear** • **Log** • **Geodesic** • **Geodesic US**
Set minimum and maximum values that you want the graph to display.	Type values in the box beside **Min** and **Max**.
Specify the number of increments on the graph	Type the number in the box beside **Inc**.
Specify the number of minor tick marks on the graph	Type the number in the box beside **Minor Ticks**. Then, click the boxes beside **Show Minor Ticks**.

Table 5.5 Axis Actions *(Continued)*

Action	Instructions
View major tick marks, major gridlines, minor gridlines, and labels	Click the boxes beside **Show Major Ticks**, **Show Minor Ticks**, **Show Major Grid**, **Show Minor Grid**, and **Show Labels.**
Add up to four reference lines to the graphs	In the boxes under **Ref Lines**, type a value for each, and up to four lines appear on your graphs at the indicated positions.
Change the orientation of the axis labels	**Horizontal** and **Vertical** describe orientations for single axes. **Perpendicular** and **Parallel** describe orientations for paired axes (for example, in Scatterplot Matrices). **Automatic** adjusts the orientation based on the length of the label text. **Angled** adjusts the orientation to be angled at a 45 degree angle.

4. Click **OK**. The properties icon (✻) now appears next to the column name in the data table's column panel, indicating that the column contains a property.

An alternate way to set the default axis properties for a column is to:

1. Create the graph.
2. Change the axis to your preferred specifications. (See "Customizing Axes and Axis Labels," p. 202.)
3. Right-click the axis and select **Save to Column Property**. The properties icon (✻) now appears next to the column name in the data table's column panel, indicating that the column contains a property.

Defining Low and High Values (DOE Coding) for Columns

When you fit a column in a model, the low and high values of the column are transformed to –1 and 1, which makes tests and parameter estimates more meaningful. This is referred to as *coding* and can be used for any continuous variable. It is the default for continuous factors generated by the DOE commands in JMP. Note that if a column has one or more limits missing, JMP substitutes the data's minimum and maximum for the high and low values.

To specify which values you want to use as the low and high values in a column

1. Double-click the column name in the data grid. The Column Info window appears.
2. Select **Coding** from the **Column Properties** drop-down menu. The Coding panel appears, as shown in Figure 5.18.

Figure 5.18 Coding Panel

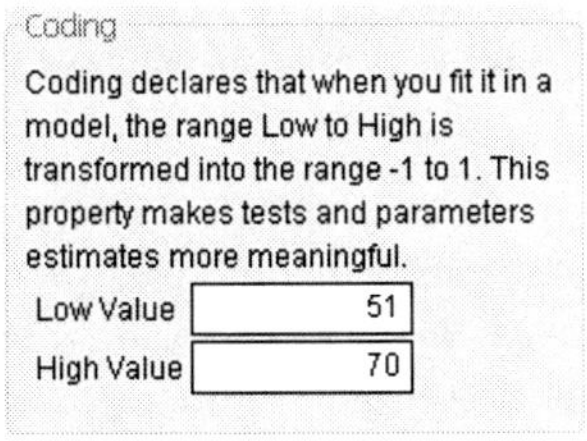

3. Type the values that you want to use as low and high values into the text boxes.
4. Click **OK**. The properties icon (✻) now appears next to the column name in the data table's column panel, indicating that the column contains a property.

Setting Columns as Factors for Mixture Experiments

You might have a column in a data table that is one of several factors that form 100% of a mixture. You can set up the column so JMP uses it to automatically generate a no-intercept model (using the Fit Model command) when you analyze the data.

After setting columns as factors for mixture experiments, the properties icon (✻) appears next to the column name in the data table's column panel, indicating that the column contains a property.

To set a column as a factor

1. Double-click the column name in the data grid. The Column Info window appears.
2. Select **Mixture** from the **Column Properties** drop-down menu. The Mixture panel appears, as shown in Figure 5.19.

Figure 5.19 Mixture Panel

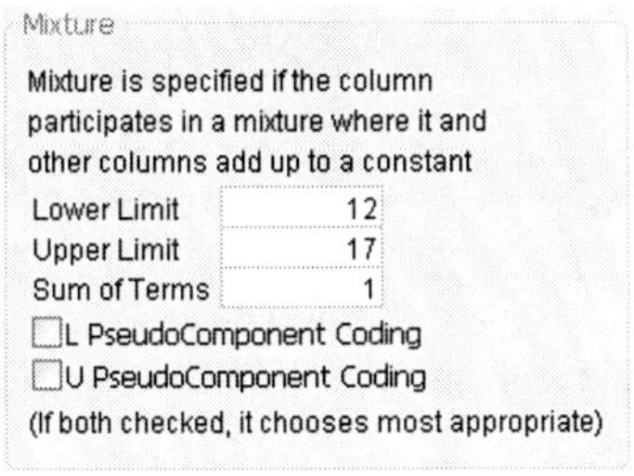

3. Enter the upper and lower limits as well as the sum of terms.
4. Check the boxes beside **L** and **U PseudoComponent Coding**, if desired. Using the example in Figure 5.19, where the mixture sum value is 1, the terms are coded as:

$X_iL = (X_i - L_i)/(1 - L)$ for the L pseudocomponent

$X_iU = (U_i - X_i)/(U - 1)$ for the U pseudocomponent

where L_i and U_i are the lower and upper bounds, L is the sum of L_i and U is the sum of U_i. If you check both in this example, the Fit Model platform uses the L coding if $(1 - L) < (U - 1)$, and the U coding otherwise.

In the Fit Model report, the main effects are labeled with the coding transformation. Crossed effects are not labeled, but coding values are used. All the features of fitting, such as the profilers and saved formulas, respect the pseudocomponent coding but present the uncoded values in the tables and plots.

5. Click **OK**. The properties icon (✻) now appears next to the column name in the data table's column panel, indicating that the column contains a property.

Specifying How Rows Appear in Analysis Reports

By default, the row data in an analysis report is ordered by the data's values. You can change the default setting to view the analyzed row data in the order in which it appears in the data table column.

To specify how rows appear

1. Double-click the column name in the data grid. The Column Info window appears.
2. Select **Row Order Levels** from the **Column Properties** drop-down menu. The Row Order Levels panel appears.
3. Select the check box beside **Row Order Levels**.
4. Click **OK**. The properties icon (✻) now appears next to the column name in the data table's column panel, indicating that the column contains a property.
5. Remember that what you have done in the above steps applies only to the selected column. To apply it to other columns, repeat the above steps for each column, or use the Standardize Attributes command, which is explained in "Standardizing Attributes and Properties Across Columns," p. 169.

Tip: To show the analyzed row data in another order (besides according to their values or their occurrence in the data table columns) use the value ordering property. See "Ordering Values in Columns," p. 153, for details. The Value Ordering property overrides the Row Order Levels property when both are evoked.

Entering Specification, Control, and Response Limits

You can save specification, control, and response limits in a column. If you do this, you can run a capability analysis, control chart analysis, or analyses displaying the prediction and contour profilers without having to re-specify roles and limits each time. Saving these limits in a column facilitates consistency from use to use. For example, you might run an analysis that uses these limits. When you come back later and change the data, you can run a new analysis on the new data using the same limits.

To enter specification, control, and response limits

1. Double-click the column name in the data grid. The Column Info window appears.

2. Select one of the following options from the **Column Properties** drop-down menu:
   - **Spec Limits** (see "Specification Limits," p. 161)
   - **Control Limits** (see "Control Limits," p. 161)
   - **Response Limits** (see "Response Limits," p. 162)
3. Click **OK**. The properties icon (✱) now appears next to the column name in the data table's column panel, indicating that the column contains a property.

Specification Limits

The **Spec Limits** option lets you enter a lower specification limit, upper specification limit, and target value for a numeric column. When you assign a column these values, JMP automatically uses them when you run a capability analysis. A capability analysis can be run after you have selected **Analyze > Distribution**. You can also select the **Show as graph reference lines** box, which draws specification limits as reference lines on a graph.

Figure 5.20 Spec Limits Panel

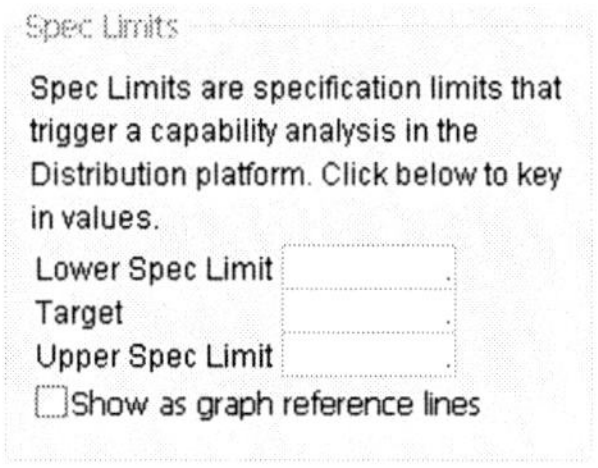

Control Limits

The **Control Limits** option lets you select a control chart type. Then you enter the values for average (**Avg**), lower control limit (**LCL**), and upper control limit (**UCL**) for each chart. Without the **Avg**, **LCL**, and **UCL** values, JMP includes a calculated value in the control chart for any of these three missing values. These values are saved with the column and automatically used when you run an analysis using **Graph > Control Chart.**

Figure 5.21 Control Limits Panel

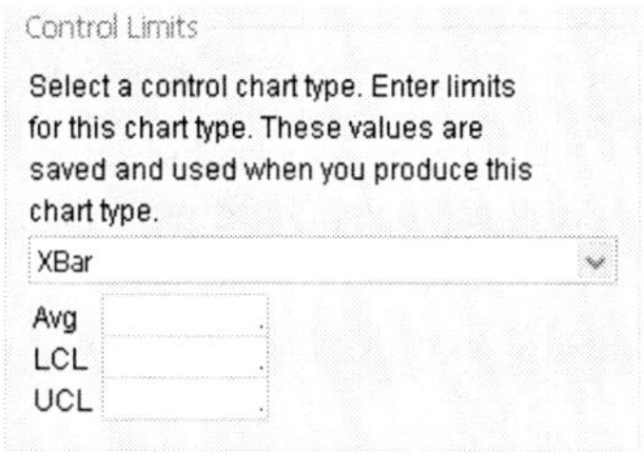

Response Limits

The **Response Limits** option lets you define response column values for use in analyses such as those displaying the prediction and contour profilers or DOE. You can perform the following actions in the Response Limits panel:

- Specify values for the lower, middle, and upper limits and desirability values.
- Specify one of the possible goals for a DOE response variable. Choose from the following: **Maximize**, **Match Target**, **Minimize**, or **None**. For example, if you are in the Prediction Profiler, and you want the desired value to be close to zero, select **Match Target.**
- If you have two responses, you can enter an importance value, which lets JMP know how to weigh one response's importance against another's.
- You can also select the **Show as graph reference lines** box, which draws specification limits as reference lines on a graph.

Figure 5.22 Defining Response Column Values

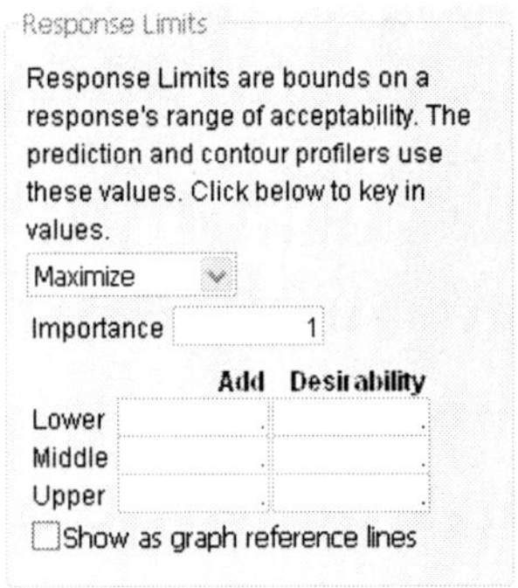

Giving Columns a Design Role

You can tell JMP how to use a factor column in a model to design an experiment. You can tell JMP to use it as a continuous, categorical, blocking, covariate, mixture, constant, signal, or noise factor.

To give a column a design role

1. Double-click the column name in the data grid. The Column Info window appears.
2. Select **Design Role** from the **Column Properties** drop-down menu. The Design Role panel appears.
3. Click the Design Role drop-down menu and select the option that specifies how you want JMP to use the factor column: **Continuous**, **Categorical**, **Blocking**, **Covariate**, **Mixture**, **Constant**, **Signal**, **Noise**, **Uncontrolled,** or **Random Block.**
4. Click **OK.** The properties icon (✱) now appears next to the column name in the data table's column panel, indicating that the column contains a property.

Identifying Factor Changes

To create a split plot design using the JMP DOE (Design of Experiments) commands, you must identify a factor as "hard," meaning that the factor would be difficult to change. Identifying a factor as "hard" can be done in the DOE design pane each time you design an experiment. (See the *Design of Experiments* for more information about creating split plot designs.) To create a split plot design every time you use a certain factor, you can save time by setting up that factor to be "hard" in all experiments.

1. Double-click the column name in the data grid. The Column Info window appears.
2. Select **Factor Changes** from the **Column Properties** drop-down menu. The Factor Changes panel appears.
3. Click the Factor Changes drop-down menu and select the option that specifies how you want JMP to use the factor column: **Easy**, **Hard**, or **Very Hard**.
4. Click **OK**. The properties icon (*) now appears next to the column name in the data table's column panel, indicating that the column contains a property.

Assigning Sigma Values to Columns

You can enter a known Sigma to be saved with the column to be used by applications, such as control charts, that require a Sigma value to complete computations. If no value is supplied, Sigma is calculated from the sample.

To assign a value to Sigma

1. Double-click the column name in the data grid. The Column Info window appears.
2. Select **Sigma** from the **Column Properties** drop-down menu. The Sigma panel appears.
3. Type the value to be used for Sigma into the text box.
4. Click **OK**. The properties icon (*) now appears next to the column name in the data table's column panel, indicating that the column contains a property.

Specifying Columns' Measuring Units

You can specify, within a column, the measurement units that were used to collect the data. For example, you might want a column to indicate that age values are measured in months or that a monetary value is in thousands of dollars.

To add a measuring unit to a column

1. Double-click the column name in the data grid. The Column Info window appears.
2. Select **Units** from the **Column Properties** drop-down menu.
3. Type the measurement that you want to use in the text box.
4. Click **OK**. The properties icon (*) now appears next to the column name in the data table's column panel, indicating that the column contains a property.

Selecting a Distribution Type for the Column

For a column that contains continuous numeric data, you can select the **Distribution** column property to produce the following results in the Distribution report:

- Whenever you run the Distribution report (**Analyze > Distribution**) for the column, JMP *automatically* estimates a fit using the specified distribution. A curve represents how the column's data is distributed.

  By default, a column has *no* Distribution column property. If you do not select the Distribution column property, then JMP does not draw the column's fit curve in the distribution nor does JMP report any information about the fit. (You could add this information to the Distribution report after you run it, of course.)

- If you set *both* the **Distribution** type *and* the **Spec Limits** column property ("Entering Specification, Control, and Response Limits," p. 160), then the Distribution report produces a **Capability Analysis** report table that is of the distribution type that you selected for the column.

Set the **Distribution** column property only when you already know how the data is distributed. For example, you might already know before you run **Analyze > Distribution** that the data has a Weibull distribution.

There are two ways to add the Distribution column property. The first way is as follows:

1. Double-click the column name in the data grid. The Column Info window appears.
2. Select **Distribution** from the **Column Properties** drop-down menu.
3. Select the type of distribution that you want to use for the column.
4. Click **OK**. The properties icon (✱) now appears next to the column name in the data table's column panel, indicating that the column contains a property.

The second way to add the distribution type to the window (along with the specification limits) is as follows:

1. From the red triangle menu in the column's report title bar, select **Fit Distribution** and select a type of distribution. JMP draws a fit curve on the histogram and adds a report table for the fitted distribution.
2. From the red triangle menu in the fitted distribution's report table, select **Spec Limits**. Enter the specification limits ("Entering Specification, Control, and Response Limits," p. 160) to generate a distribution-specific capability analysis.
3. From the same red triangle menu in the fitted distribution's report table, select **Save Spec Limits**. This saves both the specification limits and the associated distribution in the column properties for future use.

Assigning a Time Frequency to Data

The Time Series platform sometimes needs to know the frequency that data is reported (**Annual**, **Quarterly**, **Monthly**, **Weekly**, **Daily**, **Hourly**, **Minute**, **Second**). This lets JMP take things like leap years and leap days into account. If no frequency is specified, the data is treated as equally spaced numeric data.

Specify a Map Data Table

The Graph Builder platform lets you generate graphs and apply a map to geospatial data. One of your data table columns might contain the names of political boundaries (such as countries, states, provinces, counties). You can assign the column to the **Shape** zone in Graph Builder to create a map of the regions.

These political boundary files are installed with JMP. You can also define your own boundaries by creating two data tables:

- A data table named with the -XY suffix contains the boundary coordinates along with the Shape ID and Part ID for each coordinate. This information might originate as an Environmental Systems Research Institute (ESRI) shapefile.
- A data table named with the -Name suffix contains the boundary names along with the same Shape IDs from the -XY table. These boundary names are not case sensitive.

To use your own boundary files, you add a Map Role column property to the data table that you are analyzing and to the map data table. JMP shows your map on the Graph Builder graph instead of a map in the default JMP map folders. This feature is helpful when you want to store your analysis data table, reports, and map data tables together.

In the analysis data table, the Map Role property indicates the location of your -Name map data table. The property also identifies the column that contains the boundary names (for example, U.S. state names). In the -Name map data table, the Map Role column property indicates which column contains the boundary names. The columns that contain a Map Role property must contain the same boundary names, however, the column names can be different.

Figure 5.23 shows examples of how the map files and analysis data are structured.

Figure 5.23 Example of Map Data

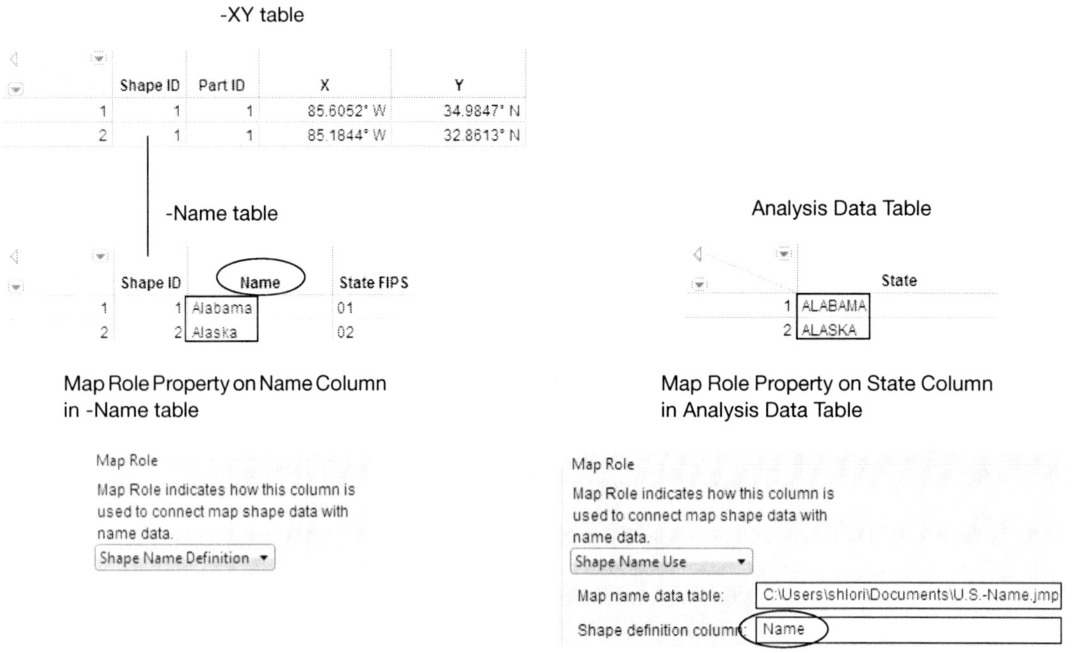

For details about the installed boundary files, see "Boundaries," p. 212. Refer to the *Basic Analysis and Graphing* book for details about where map files are stored by default, in case you do not want to use the Map Role property.

To add the Map Role property to the boundary name columns, proceed as follows:

1. Open the data table of information that you are analyzing.
2. Right-click the column and select **Column Properties > Map Role.**

   Map Role is displayed in the Column Properties box.

Figure 5.24 Map Role Column Property

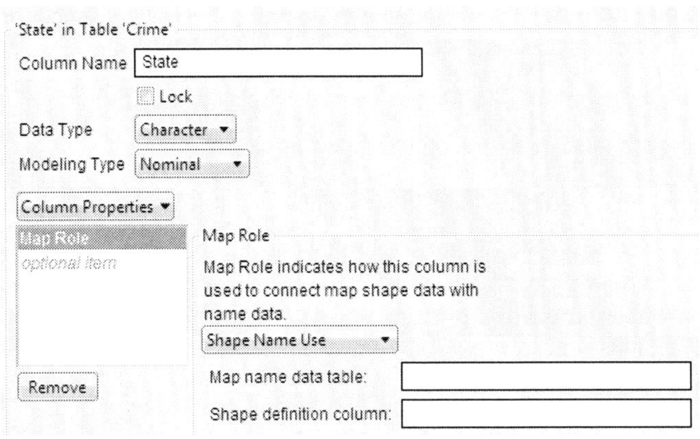

3. Select **Shape Name Use** in the Map Role area if necessary.
4. Next to **Map name data table**, enter the relative or absolute path to the -Name map data table.

   If the map data table is in the same folder, you enter only the filename. Quotes are not required when the path contains spaces.
5. Next to **Shape definition column**, enter the name of the column in the map data table whose values match those in the selected column.
6. Click **OK.**
7. Save the data table.
8. In the **-Name** data table, do the following:
   – Add the Map Role column property to the column that contains the boundary names.
   – Select **Shape Name Definition.**
   – Click **OK.**
9. Save the data table.

   When you generate a graph in Graph Builder and assign the modified column to the Shape role, your boundaries appear on the graph.

Creating Your Own Column Property

You can create your own column property and assign any name that you choose. This property is then available for JSL programming.

1. Double-click the column name in the data grid. The Column Info window appears.
2. Select **Other** from the **Column Properties** drop-down menu.
3. Type a name for the new property.
4. Enter a value for the property.

Response Probability

The Response Probability column property does not appear on the menu for selection. JMP automatically assigns this column property when the following steps are performed:

- Create a logistic regression using either the Fit Y by X platform or the Fit Model platform.
- Select the **Save Probability Formula** option in the logistic report window.

JMP assigns the Response Probability column property to the probability columns.

The Response Probability column property is used for profiling the probability columns of categorical variables. The Response Probability column property makes all of the levels of the variable appear as a single row in the Profiler report. If the Response Probability column property is not assigned, the Profiler creates separate rows for each level of the variable.

For more details, see the chapter on profiling in the *Modeling and Multivariate Methods* book.

Removing Properties

Column properties can be added or deleted at any time. To delete a property:

1. Double-click the column name in the data grid. The Column Info window appears.
2. Select the property in the list and click **Remove**.

Figure 5.25 Removing Properties

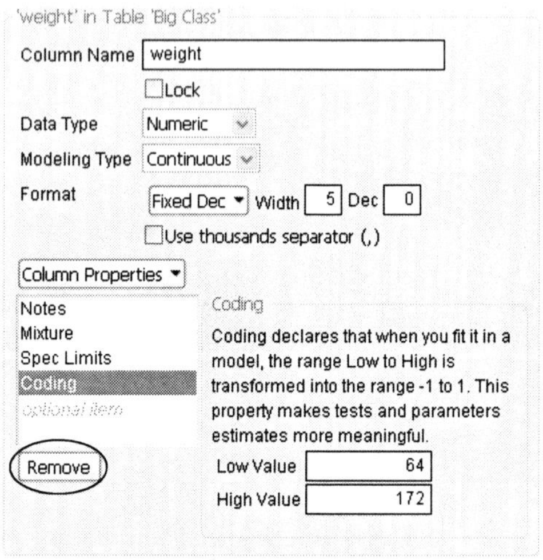

Standardizing Attributes and Properties Across Columns

A column might contain attributes (data types, modeling types, numeric formats, and so on) or properties (formulas, notes, list and range checks, and so on) that you want other columns to have. You can use the existing column to *standardize* the attributes and properties across columns. This includes both adding and deleting attributes and properties.

Adding Attributes and Properties

To apply an existing column's attributes and properties to multiple columns

1. Select the column containing the desired attributes or properties.
2. Select **Cols > Standardize Attributes.** The window in Figure 5.26 appears.

By default, the items within the Standardize Attributes panel are dimmed. To access an item:

3. Click the **Attributes** button and select the items to be duplicated across columns.

Note: The **Input Format** item is applicable only for the Date, Time, and Duration formats.

Figure 5.26 Standardizing Attributes Across Columns

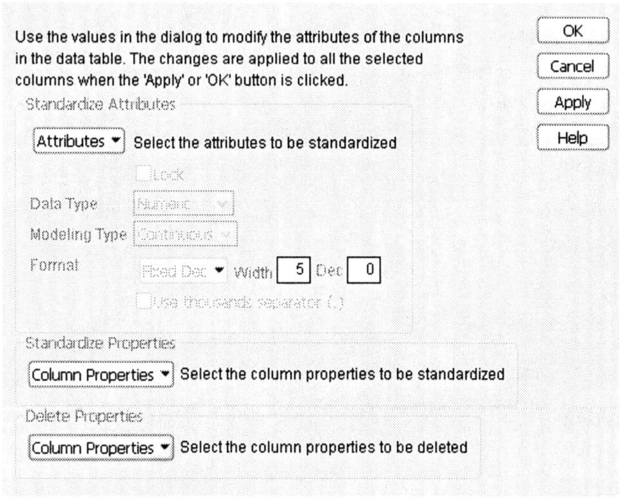

4. To change the values of any of the attributes, use the drop-down menus and text boxes in the attributes information area.
5. Click the **Column Properties** button in the Standardize Properties area and set up the properties that you want the other columns to have.
6. Click in the data table.

7. Highlight the columns in the data table to which you want to apply the attributes and properties.
8. Go back into the Standardize Columns Attributes window and click **Apply.** The attributes and properties apply to all the selected columns in the data table. You can view your results.
9. Click **OK**.

Tip: You can select all the columns that you want to standardize and then select **Cols > Standardize Attributes**. The first selected column (left-most) is used to set up the window's values. Then you might skip steps 6 and 7, because you have already selected the columns.

Deleting Properties

To delete the same properties across multiple columns

1. Select the column containing the attributes or properties that you want to delete.
2. Select **Cols > Standardize Attributes**.
3. Click the **Column Properties** button in the Delete Properties area and select the properties that you want to delete.
4. Click **OK**.

Example of Standardizing a Formula

If you have applied a formula to a column, and you want to apply that same formula to additional columns in the data table, use the **Substitute Column Reference** option. It is important to note that this option is dependent upon the location of the column that is referenced in the original formula. For example, if your original formula is based on the previous column, then any other formulas applied to additional columns are based on their previous columns.

The Blood Pressure.jmp sample data table contains blood pressure measurements taken on five subjects three times each day, over a period of three days. You want to find the log of each blood pressure (BP) column.

1. Open the Blood Pressure.jmp sample data table.

Create nine new columns, one for each existing BP column.

2. Select **Cols > Add Multiple Columns.**
3. Add nine columns.
4. Click **OK**.

Apply your original formula as follows:

5. Right-click on Column 1 and select **Formula**.
6. Select BP 8M.
7. Select **Transcendental > Log.**
8. Click **OK**.

Column 1 now contains the log of the BP 8M column. You want the rest of the empty columns to contain the log of the remainder of the BP columns.

9. In the data table, select all of the new columns that you created, including the one with the original formula (columns 1-9).
10. Select **Cols > Standardize Attributes.**
11. In the Standardize Properties panel, click on **Column Properties** and select **Formula.**
12. Select the check box next to **Substitute Column Reference.**

Figure 5.27 Completed Standardize Properties Panel

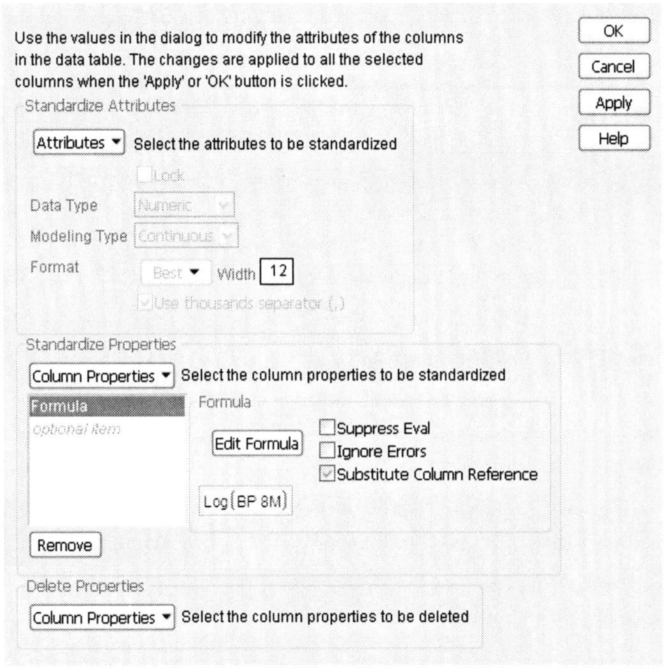

13. Click **OK.**

Now all of the new columns are populated with the log of the BP columns, in the order in which they appear. Column 1 contains the log for BP 8M, Column 2 contains the log for BP 12M, and so on. See Figure 5.28).

Figure 5.28 New Columns Populated with BP Log

BP 6F	Column 1	Column 2	Column 3	Column 4	Column 5	Column 6	Column 7	Column 8	Column 9
171	5.2072755975	5.1564669899	5.1936046592	5.1578904959	5.1841190072	5.1993379261	5.1432897243	5.179370267	5.1388226633
188	5.1504722523	5.1969409071	5.1980413843	5.1369849594	5.1864867215	5.1709668183	5.1657103661	5.2197312576	5.234328445
180	5.1972499635	5.2391005207	5.178322438	5.238787879	5.1667625357	5.203594206	5.210195218	5.2083601043	5.1936339968
183	5.1963991371	5.1745557217	5.2061014507	5.1724015063	5.1519829831	5.2128045537	5.2097555273	5.2327798951	5.2091101607
183	5.2160339977	5.193927246	5.1720568807	5.1469844063	5.1629452723	5.1680888257	5.1373223031	5.2460877197	5.2078170372
184	5.1442309913	5.2010466822	5.190272123	5.2096012721	5.1686850006	5.1599448971	5.2197014578	5.2061347434	5.2158555681
180	5.1624359599	5.1947890645	5.1939999636	5.2122053493	5.1908559374	5.1525220738	5.2158311104	5.2184801551	5.1950038697
170	5.1879960149	5.1526337052	5.2089668523	5.2358136363	5.2615147624	5.1345829767	5.1879342197	5.2184820284	5.1381696151
177	5.1805886748	5.2577570269	5.174937962	5.219947983	5.1503920678	5.1899769579	5.2298515928	5.2170237014	5.1785631176
184	5.1967689612	5.2313087654	5.2019091697	5.2032843804	5.1775361357	5.1760569244	5.2141963975	5.2118654915	5.2170798342
175	5.2029292719	5.2230611256	5.1829587628	5.1772019637	5.1737358213	5.1773153282	5.2075179631	5.2521315411	5.1647112759
184	5.1822629482	5.2142688207	5.228834261	5.1592138678	5.1632034385	5.1711384312	5.1918772486	5.2010450799	5.21531688

Compress Selected Columns

JMP lets you compress columns in a data table to minimize the size of the file and reduce the amount of memory required to analyze data. This feature is helpful when numeric columns contain many small integers or when any column contains fewer than 255 unique values. For example, compressing columns in a data table with 389 columns and 85,000 rows might decrease the file size from 250MB to 33MB, depending on the type of data.

When you compress columns, JMP verifies whether the data can be stored in a more compact form based on the data type:

- In character columns with fewer than 255 unique values, the List Check property is added to the column (shown in Figure 5.29).
- In numeric columns, data is compressed to 1-byte, 2-byte, or 4-byte integers when possible (shown in Figure 5.30). See "Using the Short-Integer Format," p. 60 for details about short integers.

  A numeric column with *non-integer* values can also be compressed if there are fewer than 255 unique values. In this case, the List Check property is added to the column.

Warning: In a column with the List Check property, you can enter only a value that is in the list. Otherwise, JMP warns that the cell contains invalid data when you try to enter the new value. For details about list checking, see "Validating Column Data," p. 148.

Figure 5.29 List Check Property Added to a Compressed Character Column

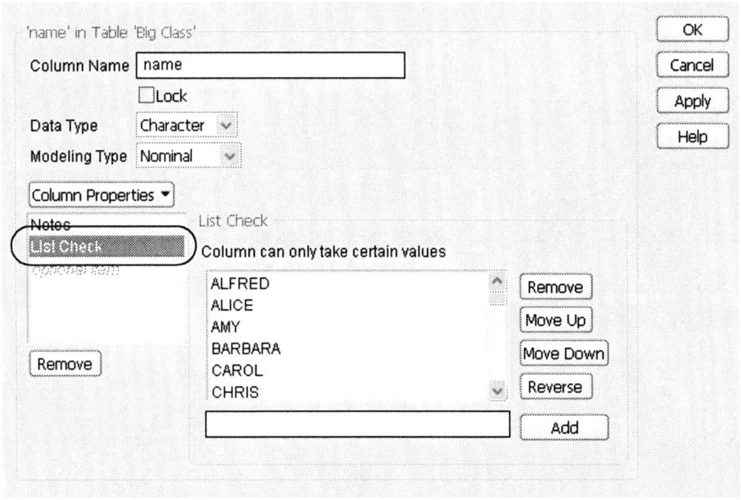

Figure 5.30 Column Info Window Showing Numeric Column Before and After Compression

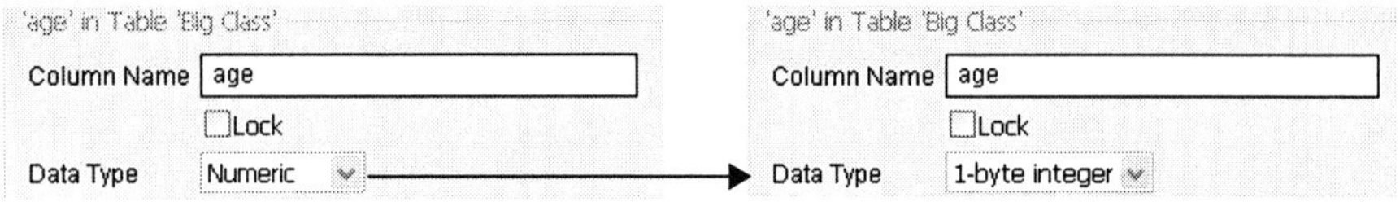

To compress columns, select one or more columns and select **Cols > Compress Selected Columns.** (Select all columns if you do not know which columns can be compressed.)

The column or columns are compressed if possible. The log shows which columns were compressed and how they were compressed. (Select **View > Log** to show the log.)

Note: To compress a numeric column manually, set your Tables preferences to allow short numeric data and then change the column's data type to **1-byte integer**, **2-byte integer**, or **4-byte integer**. See "Tables," p. 452 in the "JMP Preferences" chapter for details about this preference.

Using Row State Columns

You can create a column that contains only row state information, which stores information about whether the rows are excluded, hidden, labeled, colored, or marked. (See "Assigning Characteristics to Rows and Columns," p. 135, for details.) This column has its own data type and it does not have a modeling type because its values are not used in analyses.

Using row state columns is a two-step process. The first step is to set up the column. Then, add values to the column just as you would if it contained numeric or character data. The sections below describe how to complete these steps.

Step 1: Set Up Row State Columns

1. Create a new column by either double-clicking the top of an empty column, or by selecting **Cols > New Column**.
2. If you double-clicked to create a new column, double-click the column name again.
3. In the window that appears, click the **Data Type** menu and select **Row State**, as shown in Figure 5.31.

Figure 5.31 Select Row State for the Column's Data Type

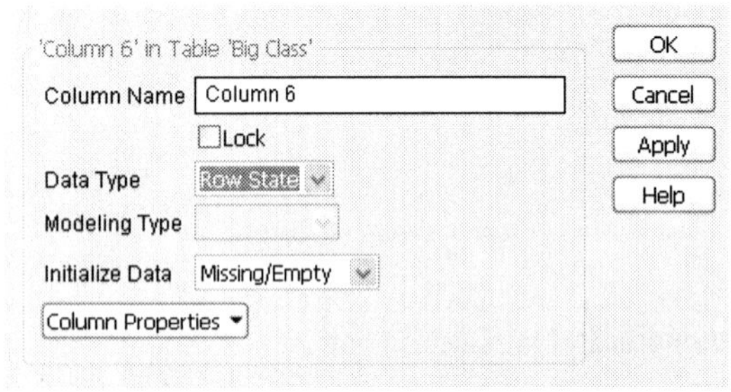

4. Click **OK**.

Step 2: Add Values to Row State Columns

After setting up the row column, populate its cells with either new information or with pre-existing row state information. There are two ways to do this: by either creating row state characteristics for the row state column or by copying existing row state information into the column. The sections below describe these two ways.

To create row state characteristics for the row state column

1. Right-click the row state column name.
2. Select **Row States Cells**. From the submenu, select a characteristic to apply to all of the rows in the column.

To populate a row state column by copying existing row state information into the column

1. To populate only certain rows in the row state column, highlight those rows. To populate on all rows in the column, highlight the row state column.
2. Click the star icon (★) beside the column name located in the column panel.
3. Select one of the following:

- **Copy from Row States** Copies the characteristics from the row number area to the row state column, replacing existing row state values in the column.
- **Add from Row States** Adds the characteristics from the row number area to the row state column, preserving existing row state values in the column.
- **Copy to Row States** Copies the characteristics from the row state column to the row number area, replacing existing active row states.
- **Add to Row States** Adds the characteristics from the row state column to the row number area, preserving existing active row states.

Permanently Highlighting Cells

You can save a highlight in a row state column just like you save other row state characteristics (hide, exclude, color, and so on). This places a "permanent" highlight on a cell. To do this:

1. Right-click a cell and select **Row States Cells > Select/Deselect**.
2. Repeat this for as many cells as you would like to highlight.
3. To remove the highlight, right-click on the cell and select **Row States Cells > Select/Deselect**.

Chapter 6

Output Reports

Using the Report Window

Reports contain both text reports and graphs. This chapter describes features common to all reports in JMP. Reports are designed to be convenient in several ways.

- Individual plots can be resized and text reports can be opened and closed to optimize screen space. (See "Resizing Plots and Graphs," p. 198, and "Showing and Hiding Parts of a Report," p. 181.)
- Plots and text reports can be printed. (See "Printing Reports," p. 184.)
- Context-sensitive help is accessible using the question mark tool, accessed from Tools > Help. After selecting it, you can click anywhere in a report. (See "Use JMP Help," p. 6.)
- All or part of a report can be copied to other applications, such as Microsoft Word. (See "Pasting Reports into Another Program," p. 185, and "Pasting Reports into Another Program," p. 123.)

Figure 6.1 Example of a JMP Report Window

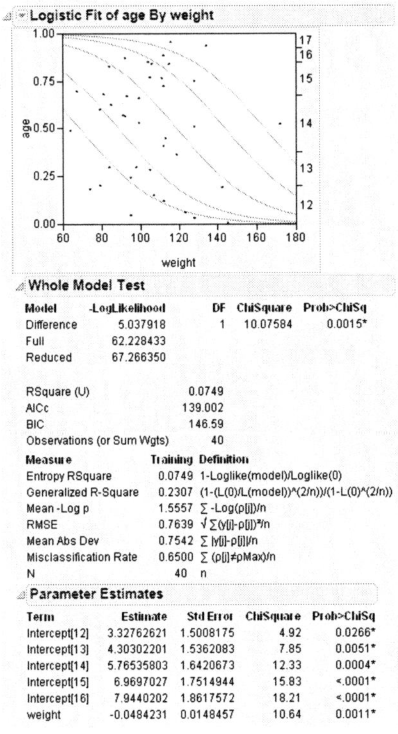

Contents

Editing Reports

JMP reports are displayed in standard windows with scroll bars and options to resize. They also have other special buttons and menus like those illustrated in Figure 6.2 and those discussed in the following sections.

Figure 6.2 Basics of the Report Window

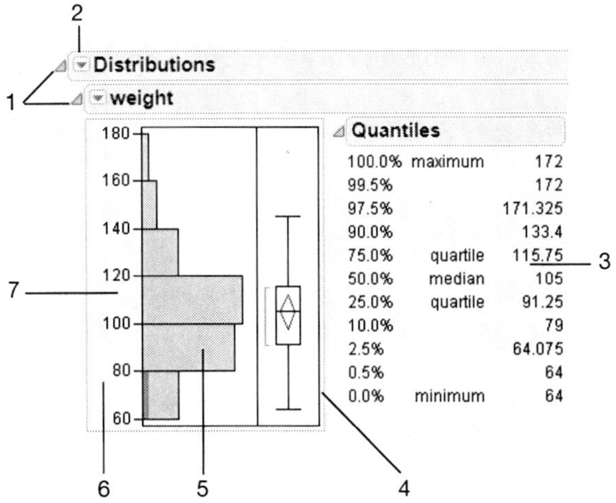

Table 6.1 Report Window Actions

Number	Action
1	Click on the disclosure buttons to hide or show sections of the report.
2	Click on the red triangle menus to access report options.
3	Right-click in the table to access formatting options.
4	Click and drag on the borders to resize graphs.
5	Right-click anywhere in the graph to access formatting options.
6	Right-click within the axis to access formatting options.
7	The arrow cursor turns into a hand when you hover over an axis. Click and drag using to scroll along the axis or to rescale the axis. See "Scrolling and Scaling Axes," p. 201.

Using the Hand Tool

Select the hand tool using the **Tools > Grabber** option. There are many functions you can use with the hand tool (also known as the grabber tool) in a report. Here are some examples of the way the hand behaves in graphs and plots:

- On histograms, for continuous variables, use the hand tool to change the number of bars or to shift the boundaries of the bars.
- In all report tables, use the hand tool to click and drag columns for rearranging.
- Use the hand tool to change the displayed range of axis values. See "Scrolling and Scaling Axes," p. 201.

Access Report Display Options

Right-click a disclosure button ◿ to show a menu that lets you rearrange the report and gives you control over report outline levels. The resulting menu has the following report formatting options:

Close Closes (hides) that section of the report. This can also be accomplished by clicking the disclosure button.

Horizontal (if available) Horizontally aligns the next lower outline levels. By default, most reports are arranged with the outline levels showing vertically with indented outline levels.

Open All Below Opens all outline levels beneath the level where this command is selected, including that level.

Close All Below Closes all outline levels beneath the level where this command is selected, including that level.

Open All Like This Opens all of the same type of reports as the one that is present in the analysis window. If you analyze several variables at a time, you often want to close many of the same type of report tables all at once. You might also want to close all of the same type of report tables at once when you select multiple options on a single analysis.

Close All Like This Closes all of the same type of reports as the one that is present in the analysis window.

Close Where No Outlines Closes all parts of the report that do not have sublevels. This command is usually used at the top level of the report outline. It is a quick way to see a nesting structure overview of a report.

Append Item Displays a submenu, which lists ways you can add structural items to the report. Items include text, outline title bars, references to other JMP files and windows, a list of all open JMP files, and URLs.

Edit Displays the submenu shown in Figure 6.3, which affect all reports at the outline level where they are used:

Select Highlights all reports for that outline level.

Deselect Deselects all selected reports for that outline level.

Journal Duplicates the report in a separate window titled Journal so you can edit it or append other reports to it. (See "Create Journals," p. 116, for details.)

Copy Picture Copies the report to the clipboard. You can then open another application and paste it.

Page Break Inserts a page break for printing purposes.

Show Tree Structure Opens a window that shows the DisplayBoxes that make up the report. This is mainly used by JSL programmers who are manipulating or reading parts of the report.

An alternative way to access these options is to hold down the ALT key and right-click the disclosure button ⊿. This displays a window, as shown in Figure 6.3, with check boxes for commands and options so that you can select multiple actions at the same time. You can also do the same for the menu under a red triangle icon.

Figure 6.3 Menu Items in a Window

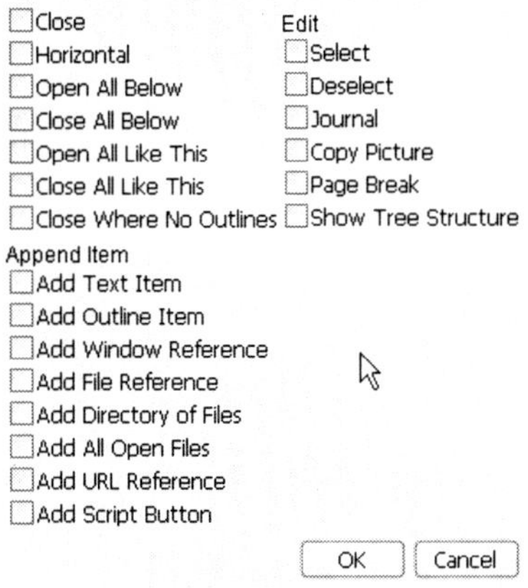

Showing and Hiding Parts of a Report

JMP reports are organized in a hierarchical outline. Each level of the outline has a triangle-shaped disclosure button ⊿. Click the disclosure button to open and close that section of the report. In Figure 6.4, the report on the left has the Whole Model Test section open, while the report on the right has that section closed.

Figure 6.4 Reports With a Section Open and Closed

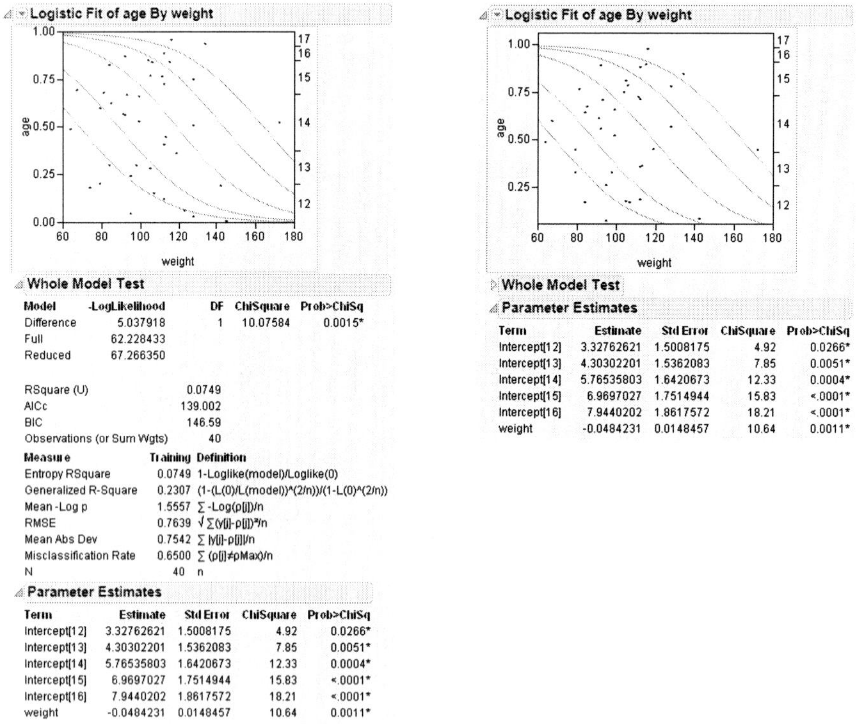

Renaming a Report

Figure 6.5 illustrates how to change the title that appears on the top of a report.

Figure 6.5 Renaming a Report

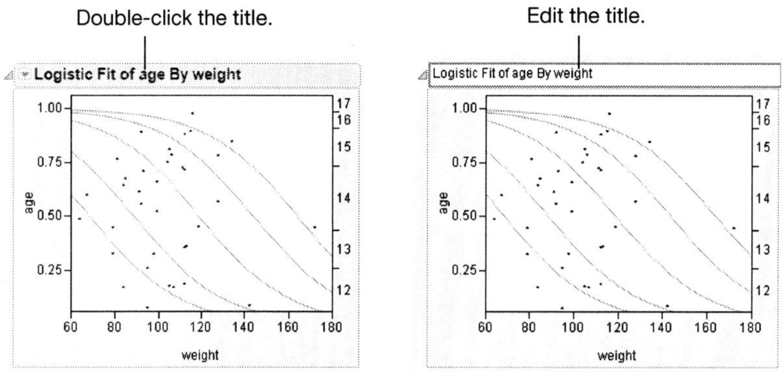

Increasing Font Sizes

On Windows, change the font size that JMP uses in reports and data tables by selecting **Window > Font Sizes**. Then choose from one of the submenu items:

Increase Font Size Increases the font size. Select again to increase the font size again.

Decrease Font Size Decreases the font size. Select again to decrease the font size again.

Editing Data Table Rows Using the Row Editor

Note: You cannot edit text in a locked column or a locked data table. See "Locking Columns," p. 147, for details about unlocking a column. See "Locking Tables," p. 80, for details about unlocking a data table.

To browse or edit values one row at a time from the report graph:

1. Highlight a point in the graph.
2. Right-click anywhere in the graph.
3. Select **Row Editor**.

   When the row editor appears, it displays information about the selected row.

Figure 6.6 The Row Editor

4. Edit the data.
5. Click the arrow buttons to browse through selected rows or the entire data set. Table 6.2 describes the function of each button and the corresponding red triangle option.

Table 6.2 Description of Row Editor Buttons

Button	Function	Red Triangle Option
	Shows information for the previous row.	**Prev**
	Shows information for the previously selected row.	**Prev Selected**
	Makes the currently highlighted row flash.	**Blink**
	Shows information for the next row.	**Next**
	Shows information for the next selected row.	**Next Selected**

Table 6.2 Description of Row Editor Buttons *(Continued)*

Button	Function	Red Triangle Option
	Finds rows based on criteria that you specify. See "Selecting Cells That Contain Specific Values," p. 91.	**Find**
	Creates a new row in the data table. Enter the values for the row.	**New Row**
none	Saves the data table.	**Save**

Note: You can also access the row editor from a data table by selecting **Rows > Row Editor.**

Understanding the p-value Indicator

In output reports, JMP places an asterisk next to p-values that are less than 0.05.

Figure 6.7 p-values

Whole Model Test

Model	-LogLikelihood	DF	ChiSquare	Prob>ChiSq
Difference	5.037918	1	10.07584	0.0015*
Full	62.228433			
Reduced	67.266350			

RSquare (U)	0.0749
AICc	139.002
BIC	146.59
Observations (or Sum Wgts)	40

Measure	Training	Definition
Entropy RSquare	0.0749	1-Loglike(model)/Loglike(0)
Generalized R-Square	0.2307	(1-(L(0)/L(model))^(2/n))/(1-L(0)^(2/n))
Mean -Log p	1.5557	∑ -Log(p[j])/n
RMSE	0.7639	√ ∑(y[j]-p[j])²/n
Mean Abs Dev	0.7542	∑ |y[j]-p[j]|/n
Misclassification Rate	0.6500	∑ (p[j]≠pMax)/n
N	40	n

Parameter Estimates

Term	Estimate	Std Error	ChiSquare	Prob>ChiSq
Intercept[12]	3.32762621	1.5008175	4.92	0.0266*
Intercept[13]	4.30302201	1.5362083	7.85	0.0051*
Intercept[14]	5.76535803	1.6420673	12.33	0.0004*
Intercept[15]	6.9697027	1.7514944	15.83	<.0001*
Intercept[16]	7.9440202	1.8617572	18.21	<.0001*
weight	-0.0484231	0.0148457	10.64	0.0011*

Printing Reports

To print the report in the active window, select **File > Print**. This command displays the standard window for printing. The appearance of the window depends on your operating system and printer driver.

To insert a page break for printing purposes:

1. Right-click the disclosure button in the report window.
2. Select **Edit > Page Break**.

Pasting Reports into Another Program

When you need to use JMP reports or data tables in another program, you can copy and paste or drag and drop parts of a report or table into another program, such as Microsoft Word or PowerPoint. Then save the document in that application.

1. Click the selection tool.
2. Click and drag (or hold down the SHIFT key and click) to select items in a report window or data table. Clicking near the edge of the report window selects the entire report.
3. Click the selected items and drag them from JMP to the other program. Or, copy the selected items in JMP and paste them into the other program. When you paste an element into another application, the format used depends on the application into which you paste. If the application has a **Paste Special** command and you use it, you can select a format to use, such as text (.rtf), unformatted text (.txt), bitmap (.bmp), or enhanced picture (.emf).

To copy all text (no graphs) from the active report window as unformatted text, select **Edit > Copy As Text**. On the Macintosh, press the SHIFT key and select **Edit > Copy As Text**. To copy only the graph (no text), right-click the graph and select **Edit > Copy Picture**.

Adding Options and Working with Analyses

How to Access Analysis Options

Click the red triangle icon in a report to access a list of options that apply for that particular report, as shown in Figure 6.8.

Figure 6.8 The Red Triangle Icon and Options

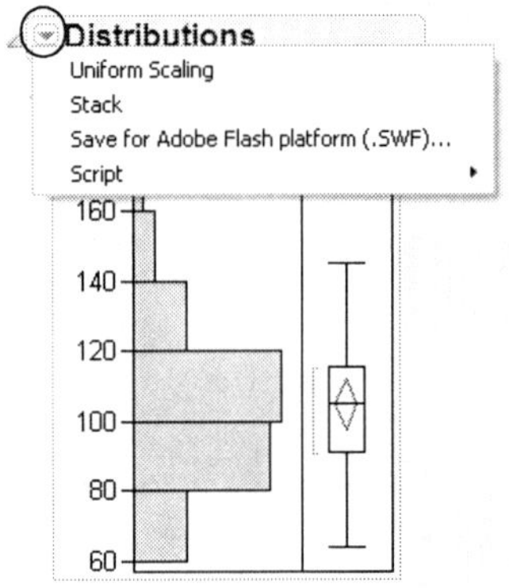

In addition to clicking the red triangle icon, you can also:

- Select multiple actions at the same time. Hold down the ALT key and click the red triangle icon. A panel of all commands and options appears with check boxes.
- Apply a command to all similar reports in the report window. Hold down the CTRL key and click the red triangle icon. For example, in a Oneway analysis, if you hold down the CTRL key, click the icon, and select **Means/Anova/t Test**, an analysis of variance is performed for all Oneway analyses in the active report window.

The red triangle options applicable to each platform in the **Analyze** and **Graph** menus are described in the following books:

- *Basic Analysis and Graphing*
- *Modeling and Multivariate Methods*
- *Quality and Reliability Methods*

Rerunning An Analysis

You can redo an analysis and get its report without having to return to the main menu. For example, if you run a bivariate fit, and then alter a value or two in the data table, you could go back to the original report and rerun the analysis from there.

To redo an analysis from the report window:

1. Click the red triangle icon in the report title.
2. Select **Script > Redo Analysis**.

Note: For details about the **Relaunch Analysis** and **Automatic Recalc** options, see the *Basic Analysis and Graphing* book.

Saving Your Steps as a Script

After completing an analysis and receiving a report, you might want to save the process you used to arrive at the report as a JSL script. You can save the script inside the data table, inside the report, or in a separate script window. To do this, click the red triangle icon in the report title and select **Script.** Then select the **Save Script** option that you prefer. See “Script Menus,” p. 11.

Non-English versions of JMP can save scripts in either English or your local language. See “General,” p. 445 in the “JMP Preferences” chapter.

Formatting Report Tables

There are many ways you can format your report to meet your needs. Right-click on a report to see formatting options. The sections below detail how to make changes to the textual portions of your reports.

Figure 6.9 Right-Click Report Menu

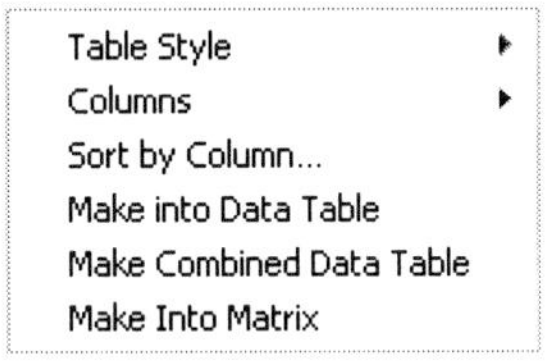

Reordering Rows (Sorting)

To sort, or rearrange, the rows found in a report table:

1. Right-click anywhere in a report table.
2. Select **Sort by Column.**
3. From the window that appears, highlight a column to use as a sort variable.
4. (Optional) To display the sorted column in ascending order (instead of the default, which is descending order), click the box beside **Ascending**.

Showing and Hiding Columns

Report tables often have many columns. You can specify which columns you want to show or hide, as follows:

1. Right-click anywhere in a report table.
2. Select **Columns.**
3. Select a column to show or hide. Columns with checks beside their names appear in the table.

Note: Columns whose names begin with a tilde (~), such as **~Bias**, are not applicable to the analysis that you ran and do not appear in the table, even if you place checks next to their names.

Adding Table Borders

By default, a report table has no borders to separate rows and columns. You can tailor the appearance and content of the tables. For example, see Figure 6.10.

Figure 6.10 Default Table and Table with Bordered Style

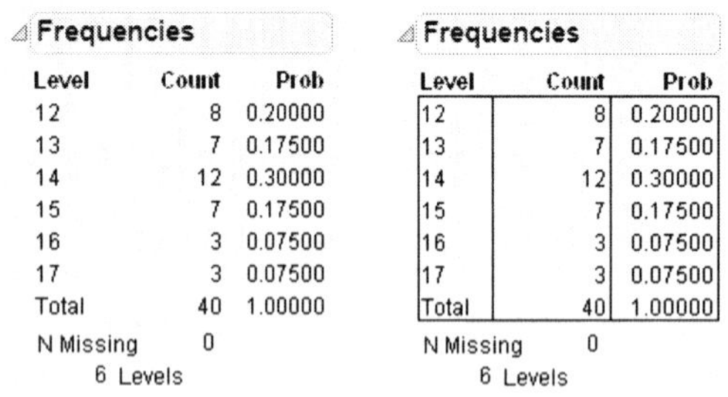

Frequencies

Level	Count	Prob
12	8	0.20000
13	7	0.17500
14	12	0.30000
15	7	0.17500
16	3	0.07500
17	3	0.07500
Total	40	1.00000

N Missing 0
6 Levels

Frequencies

Level	Count	Prob
12	8	0.20000
13	7	0.17500
14	12	0.30000
15	7	0.17500
16	3	0.07500
17	3	0.07500
Total	40	1.00000

N Missing 0
6 Levels

To add borders:

1. Right-click anywhere in a report table.
2. Select **Table Style**.
3. Select one of the options to enhance the appearance of the table:

   Plain Contains no divider lines or borders.

   Bordered Contains a border around the table and divider lines between columns.

   Embossed On Windows, adds a three-dimensional effect to the border and divider lines. On the Macintosh, changes the border from black to gray.

   Beveled On Windows, adds a three-dimensional effect and causes the corners of borders to be rounded instead of square. On the Macintosh, shades the table background gray.

   Heading Separator Adds a thin line under the boldface table headings. If there are no headings in the table, the **Heading Separator** setting appears to have no effect.

Note: You can also change the table style using the Report table style option in **File > Preferences > Reports.**

Changing Numeric Formats and Field Widths

By default, JMP picks what it considers to be the best format for numbers both in reports and in the data table. For long decimals, this format is as wide as will fit in the column. It also truncates trailing zeros. To change this in a report table:

1. Double-click anywhere in a report's numeric column, as shown in Figure 6.11. The Column Numeric Format window appears.

Figure 6.11 Change the Table's Format

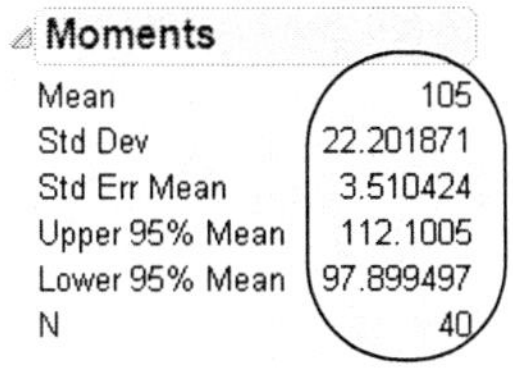

2. Select a format from the **Format** drop-down menu.
3. If you select **Fixed Dec**, **Date**, **Time**, or **Duration**, other fields appear. Type in the number of decimals, select the date format, or select the time format from the next box.
4. In the **Width** box, type the number of characters that you want the column in the report to accommodate. Be sure to type a number that is the largest number of digits that you think a value in the column could have.
5. (Appears only for selected formats.) Select **Use thousands separator (,)** to add commas to values that equal a thousand or more. You must allow spaces for these values in the **Width** box, or else they might not appear.
6. (Optional) Select the **Conditional Highlighting** that you want to use, if any. See "Conditional Highlighting," p. 189.
7. Click **OK**, and the column in the analysis table formats accordingly.

Conditional Highlighting

Note: You can control the **Conditional Highlighting** option using the **Show conditional formatting** preference. See "Reports," p. 449.

The **Conditional Highlighting** option conditionally formats the color of the text that represents the values. You can choose from the following selections:

None Applies no conditional formatting to the selected values.

Correlation Applies the following color rules to correlation values:

- Values of -1 to 0 range on a gradient from red to gray.
- Values of 0 to 1 range on a gradient from gray to blue.

PValue Applies the following color rules to p-values:

- Values greater than 0.01 and less than 0.05 are red.
- Values less than 0.01 are orange.

Factor Pattern Applies the following rules to factor pattern values:

- Values between -0.4 and 0.4 range on a gradient of transparency.
- Values close to 0 are the most transparent, with a value of 0 being invisible.
- Values below -0.4 and above 4 appear as normal (black).

For additional information about factor patterns, see the *Modeling and Multivariate Methods* book.

Changing Outline Titles and Column Headings

You can change the name of a outline title and a column heading as follows:

1. Double-click the outline title or column heading. It is highlighted, as shown in Figure 6.12.

Figure 6.12 Double-Click an Outline Title (Left) or a Column Heading (Right)

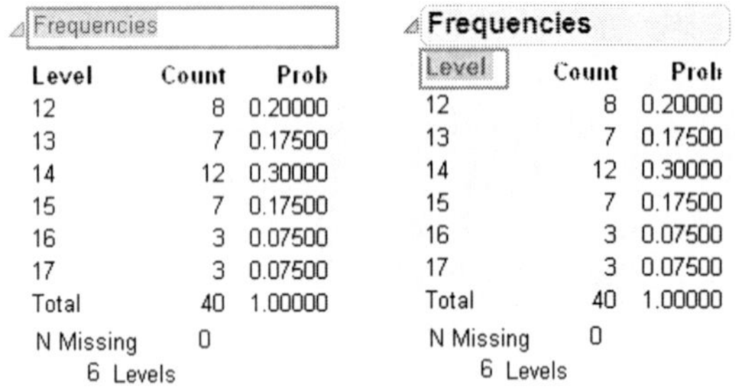

2. Click it again. The cursor becomes a flashing text insertion bar.
3. Edit its contents.

Turning a Report Table Into a Data Table

You can create a JMP data table from a report table, as follows:

1. Right-click anywhere in a report table.
2. Select **Make into Data Table** to create a data table from the report table.

   Select **Make Combined Table** to search the report for other tables like the one you clicked and combine them into a single data table.

Turning a Report Table Into a Matrix

You can create a JMP matrix from a report table. For example, when working with JMP Scripting Language (JSL), you might want to access a report's table that has been stored into a JSL variable. Or, you might want to store a report table's values into a table property as either a table property or as a JSL assignment, which is stored within the data table and is accessible via a script or the formula editor.

To store a table in matrix form into a global variable, into a table property, or into a table property as an assignment:

1. Right-click anywhere in a report table.
2. Select **Make into Matrix**.
3. In the window that appears, tell JMP how you want to store the table.
4. (Optional) Rename the variable or property by typing a new name into the box beside **Name**.

Selecting Points in Plots

To select a point in a plot, click the point with the arrow cursor. This selects the point as well as the corresponding row in the current data table. To select multiple points, hold down the SHIFT key while you select points. A point's label appears when you place the cursor over the point with or without clicking.

Selecting Rows and Columns in Plots, Charts, and Graphs

All graphs and plots that represent the same data table are linked to each other and to the corresponding data table. When you click points in plots or bars of a graph, the corresponding rows highlight in the data table. The example in Figure 6.13 shows a histogram with the SPEEDYTYPE bar highlighted, and the corresponding rows highlighted in the table. You can also extend the selection of bars in a histogram by holding down the SHIFT key and then making your selection.

Figure 6.13 Highlighting Rows In a Histogram

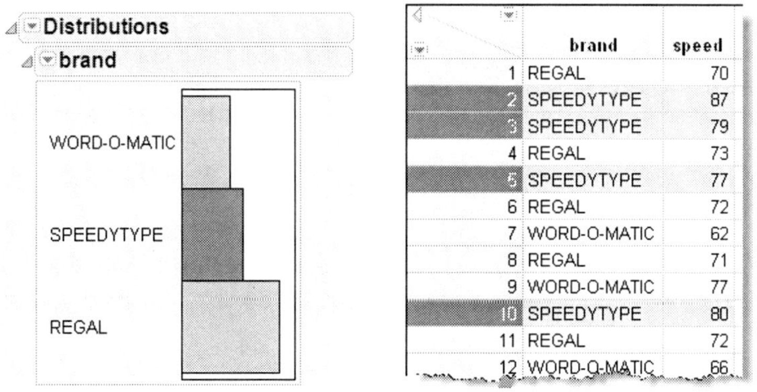

Selecting a Rectangular Area of Points

You can select all points that fall in a rectangular area using the arrow cursor. Click and drag the arrow to highlight points. Alternatively, you can use the brush tool. As you move the brush over the graph, points that fall within the rectangle are selected. Any points marked in the data table as hidden are not selected. (See "Hiding Rows and Columns," p. 136.)

To select points using the brush tool:

1. Click the brush tool  in the toolbar.
2. Click and hold the cursor (now brush-shaped) in a plot. A rectangle appears.
3. Move the rectangle over points. As it passes over them, they appear larger and are highlighted both in the plot and in the active data table.

Note: To keep all points selected as you move the brush-shaped cursor over points, press the SHIFT key before you click in the plot.

Figure 6.14 Using the Brush Tool

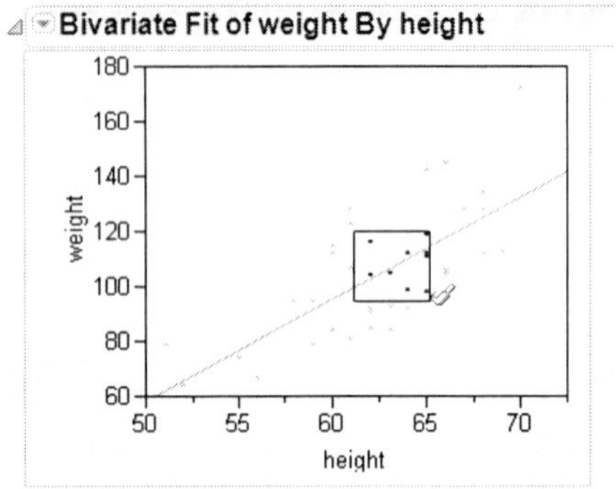

4. Release the mouse. The points within the rectangle remain selected.

You can also use the brush tool to change the size of the selection rectangle. Hold down the ALT key before you click in the plot. This shape acts like a slicing tool that can traverse and highlight slices of points across either axis.

Selecting an Irregular-Shaped Area of Points

You can use the lasso tool to select points that fall in an irregular-shaped area. Any points marked in the data table as hidden are not selected. (See "Hiding Rows and Columns," p. 136.)

To select points within an irregular-shaped area:

1. Click the lasso tool  in the toolbar.
2. Click and hold the cursor (now lasso-shaped) in a plot.

Note: To keep all points selected as you drag the lasso around several sets of points, press the SHIFT key before you click in the plot.

---

3. Drag the lasso around any set of points.

Figure 6.15 Using the Lasso Tool

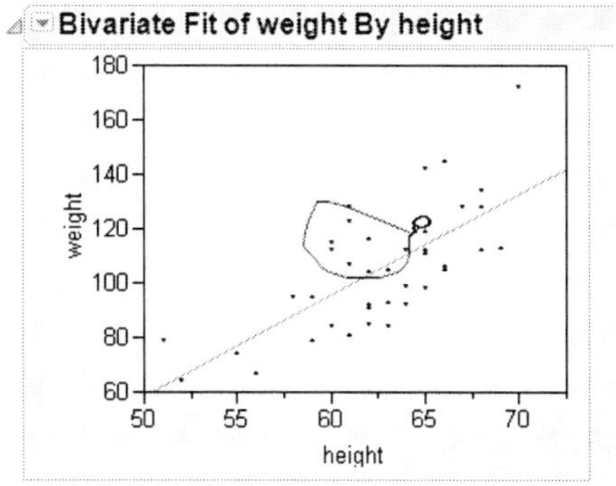

4. Release the mouse. JMP automatically closes the lasso and highlights the points within the enclosed area.

Using Markers

Markers are points on a graph that represent data. Once they are changed from their default setting, they also appear next to rows in the data table. The following sections show you how to change marker shape, size, color, and so on.

Changing Marker Shape

You can assign a character from the JMP markers palette to replace the standard points in scatterplots. These markers also appear next to row numbers in the data table.

1. Highlight one or more markers whose shape you would like to change.
2. Right-click anywhere in the graph. ((To change only the color of a box plot, right-click the box plot area.)
3. Select **Row Markers.**
4. Select a marker shape from the options that appear, or click **Custom** to enter a character to use as a marker.

Changing Marker Colors

You can assign any color to highlighted rows. When you do this, the points in scatterplots appear in the color that you select from the colors palette. The active color assigned to a row appears next to the row number in the data grid.

To change the color of markers (points) on a graph:

1. Highlight one or more markers whose color you would like to change.
2. Right-click on the graph. (If you want to change the background color of the box plots, for example, right-click the box plot area.)
3. Select **Row Colors**.
4. Select one of the colors, or click **Custom** to apply a custom color.

Changing Marker Size

To increase or decrease the size of markers (points) on a graph:

1. Right-click anywhere in a graph. Hold down the CTRL key and right-click to broadcast the command and apply it to all plots of the same type located in the same window. In a histogram, right-click the box plot area on the right.
2. Select **Marker Size**.
3. Select one of the marker sizes listed. **Preferred Size** is the size that JMP estimates to be the best size for the graph.

Changing the Marker Drawing Mode and Transparency

When working with a large number of markers on a graph, the markers can appear crowded. If this is the case, you might need to alter the transparency to gain a better view. Altering the transparency might also affect the marker drawing mode, which is the mode JMP uses when it refreshes a report window. As it draws markers on a plot, it uses one of two speeds: normal or fast.

To change the marker drawing speed:

1. Right-click anywhere in a graph. In a histogram, right-click the box plot area on the right.
2. Select **Marker Drawing Mode**, and then select either **Normal** or **Fast**.

   Normal If JMP is in normal drawing mode and the number of markers in a graph are more than the specified threshold number, JMP automatically switches to fast mode. See "Reports," p. 449 in the "JMP Preferences" chapter, for details about setting the marker threshold.

   Fast Graphs displaying a large number of markers appear faster if you set the marker drawing speed to **Fast**. Note that when the drawing speed is set to **Fast**, marker size reverts to Preferred Size, and marker transparency settings revert to the default opaqueness.

   Outlined see "Adding Outlines Around Markers," p. 194.

Adding Outlines Around Markers

You can add a black outline, or frame, to markers in a plot. Outlined markers are available at the medium, larger, XL, XXL, and XXXL marker size. (See "Changing Marker Size," p. 194, for details.)

To add outlines:

1. Right-click a plot or graph.

2. Select **Marker Drawing Mode**.
3. Select **Outlined**.

To use an outline effectively, it is best if your marker is a color other than black.

To change marker colors:

1. Highlight the markers whose color you want to change.
2. Right-click anywhere in the graph.
3. Select **Row Colors**.
4. Select a marker color from the options that appear.

Marker Selection Modes

When you select markers on a graph, only the selected markers are highlighted. You can change the way markers are highlighted on the current graph. The options are applied to the top two triangles in the following figure.

Figure 6.16 Examples of Highlighted Triangular Markers

To change the marker highlighting on the current graph:

1. Right-click anywhere in a graph and select **Marker Selection Mode**.
2. Select one of the following options:

   Unselected Faded only the selected markers are highlighted. Everything else is dimmed.

   Selected Larger the selected markers are larger than the deselected markers.

   Selected Haloed the edges of the selected markers are outlined in blue.

   Selected Outlined the selected markers are outlined in black.

For information about setting the default marker selection mode, see “Reports,” p. 449 in the “JMP Preferences” chapter.

Specifying Marker Transparency

You can change the transparency of markers (points) on a graph. For example, this enables you to control the visibility of overlapping points.

Note: When the drawing speed is set to **Fast**, marker transparency settings revert to the default opaqueness and marker size reverts to **Preferred Size**.

To adjust markers' transparency:

1. In a graph, right-click anywhere and select **Transparency.** In a histogram, right-click the box plot area on the right and select **Transparency.**
2. Type the level of transparency that you want the markers (points) to have on the graph, and click **OK.**

   A value of 1 indicates total opaqueness, while 0 indicates invisibility. Values between 1 and 0 are semi-transparent.

Excluding and Hiding Markers

Using the **Exclude/Unexclude** command, you can exclude highlighted rows from statistical analyses. Data remain excluded until you select **Rows > Exclude/Unexclude** for those highlighted rows.

Note: Excluded data are not automatically hidden in plots even though they are excluded from calculations in text reports and graphs.

Using the **Hide/Unhide** command, you can suppress (hide) the appearance of highlighted points in scatterplots. For example, you can exclude points from analysis and then hide those same points in scatterplots. The data remain hidden until you select **Rows > Hide/Unhide** for highlighted hidden rows.

Note: Hidden points are not automatically excluded from statistical computations that affect text reports and graphs, even though they are not displayed in the plots. To exclude hidden observations from analyses, you must highlight them and select **Rows > Exclude/Unexclude** characteristic.

To exclude or hide markers (points) from analyses:

1. Highlight the marker(s) you would like to exclude or hide.
2. Right-click anywhere in a graph.
3. Select **Row Exclude** or **Row Hide.**

Adding Labels to Markers

When you position the arrow cursor over a point in a plot, the point's label appears. By default, the label is the row number. There are three ways you can customize the label:

- You can change the label so it displays values found in one or more columns instead of the row number.
- You can enable the label to appear always, not just when you position the cursor over points.
- You can click the label with the Arrow tool and drag it to a new location. If a label is dragged a certain distance away from the marker, then a tail is added connecting the label to its point.

To change the label so it displays values found in one or more columns instead of the row number:

1. In the data table, highlight the column(s) whose values you want to appear as the label in plots.
2. Select **Label/Unlabel** from one of the following places:
   - the **Cols** menu
   - the red triangle icon in the columns panel
   - the top red triangle icon in the upper left corner of the data grid

A label or yellow tag icon beside the column name in the columns panel indicates that points on plots are identified by the column value. If there are multiple labeled columns, their values appear on plots separated by a comma. Data remain labeled until you highlight the column and select **Label/Unlabel** again.

To enable the label to appear always, not just when you position the cursor over points:

1. Highlight the point(s) whose label you want to always appear in plots.
2. Right-click anywhere in a graph. In a histogram, right-click the box plot area on the right.
3. Select **Row Label**.

A label or yellow tag icon beside the row number in the data table indicates that points on plots corresponding to the row appear with a label.

Changing Marker Shape or Colors Based On Values

In some plots, you can change marker shapes or colors based on the values of points by adding a *row legend.* It is called a row legend because JMP automatically inserts a legend using row color or row marker settings. When you assign markers or colors in this way, it assigns the characteristic(s) to all points in a graph, regardless of what points you have selected. All previous marker and color settings are overwritten.

To add shapes or colors based on column values:

1. Right-click anywhere in a graph. In a histogram, right-click the box plot area on the right.
2. Select **Row Legend**.
3. In the window that appears (Figure 6.17), highlight the column whose values you want to color or mark. A preview of the legend is shown on the right.

Figure 6.17 Adding a Row Legend

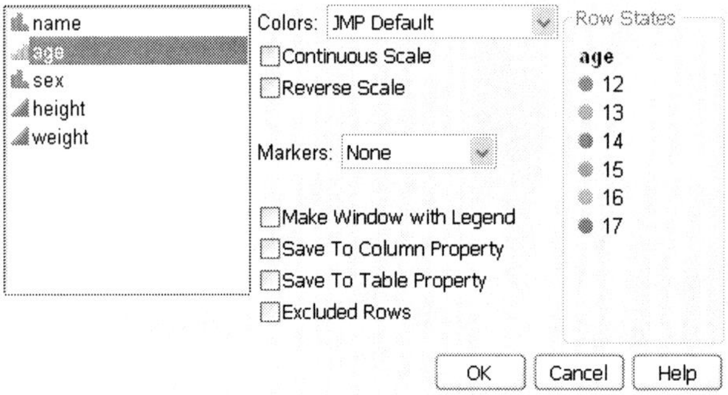

4. Refine your row legend using the following options:

   Colors Lets you choose among several pre-defined color schemes.

   Continuous Scale Assigns colors on a spectrum that corresponds to the ascending or descending order of the values. Use this option when the highlighted column contains continuous values.

   Reverse Scale Reverses the scale of colors.

   Markers Lets you choose among several marker schemes.

   Make Window with Legend Creates a separate legend window that tells you what colors and shapes correspond to which value.

   Save Column Property Adds a column property that stores the selected color theme.

   Save Table Property Adds a table property that preserves the selected color and marker configuration.

   Excluded Rows Assigns colors or markers to rows that are excluded.

Removing the Legend

After you make the legend, right-click it to change the colors and the markers or to remove the legend. To remove colors or markers, select **Rows > Clear Row States** in the data table.

Altering Plot and Chart Appearances

There are many ways that you can format your report to meet your needs. The sections below detail how to make changes to the graphical portions of your output reports.

Resizing Plots and Graphs

There are two main ways to resize plots and graphs: using the click and drag method or resizing it according to pixel size.

Note: You can also change the default size of a graph using the Graph Height option in **File > Preferences > Reports**.

Using Click and Drag

To resize a plot or graph using the click and drag method:

1. Place the cursor on the right edge, bottom edge, or lower right corner of the plot frame. The cursor changes to a small double-arrow pointer.
2. Click and drag to change the size of the plot. When you resize, the height and width of all plots in that frame adjust independently of other frames in the same report window. Table 6.3 describes how to adjust the plot.

Table 6.3 Resizing Actions

Action	Instructions
Adjust the plot frame but preserve the proportions (aspect ratio)	Hold down the SHIFT key and click and drag the corner of the frame.
Adjust a plot in 8-pixel increments	Hold down the ALT key and click and drag the corner of the frame.
Adjust all plots of the same type simultaneously	Hold down the CTRL key and click and drag the corner of one of the plots. If you do this for one scatterplot, the action is broadcast to all scatterplots in the window, and they resize together. Any other types of plots remain unchanged.

Specifying Size in Pixels

To resize a plot or graph to a specific pixel size:

1. Right-click the plot or graph.
2. Select **Size/Scale** > **Frame Size**.
3. Enter the number of pixels for the frame's height and width.

Note: For details about the other options in the **Size/Scale** menu, see "Scrolling and Scaling Axes," p. 201.

Zooming In and Out

The magnifier lets you automatically zoom in on any area of a plot. When you click the magnifier, the point or area where you click becomes the center of a new view of the data. The scale of the new view enlarges, giving you a closer look at interesting points or patterns. You can perform any of the following actions:

- Click and drag the magnifier to focus in on a particular region of the plot.
- On a ternary plot, drag the magnifier to zoom the triangular axes.
- Zoom repeatedly to look closer at the data.
- Hold down the CTRL key and click to return to your previous state before the last zoom.
- Double-click or hold down the ALT key and click the magnifier to restore the original plot.

Example of Zooming with the Magnifier Tool

1. Open the Pollen.jmp sample data table.
2. In the Tables panel, click on the red triangle next to Bivariate and select **Run Script**.

In this example, the data were devised with a hidden pattern (message) buried in over 3,800 points. Now you see the whole table as the dense cloud shown to the left in Figure 6.18. Clicking once with the magnifier produces the plot in the middle, and clicking twice more reveals the surprise message shown in the plot on the right.

Figure 6.18 Example of Zooming with the Magnifier Tool

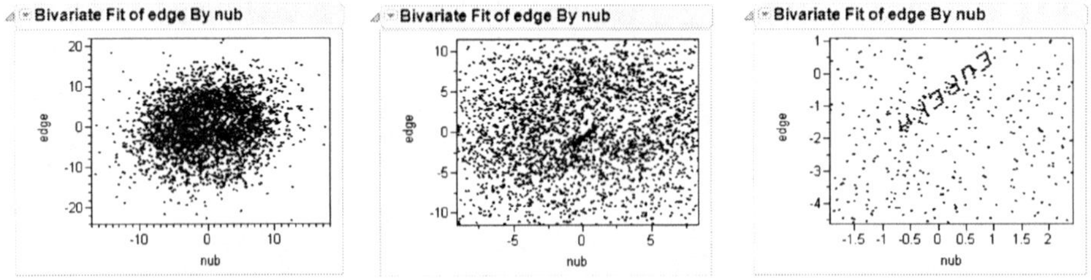

Changing Line Widths

After fitting a line to a graph, or producing a graph with a line already present, you can adjust the width of the line:

1. Right-click anywhere in a graph.
2. Select **Line Width Scale.**
3. Select to increase the current line width one to three times its default width. Or, select **Other** and specify a larger or smaller number. Select **Scale with Font** to increase the line size as you increase the display font size using **Window > Font Sizes** (Windows) and **View > Make Text Bigger/Smaller** (Macintosh).

Changing the Background Color in a Graph

To change the background color in a graph, follow these steps:

1. Right-click anywhere in a graph. (To change only the color of a box plot, right-click the box plot area.)
2. Select **Background Color.**

3. Select one of the predefined colors, or create your own color.
4. Click **OK**.

Changing the Color of Histogram Bars

To change the color of histogram bars, follow these steps:

1. Right-click anywhere in a histogram and select **Histogram Color.**
2. Click a color, or click **Other** and create your own color.

See your operating system documentation for details about creating your own colors.

Displaying Coordinates and Temporary Reference Lines

You can measure points and distances in graphs, or easily find the exact value, or coordinates, of points and distances on plots and graphs. To do this, click the crosshairs tool and click and hold anywhere on a graph. The coordinate values appear where the crosshairs intersect the vertical and horizontal axis as you drag the crosshairs within a plot.

Figure 6.19 Using the Crosshairs Tool

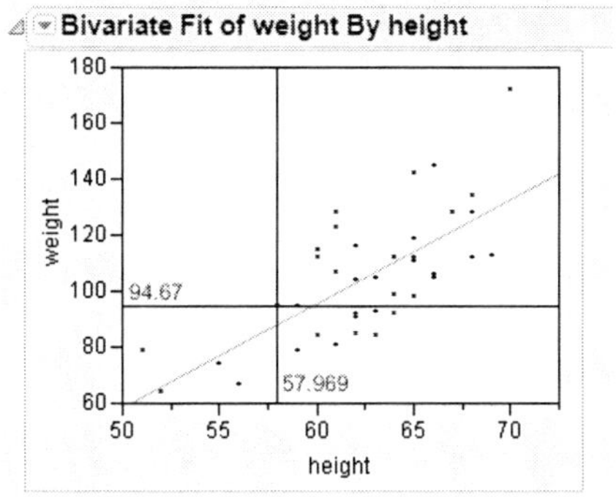

On a fitted line or curve, the crosshairs identify the response value for any predicted value. On a ternary plot, this tool displays triangular crosshair lines.

Scrolling and Scaling Axes

The hand tool (also known as the grabber tool) ( ) provides a way to change the axes and view of a plot:

- On a *y*-axis, dragging scrolls the *y*-axis; dragging or scales the *y*-axis.
- On an *x*-axis, dragging scrolls the *x*-axis; or scales the *x*-axis.

You can also right-click in a plot or graph, and select **Size/Scale**. Choose one of the following options:

- To adjust the scale of the X axis, select **X Axis**.
- To adjust the scale of the Y axis, select **Y Axis** or **Right Y Axis**.
- For details about the **Frame Size** option, see "Specifying Size in Pixels," p. 199.
- Select **Size to Isometric** when the *x*- and *y*-axes are measured in the same units and you want distances on the graph to be represented accurately regardless of direction.

Customizing Axes and Axis Labels

Double-click a numeric axis to customize it using the Axis Specification window. Or, right-click the axis area and select **Axis Settings** to access the window.

Customization features in the window depend on the data type of the axis and the specific platform JMP uses to create the plot or chart. Figure 6.20 shows a typical Axis Specification window for numeric axes.

Figure 6.20 The Axis Specification Window for a Numeric Axis

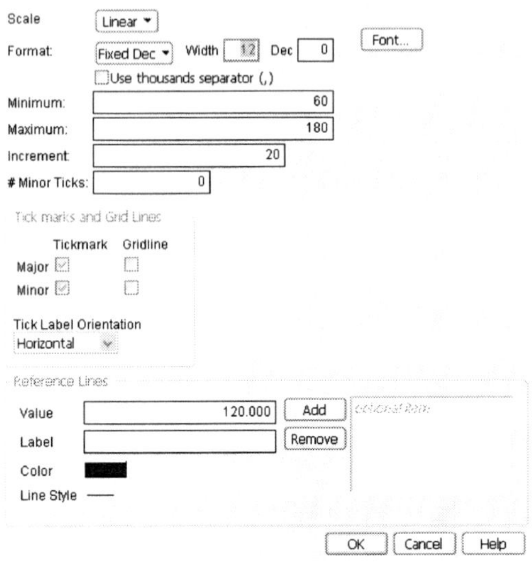

The following sections describe how to use the axis specification window and what tasks you can accomplish with it.

Establishing the Minimum and Maximum Axis Values

For plots and charts that contain a numeric axis area, you can change the minimum and maximum values that you want the graph to display.

To change the minimum and maximum axis values while viewing a graph:

1. Double-click a numeric axis. Or, right-click and select **Axis Settings**. The axis specification window appears.
2. Type in a value for the minimum and maximum values that you want the graph to display.

Note: To restore the default minimum and maximum axis settings of a numeric axis, right-click a numeric axis and select **Revert Axis**.

To set a default minimum and maximum axis value for a variable, which avoids making this change every time you run an analysis, see "Changing Columns' Default Axis Settings," p. 157.

The example on the right in Figure 6.21 is an enlargement of the point cluster that shows between 80 and 140 in the plot to the left. The enlarged plot is obtained by reassigning the maximum and minimum axis values and changing the number of minor tick marks to 1. (See "Adding Minor Tick Marks," p. 208, for details.)

Figure 6.21 Rescale Axis to Enlarge a Plot Section

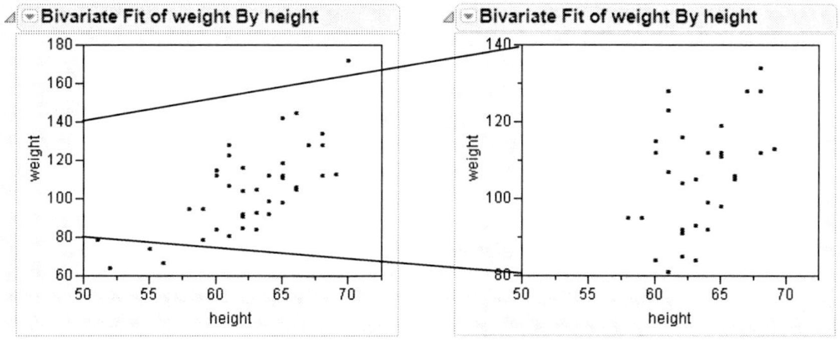

Changing the Axis Scale Type

When viewing a graph with a numeric axis, you can change the axis scale as follows:

1. Double-click a numeric axis. Or, right-click and select **Axis Settings**. The Axis Specification window appears.
2. Select a scale type:
   - **Linear**
   - **Log**
   - **Geodesic**
   - **Geodesic US**

Note: Specific platforms might use other scale types that are fixed and cannot be changed.

To set a default scale type for a variable, which avoids making this change every time you run an analysis, see "Changing Columns' Default Axis Settings," p. 157.

Changing the Orientation for Tick Labels

You can set the orientation for tick labels as follows:

1. Double-click a numeric axis. Or, right-click and select **Axis Settings.** The Axis Specification window appears.
2. From the Tick Label Orientation menu, select one of the options:

   Automatic Tick labels are oriented automatically to be readable, depending on the tick increment and length of the labels.

   Horizontal Baseline at the bottom.

   Vertical Vertical with the baseline on the right.

   Perpendicular (*Recommended in place of Horizontal*) Horizontal for vertical axes and vertical for horizontal axes.

   Parallel (*Recommended in place of Vertical*) Vertical for vertical axes and horizontal for horizontal axes.

   Angled Angled at about 45 degrees.

Changing the Axis Increments

While viewing a graph, you can change the axis increments:

1. Double-click a numeric a xis. Or, right-click a numeric axis and select **Axis Settings**. The Axis Specification window appears.
2. In the text box beside **Increment**, type the number of increments that you want displayed.
3. If the format of the axis is **Date** or **Time** (as shown in the **Format** drop-down menu), another drop-down menu appears beside **Increment**. (See Figure 6.22.) From that box, select which format you want the increments to take.

Figure 6.22 Changing Axis Increments

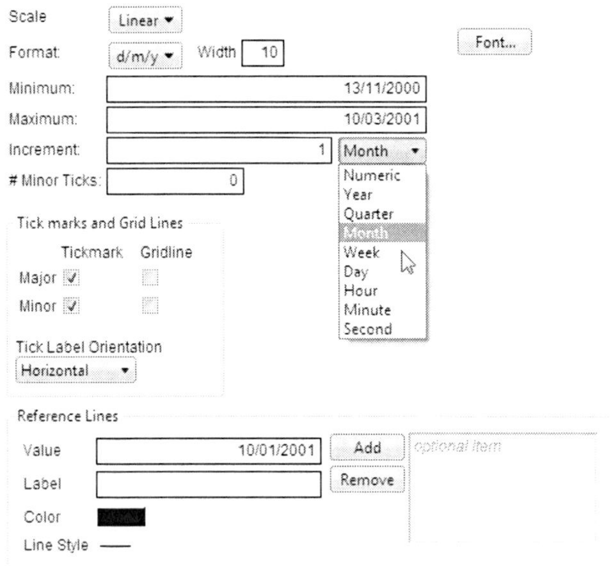

To set a default axis increment for a variable, which avoids making this change every time you run an analysis, see "Changing Columns' Default Axis Settings," p. 157.

Adding and Removing Axis Labels

You can add or remove labels in a numeric axis. To add an axis label:

1. Right-click a numeric axis and select **Add Axis Label**.
2. Type a name for the axis label. The axis area enlarges to hold the number of label lines that you enter.

This command can be used multiple times to add multiple labels. To edit the label after it has been added to the axis, click it and it will turn into an edit box.

To remove an axis label, right-click a numeric axis and select **Remove Axis Label**. The last label added is removed.

Changing the Numeric Format of an Axis

For plots and charts that contain a numeric axis area, you can change the format of the axis. To change the numeric format while viewing a graph:

1. Double-click a numeric axis. Or, right-click and select **Axis Settings**. The Axis Specification window appears.
2. In the box beside **Format**, use the drop-down menu to select an option. See "Numeric Format Options," p. 61, for details.

3. If you selected:
   - **Date** or **Time** from the **Format** drop-down menu in the previous step, use the drop-down menu to the right of the selection to select date increments for tick marks. See "Numeric Format Options," p. 61, for descriptions of options. You also need to specify the format of the increments, as described in "Changing the Axis Increments," p. 204.
   - **Fixed Dec** from the **Format** drop-down menu in the previous step, a text box appears beside the **Format** box. Type the number of decimal places that you want JMP to display.

Note: When you change the numeric format of an axis, you do not change the numeric format of the way the values appear in the corresponding data table. To change the way a date or time appears in a data table, see "Choosing Numeric Formats," p. 60.

Selecting a date interval from the date increment drop-down menu divides the JMP date (number of seconds) into the appropriate units. This gives the plot scale that you want for your data. The date axis must be a column with a JMP date value and appear in the Axis Specification window in the date format found in the Column Info window. However, you can use the Axis Specification window to format the date any way that you want it to be displayed in the plot.

Changing Axis Label Font

You can modify the axis label font on any axis type. When you modify it, your change only applies to the active graph. To set the default axis label font, see "Fonts," p. 463 in the "JMP Preferences" chapter.

To change the current font type and size:

1. Right-click an axis label.
2. Select **Font**.
3. Make your selections in the window.

To change the font color:

1. Right-click an axis label.
2. Select **Font Color.**
3. Select a color for the text from the color palette provided.

Rotating Axis Labels

You can modify the axis label on any axis type. To rotate an axis label:

1. Right-click an axis label.
2. Select **Rotate Text.**
3. Select which direction to rotate the text: **Horizontal**, **Left**, or **Right**.

To set a default axis label position for a variable, which avoids making this change every time you run an analysis, see "Changing Columns' Default Axis Settings," p. 157.

Copying and Pasting Graph Contents

After customizing a graph by adding elements such as a fitted line, you can copy and paste the contents from one graph to another compatible graph:

1. Right-click the graph that you have customized.
2. Select **Edit > Copy Frame Contents.**
3. Right-click the graph to which you would like to copy the settings.
4. Select **Edit > Paste Frame Contents.**

Copying and Pasting Axis Settings

After customizing an axis (as described in "Customizing Axes and Axis Labels," p. 202), you can copy and paste your new settings to another axis:

1. Right-click the axis that you have customized.
2. Select **Edit > Copy Axis Settings.**
3. Right-click the axis to which you would like to copy the settings.
4. Select **Edit > Paste Axis Settings.**

Changing the Order of Values

Data in a JMP report might not appear in the order that you prefer. To give data a specific order so it appears that way in a report, assign the column the Value Ordering property before running the analysis, as described in "Ordering Values in Columns," p. 153.

If your values include any of the following, they automatically appear in the appropriate order in reports:

- January, February, March, April, May, June, July, August, September, October, November, December
- Jan, Feb, Mar, Apr, May, Jun, Jul, Aug, Sep, Oct, Nov, Dec
- Sunday, Monday, Tuesday, Wednesday, Thursday, Friday, Saturday
- Very Low, Low, Medium Low, Medium, Medium High, High, Very High
- Strongly Disagree, Disagree, Neutral, Indifferent, Agree, Strongly Agree
- Failing, Unacceptable, Very Poor, Poor, Bad, Acceptable, Average, Good, Better, Very Good, Excellent, Best

Customizing Tick Marks and Tick Mark Labels

On a numeric axis, you can add and remove tick marks, add and remove gridlines, and add minor tick marks. You can also modify font properties, rotate the axis labels, or add an outline box to nominal or ordinal axes' tick mark labels. The following sections describe how to customize them.

Adding Tick Marks and Gridlines

For plots and charts that contain a numeric axis area, you can show or hide tick marks and gridlines. To do this, double-click a numeric axis. Or, right-click a numeric axis and select **Axis Settings**. The Axis Specification window appears, as shown in Figure 6.23. Use the table below to show or hide tick marks and gridlines.

Figure 6.23 Show or Hide Tick Marks and Gridlines

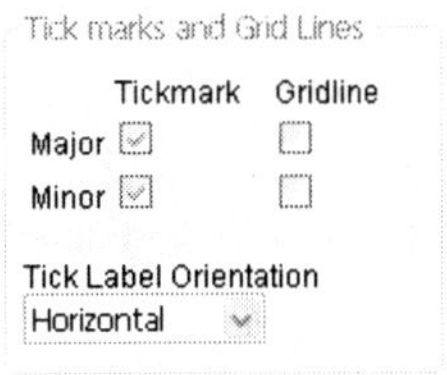

Below the check boxes, the orientation of the Tick Labels can be set by a drop-down list. **Horizontal**, **Vertical**, and **Angled** refer to single axes, while **Parallel** and **Perpendicular** refer to paired axes, like in Multivariate plots.

To set default tick marks for a variable, which avoids making this change every time you run an analysis, see "Changing Columns' Default Axis Settings," p. 157.

Adding Minor Tick Marks

To add tick marks to a numeric axis, or to change the number of minor tick marks that appear on a numeric axis, proceed as follows:

1. Double-click the tick mark. Or right-click it and select **Axis Settings**. The Axis Specification window appears.
2. In the box beside **# Minor Ticks**, type the number of minor tick marks that you want to appear between major tick marks.
3. Click the box to the right of **Minor** and below **Tickmark** to indicate that you want the tick marks to appear on the axis.

To set default minor tick marks for a variable, which avoids making this change every time you run an analysis, see "Changing Columns' Default Axis Settings," p. 157.

Adjusting the Field Width of Labels

On plots and charts that contain a numeric axis area, you can adjust the width of the tick mark labels. Then they show large values with many decimal places or only a small amount of space or truncated values. If the field width is set too small, your tick mark labels appear as an ellipsis (...).

To change the field width while viewing a graph:

1. Double-click a numeric axis. Or, right-click and select **Axis Settings**. The Axis Specification window appears.
2. In the **Width** text box, type the approximate number of characters in the widest tick mark label. The maximum field width is 40 for numeric values. There is no limit for character values.

Note that increasing or decreasing the field width of tick mark labels does not change the field width for the cells containing the values in the corresponding data table. To change the field width of cells in a data table, see the section "Choosing Numeric Formats," p. 60.

Changing Tick Mark Label Font Types and Sizes

You can modify a numeric axis tick label's font type, size, and style (bold and italic). To change font type and size:

1. Double-click the tick label. Or, right-click it and select **Axis Settings**. The Axis Specification window appears.
2. Select **Font**, and make your selections.

Note: To make global changes to all types of axes' font types and sizes, select **File > Preferences** (**JMP > Preferences** on the Macintosh). Click the **Fonts** tab, and then click **Axis.**

Rotating Tick Mark Labels

To rotate tick mark labels vertically or horizontally on a *numeric* axis:

1. Double-click the tick label. Or, right-click it and select **Axis Settings**. The Axis Specification window appears.
2. Modify the **Tick Label Orientation.**

To rotate tick marks on a *nominal* axis, right-click a tick label and select **Rotated Tick Labels.**

To rotate tick mark labels back, complete the steps again.

Extending Divider Lines and Frames for Categorical Axes

Extending the vertical divider line(s) between tick labels is useful when there are many levels of a nominal or ordinal (categorical) variable.

To extend the divider line to the *x*-axis labels:

1. Right-click a nominal or ordinal axis.
2. Select **Tick Marks > Divider Lines** to add the lines, or **Lower Frame** to add a frame around the axis area.

Figure 6.24 Divider Lines

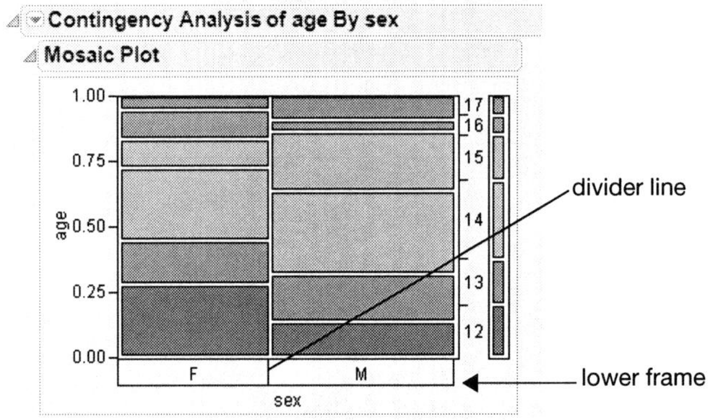

Adding Reference Lines

To add references lines to graphs, proceed as follows:

1. Double-click a numeric axis. Or, right-click a numeric axis and select **Axis Settings**. The Axis Specification window appears.
2. In the text box to the left of the **Add** button, type the value to which you want the reference line to correspond. This is the position on the graph at which the line is placed.
3. Further customize the reference line by choosing either of these options:

Table 6.4 Reference Line Actions

Action	Instructions
Give the line a color	Click the **Color** bar and select a color.
Use a stippled line	Click the line next to **Line Style** and select one of the line types.

4. Click the **Add** button. The value moves into the box to the right of the **Add** button, indicating that it will be placed on the graph.
5. To add more lines, repeat the above steps.

To set a default reference line for a variable, which avoids making this change every time you run an analysis, see "Changing Columns' Default Axis Settings," p. 157.

Add Geographical Images and Boundaries

Adding map images and political boundaries to graphs provides visual context to geospatial data. For example, you can add a map to the graph that displays an image of the U.S. Another option is displaying the boundaries for each state (when data includes the latitudes and longitudes for the U.S.).

The data should have latitudinal and longitudinal coordinates. Otherwise, the map has no meaning in the context of the data. The x and y axes also have range requirements based on the type of map. These requirements are described in the following sections.

Two tools are especially helpful when you are viewing a map:

- The grabber tool ( ) lets you scroll horizontally and vertically through a map.
- The magnifier tool ( ) lets you zoom in and out.

The following sections describe the map images and boundaries you can add to a graph.

Earth Images Installed with JMP

JMP provides two levels of earth imagery, simple and detailed. Both maps show features such as bodies of water and terrain. However, detailed maps show more precise terrain. And with detailed maps, you can zoom in farther, and the map features remain clear.

The maps wrap horizontally, so you continue to see map details as you scroll from left to right. The maps do not wrap vertically. Beyond the -90 and 90 y-axis range, a plain background appears instead of the map.

Figure 6.25 Examples of Simple and Detailed Maps

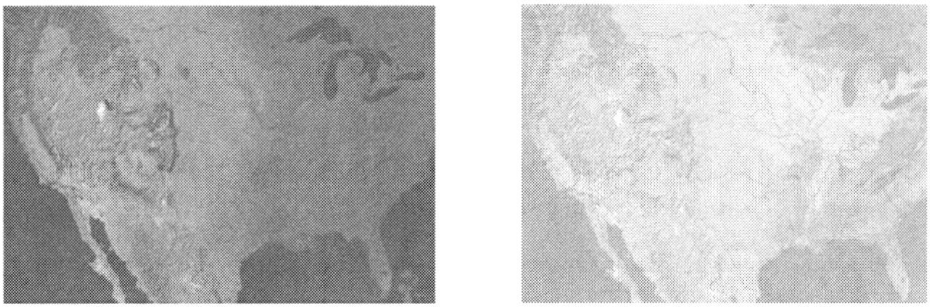

Simple Earth image   Detailed Earth image

Maps from the Internet

The National Aeronautics and Space Administration (NASA) map and other Web sites provide map image data using a protocol called Web Map Service (WMS). These maps have the advantage of showing the most up-to-date geographical information. However, the display of the maps can be slow depending on the response time of the server. An Internet connection is also required.

Unlike with simple and detailed maps, WMS maps do not wrap. You can scroll horizontally and vertically. However, beyond the -180 to 180 (x axis) and -90 to 90 (y axis) ranges, a plain background appears instead of the map.

Figure 6.26 Examples of NASA and WMS maps

NASA image WMS image from the MetaCarta™ Server

Custom Map Image

There are two ways to use your own map images:

- Assign the Map Role column property in a data table and specify the path to the map. See "Specify a Map Data Table," p. 165 in the "Properties and Characteristics of Data" chapter for details.
- You can create a map image in a graphics program and drag and drop the image onto a report. See "Drag and Drop an Image into a Graph," p. 214 for details.

Boundaries

JMP can display boundaries (such as U.S. state or French region boundaries). These boundaries can be displayed alone on a graph or combined with image data.

Several boundaries are installed with JMP. Alternatively, you can create your own boundaries from Economic and Social Research Institute (ESRI) shapefiles or from scratch. See the *Basic Analysis and Graphing* book for details.

Figure 6.27 Example of U.S. State Boundaries

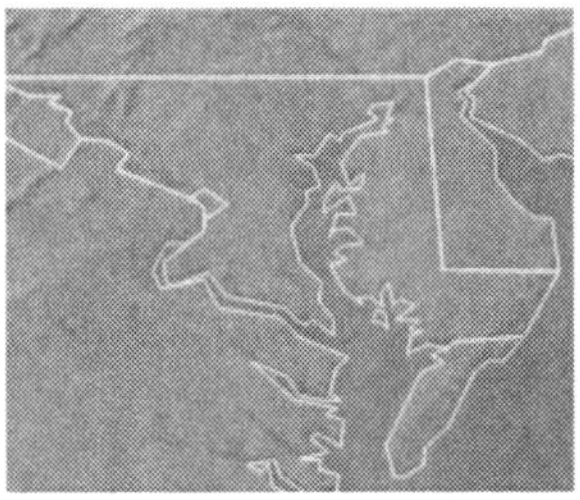

Add a Background Map and Boundaries

To add a background map and boundaries, proceed as follows:

1. Right-click a blank area on the graph and select **Background Map** (or select **Graph > Background Map** in Graph Builder).

   The Set Background Map window appears.

Figure 6.28 Example of Background Map Options

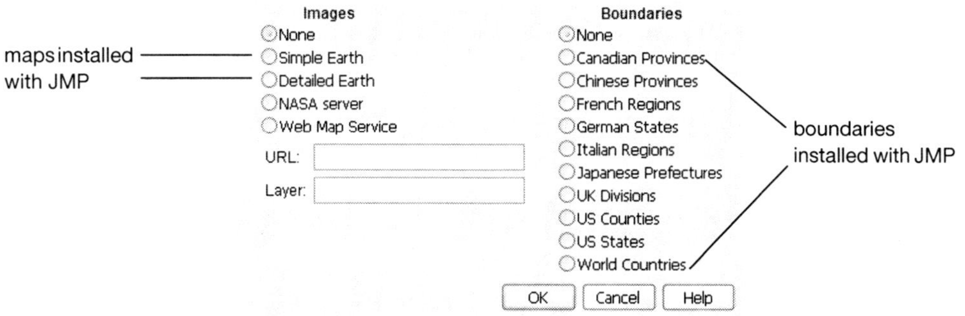

2. To display a background, do one of the following:
   - Select **NASA server**, **Simple Earth**, or **Detailed Earth** in the Images column.
   - Select **Web Map Service** and paste a WMS URL next to URL. Type the layer identifier next to Layer.
3. To display geographic borders on the map, select an option in the Boundaries column. (If you installed your own boundary shapefiles, they are also listed in this column.)
4. Click **OK**.

   If the NASA or WMS map does not appear after you add it, the map server might not be available. View the error log to verify the problem. (Select **View > Log** to open the log. Alternatively, you can set the log to open any time text is added to it. See "Open the JMP Log window," p. 461 for details.)

Table 6.5 Descriptions of Background Map Options

Images	
None	Removes the background map that you selected in the Images column.
NASA Server	Shows a map from the NASA server.
Simple Earth	Shows a map of basic terrain.
Detailed Earth	Shows a high-resolution map with detailed terrain.
Web Map Service	Shows the map from the URL (Uniform Resource Locator) and the layer that you specify.

Table 6.5 Descriptions of Background Map Options *(Continued)*

Boundaries	
None	Removes the boundaries that you selected in the Boundaries column.
Boundaries for various regions	Shows borders for the map regions, such as Canadian provinces, U.S. counties, U.S. States, and world countries. The maps that you created from ESRI shapefiles are also listed here.

Customize the Map Image

After you add a map image, you can change certain display properties, such as adding state boundaries and setting the transparency level.

- To change the type of map or the boundaries that appear on the graph, right-click on the map, select **Background Map**, and then modify the options. (In Graph Builder, right-click on the map, select **Graph > Background Map,** and then modify the options.)
- To change the map transparency level, right-click the map, select **Customize > Background Map**, and then type a value between 0 and 1 (where 1 is completely opaque). Note that the other Transparency options (on the right-click and right-click **Graph** menu) do not affect map images.

Drag and Drop an Image into a Graph

To add an image that you created, generate your report and then drag and drop the image from your file system onto the graph in the report. After you add the image, JMP provides several options such as resizing, formatting, and rotating an image. Right-click the image and select **Image** to see these options.

The following example shows a bivariate plot of wind speeds in the Chicago area. The plot on the left includes arrows to illustrate the wind direction and speed. A map image was dropped onto the plot and resized to line up the markers with the stations that provided the wind data (each dot representing a station).

Figure 6.29 Example of a Custom Map Image

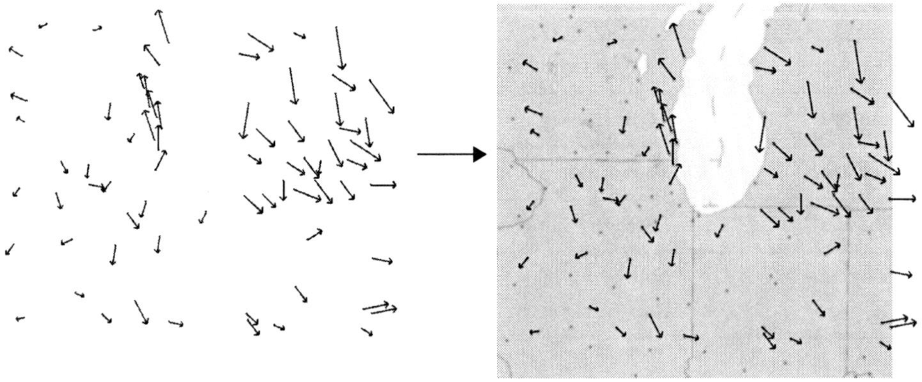

Note: Repeated changes to images can degrade image quality. If you are not happy with the results of sizing, scaling, or applying filters to the image, remove the image and start again.

Table 6.6 Descriptions of Image Options

Lock	Locks the image in place so it cannot be moved.
Size/Scale	• **Fill Graph**: Resizes the image proportionately to fit the graph. • **Specify Size**: Resizes the image according to the values that you enter. (The units for these values are the same as your graph axes.) • **Crop**: Crops the image according to the values that you enter. (The units for these values are the same as your graph axes.) For example, the left edge might be positioned at 50. You type 60 next to Left, and the portion of the image between 50 and 60 is removed from the image.
Flip	• **Flip vertical**: Turns the image upside down. • **Flip horizontal**: Flips the image left to right. • **Flip both**: Flips the image both horizontally and vertically.
Rotate	Rotates the image the specified number of degrees. Enter a negative value to rotate the image counterclockwise.
Transparency	Changes the marker transparency level. Type a value between 0 and 1 (where 1 is completely opaque).

Table 6.6 Descriptions of Image Options *(Continued)*

Filter	Provides filters found in many graphic editing programs to change the appearance of the image. Select a filter repeatedly to increase its effects on the image. • **Contrast**: Optimizes the light and dark colors. Larger values lighten the image. • **Despeckle**: Removes pixels that do not blend with surrounding pixels. For example, a black pixel surrounded by white pixels is converted to a white pixel. • **Edge**: Darkens everything but the outlines of objects. • **Enhance**: Reduces the contrast between pixels in a noisy image. • **Gamma**: Balances the brightness of an image and the red, green, and blue (RGB) ratios. Larger values create a lighter image. • **Gaussian Blur**: Blurs pixels by the specified radius. Larger radii create a smoother image. (In JSL, you can also specify the sigma value. Larger sigma values create a smoother image.) • **Median**: Replaces each pixel color value with the median value of the surrounding pixels. • **Negate**: Converts each pixel to its complementary color (such as pink to green and white to black). • **Normalize**: Removes a percentage of the top and bottom color values. The color values are then stretched to fill the remaining image. This process increases the intensity of the colors. • **Reduce Noise**: Finds the minimum and maximum color values and replaces them with values more consistent with the surrounding pixels. Larger values create a smoother image. • *Sharpen*: Makes the edges of pixels more distinct.
Remove	Removes the image from the report.

Extract Data from an Image

JMP provides the ability to extract information from images into a data table and then analyze that information. Researchers at WildTrack.org analyze digital footprint photos in JMP to track endangered species. They drag and drop a footprint image into a JMP report and draw data points to capture the size and shape of the print. A JMP Scripting Language (JSL) script extracts those measurements into a data table. At that point, the researchers can analyze the data and determine whether the footprint is from a new animal. This method helps them track populations of endangered species in specific regions of the world.

Figure 6.30 Example of Extracting and Analyzing Data

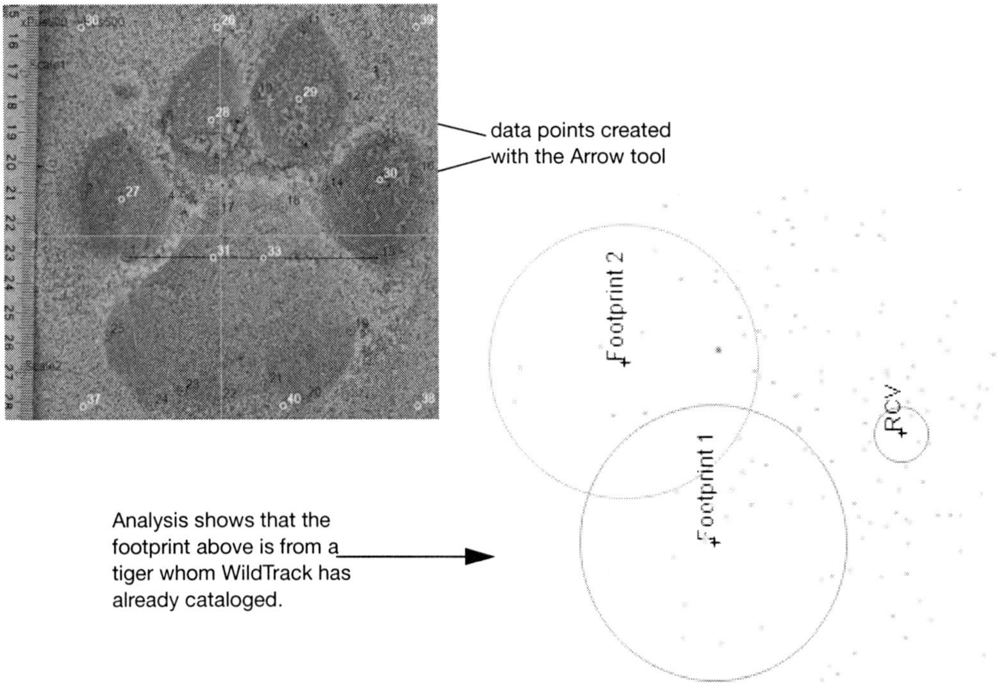

The details for implementing this feature are beyond the scope of this book, because the needs of the user vary widely. For details about writing JSL scripts, see the *Scripting Guide*.

Adding Elements to a Report

You can add text notes, lines, polygons, ovals, and rectangles to a report using graphics tools found in the toolbar.

Annotations

You can add editable text notes to a JMP report using the annotate tool. To add an annotation:

1. Select the annotate tool from the **Tools** menu or toolbar.
2. Click the location in the window where you want to add the annotation. Or, click and drag to size the annotation note. A white editable text box is displayed.

Figure 6.31 Annotation Box

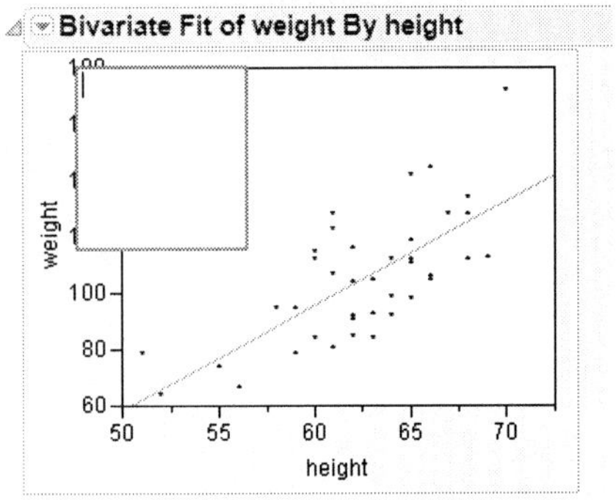

3. Enter text.
4. Click outside the annotation. The annotation turns yellow.

Figure 6.32 Annotation

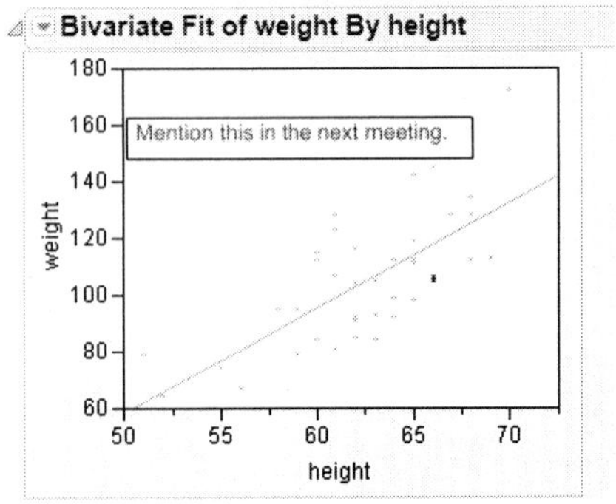

5. (Optional) Right-click the annotation to access the following options:

Background Color Provides you with a color palette, from which you can select the background color for the annotation.

Text Color Provides you with a color palette, from which you can select the color for the annotation's text. The color of the font also defines the color of the annotation outline. If you select the same color for both the background and the font, the font does not show except for black and white. A black font on a black background changes to white, and a white font on a white background appears black.

Font Lets you change the current font type, style, and size. To change the default font, see "Fonts," p. 463 in the "JMP Preferences" chapter.

Tag Line Attaches a line to the annotation that points to a place in the text, as shown below. To move the line with the annotation to a new position, press the CTRL key and drag the annotation.

Figure 6.33 Tag Line

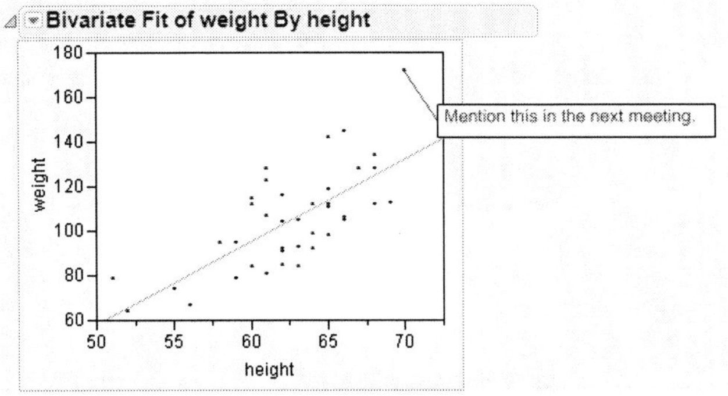

Filled Removes the background color from the annotation so it looks transparent. A transparent note is handy for putting titles and footnotes on a graph.

Editable Makes the annotation editable by double-clicking on it.

Reanchor Reanchors the annotation.

Delete Deletes the entire annotation.

Note: When adding multiple annotations, press the SHIFT key when selecting the annotation tool for the first time. This causes subsequent clicks to add an annotation, and you do not have to select the annotate tool from the toolbar before the addition of each annotation.

Once you have added an annotation, you can do the following:

Table 6.7 Using Annotations

Action	Instructions
Add to or edit an annotation	Click inside the text box.
Move an annotation	Click inside the annotation box and drag it. When an annotation is moved, it becomes selected, as indicated by a double blue line with handles around the perimeter.
Resize an annotation	Place the cursor on the handle of a selected note (showing in the middle of the edges and in the corners). The cursor appears as a single crossed arrow; drag to resize the annotation.
Delete an annotation	Highlight the annotation by clicking the handle of a note. Then press the Delete (or Backspace) key.

Adding Shapes

You can add editable lines, polygons, and simple shapes (ovals or rectangles) to a JMP report using the drawing tools. The following sections describe how each of these tools can be used.

Note: Each graphics tool remembers the most recent options chosen. This is useful if you need many annotations or other graphics with the same characteristics. For example, suppose you want many thick green lines with an arrow on one end. Create a line the way you want it, set the options, and subsequent lines appear with those options in effect. The options persist until you change them.

Adding a Line

To add a line to a report window:

1. Click the line tool in the tool palette.
2. Click and drag where you want to insert the line. The line appears selected, showing handles on both ends, such as is shown in Figure 6.34.

Figure 6.34 Line Tool

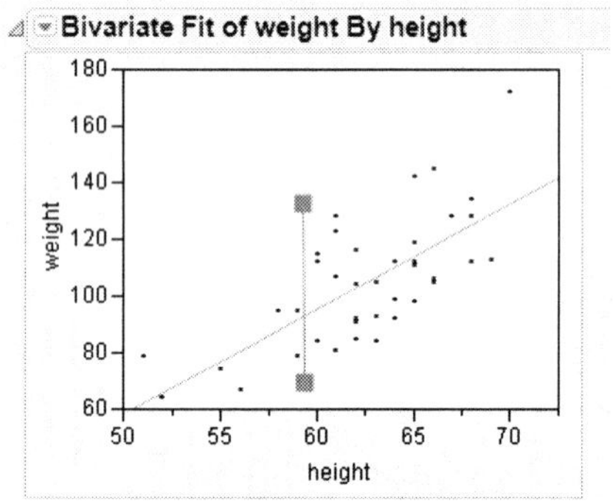

3. Click and drag the line to move it.
4. Click a handle and drag to rotate the line.
5. Right-click a line for a menu of options to tailor the appearance of the line, as follows:

   Point to and Point from Places arrows on either end (or both ends) of the line.

   Thick Alternately displays the width of a line as thick or thin. A line is thin by default.

   Dashed Alternately displays a line as dashed or solid. A line is solid by default.

   Color Displays the JMP color palette to change the color of the shape.

   Reanchor Reanchors the shape.

   Delete Removes the shape from the report surface. You can also remove the shape by selecting it and then pressing the Delete (or Backspace) key.

Adding a Polygon or Spline

To add a polygon (also known as a spline) to a report window:

1. Click the polygon tool in the tool palette.
2. Click to create the beginning point for the first side of a polygon.
3. Click again at the location where you want to complete the first side and begin an adjacent side. A square selection box with handles appears around the polygon area.
4. Click a third time to complete the second side.
5. Continue this process until the polygon is the way you want it. Each time a side is complete, the selection box adjusts to encompass the polygon sides.
6. Double-click to release the polygon tool.

Figure 6.35 Polygon Tool

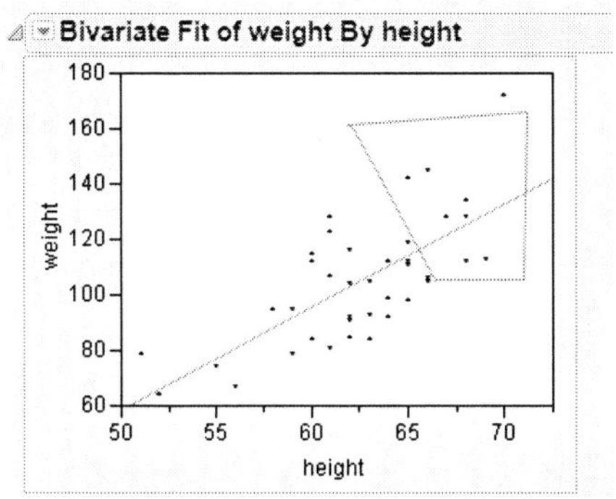

Once you have added a polygon, you can perform the following actions:

Table 6.8 Working with Polygons

Action	Instructions
Select or deselect the polygon	Click the edge of a completed polygon.
Resize the polygon	Select it and drag one of the selection box's handles.
Move the polygon	Click between the box's handles and drag the selection box.
Change the number of sides of the polygon	Click and drag the sides to form the new shape.

Right-click a polygon for a menu of options to tailor its appearance, as follows:

Filled Alternately fills or empties the area of the shape.

Raised Displays thick shaded lines around the shape. If the shape is also filled, the lower edge of the figure appears raised, giving it a three-dimensional look.

Smooth Smooths the vertices of a polygon to produce a Bezier curve. The smoothed figure is reshaped and resized the same way as the polygon, and can be filled and raised.

Closed Alternately opens or closes the last segment of a polygon.

Color Displays the JMP color palette to change the color of the shape's sides, and its fill color when the **Filled** option is in effect.

Reanchor Reanchors the shape.

Delete Removes the shape from the report. You can also remove the shape by selecting it and then pressing the Delete (or Backspace) key.

Adding an Oval or Rectangle

To add an oval or rectangle to a report window:

1. Click the simple shape tool  in the tool palette.
2. Click and drag where you want to insert the shape. An oval appears with a selection box around it.
3. (Optional) To turn the oval into a rectangle, right-click on the oval and select **Shape > Rectangle**.
4. Double-click to release the simple shape tool.

Figure 6.36 Simple Shape Tool

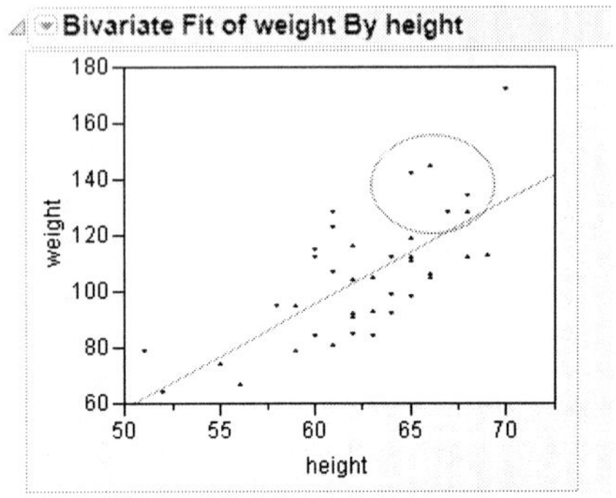

Once you have added the shape, you can do the following:

- Select the shape and drag one of the selection box handles to reshape or resize.
- Click and drag an edge of the selection box (located between the handles) to move it.
- Select the shape, and then right-click it for a menu of options to tailor its appearance, as follows:

  Filled Alternately fills or empties the area of the shape.

  Raised Displays thick shaded lines around the shape. If the shape is also filled, the lower edge of the figure appears raised, giving it a three-dimensional look.

Shape Displays a submenu whose options transform the shape into either an oval or a rectangle when selected.

Color Displays the JMP color palette to change the color of the shape's sides and its fill color when the **Filled** option is in effect.

Reanchor Reanchors the shape

Delete Removes the shape from the report. You can also remove the shape by selecting it and then pressing the Delete (or Backspace) key.

Adding Graphics

To enhance your graphs with logos, pictures, or any other type of graphic, you can paste it into a report in .bmp, .jpeg, or .png format. You can also drag and drop graphics into reports.

To add graphics:

1. Open the graphic file and select the graphic. Copy it to the computer's clipboard.
2. Right-click inside a graph.
3. Select **Edit > Paste Background Image**. The graphic is inserted at the point in the graph that you right-clicked.

You can also add the graphic to the end of a report window: copy the graphic onto your computer 's clipboard, making sure the report is the active window and selecting **Edit > Paste**.

Customize Graphical Elements

Graphs consist of markers, lines, text, and other graphical elements that you can customize. For example, in a Scatterplot Matrix graph, you might want to highlight data points in one of the bivariate graphs with a pink solid marker. In a Contour Plot graph, you can increase the width or transparency of the contour lines.

Example of Customizing a Contour Plot

1. Open the Little Pond.jmp sample data table.
2. In the Tables panel, click on the red triangle next to Contour Plot and select **Run Script**.
3. Click on the red triangle next to Contour Plot for Z and deselect **Fill Areas**.

Figure 6.37 Contour Plot Before Customization

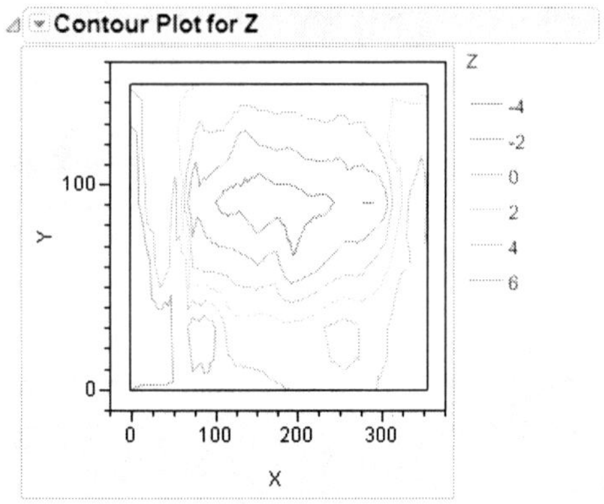

4. Right-click on the plot and select **Customize**.

   The Customize Graph window appears.

Figure 6.38 Customize Graph Window

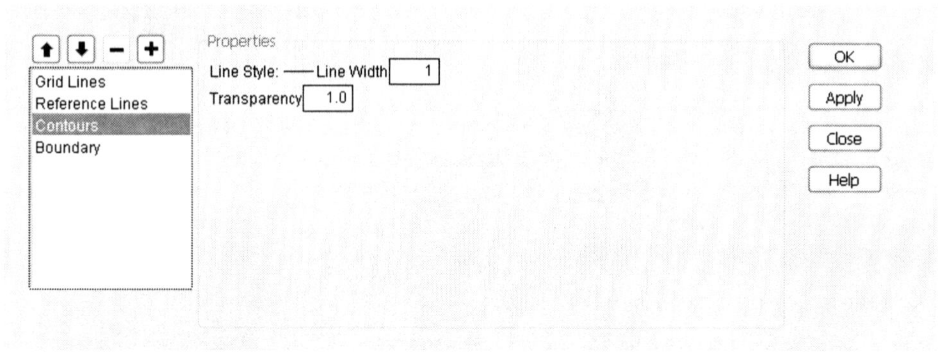

5. Click on **Contours.**
6. Change the Line Width to 3.
7. Click on **Boundary.**
8. Change the Line Width to 2.
9. Click on the color next to Line Color and select a shade of green.

Figure 6.39 Customized Contour Plot

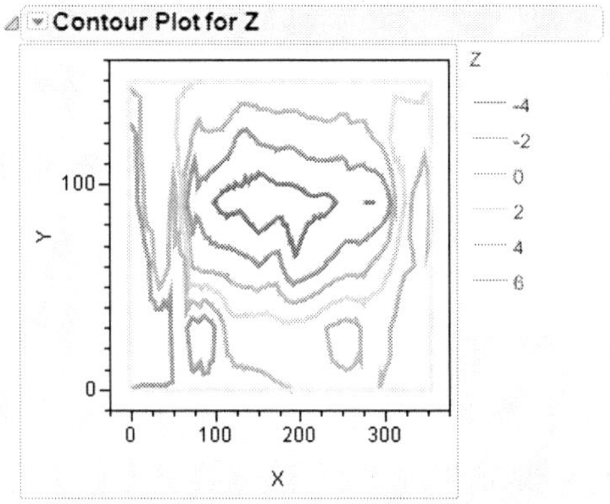

The graphical elements that you can customize differ on each graph. For example, in a Control Chart graph, there are three line elements (**Center Line**, **Limits**, and **Connect Line**). In other graphs, the line element might be named **Line** or **Custom**.

To Customize Graphical Elements in the Current Graph

1. Right-click the graph and select **Customize**.
2. Select the element that you want to change, and then modify the properties.

   Your changes are immediately shown on the graph.
3. Click **OK** to save your changes to the current graph.

Each element can include properties for line color, marker style and size, text style, and so on. Here are some common properties:

Line Color Changes the color of the line. Click to select any color in the window. Right-click for to choose from more colors.

Line Style Changes the style of the line. Click to select one of five different styles.

Line Width Changes the width of the line. Click in the box and enter the desired line width in points.

Marker Changes the marker shape or size. Click to change the shape. Right-click to change the size. For more information about markers, refer to "Using Markers," p. 193.

Line of Fit Changes the color, width, or style of the line.

Text Color Changes the color of the text.

Font Changes the font, style, and point size of the text.

Text Style Changes the text alignment to centered, left-aligned, or right-justified. The **Fill** option applies the selected color to the background.

Fill Color Changes the color of objects such as box plots and Fit Confidence regions.

Transparency Changes the marker or label transparency. Enter the level of transparency to draw markers (points) on the graph. The degrees of opacity ranges from 0 (clear) to 1 (opaque). For more information about changing the transparency of markers, see "Specifying Marker Transparency," p. 195.

Create Scripts for Graphical Elements

In addition to customizing graphical elements, you can write JSL scripts that add elements. These scripts run when you display the graph. You can write the scripts from scratch, or you can select from the following lists of commands and scripts, accessed by clicking on the **Add a new script button** :

- Use the **Templates** list to insert a single JSL command. For example, the **Polygon** option inserts the Polygon command. The text enclosed in underscores are placeholders for point values, which you replace with your own values.

  ```
  Polygon([_x0_, _x1_, ... ], [_y0_, _y1_, ...]);
  ```

- Use the **Samples** list to insert a script that creates elements such as bubble plots and sine waves. In this list, the **Polygon** option shows examples of the Transparency, Fill Color, and Polygon values, which you replace with your own values.

Figure 6.40 The Polygon Sample Script

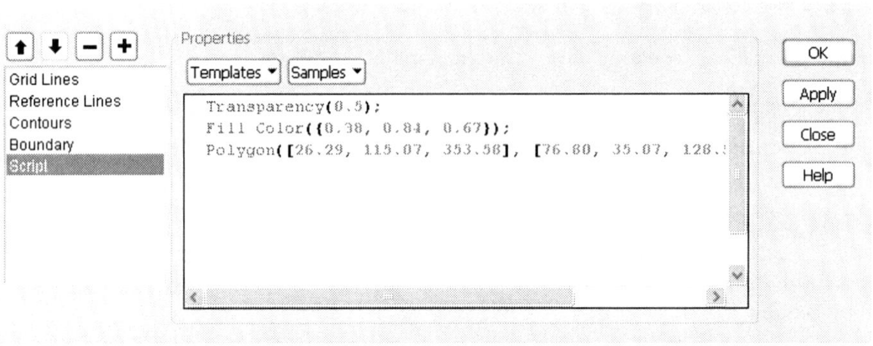

In some graphs, you can view the JSL that creates graphical elements. Figure 6.41 shows the Group Label script for a Discriminant Analysis graph. The script defines markers and text for group labels. The commands included in these built-in scripts cannot be modified or deleted. You can insert commands from the Templates or Samples list, but you cannot click in the window and type new commands.

Figure 6.41 Example of JSL that Creates Group Labels

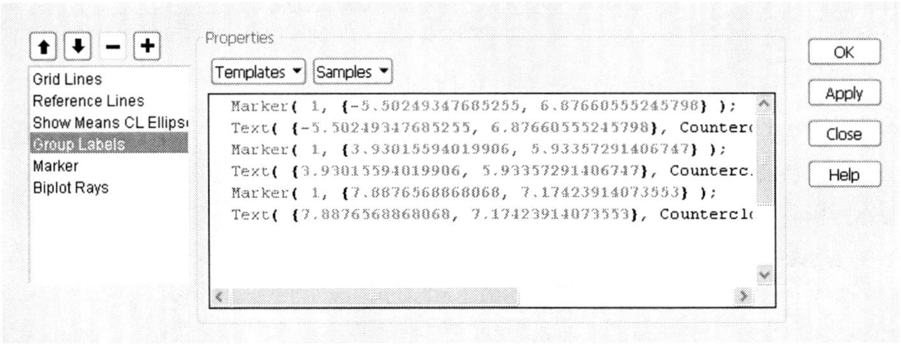

For information about JSL, see the *Scripting Guide* book.

To Create a Graphics Script

1. Right-click on the graph and select **Customize.**
2. Click the **Add a new script** button (**+**) to create a new script.

   The default name, Script, is highlighted.
3. With the default name highlighted, type a more descriptive name, and then press ENTER. (If you already moved the cursor and the name is no longer highlighted, double-click Script, and then type the new name.)
4. Do one or more of the following:
   – Enter JSL in the Properties window.
   – Select one or more JSL commands from the **Templates** list, and modify the placeholder text. For example, change the **Pen Color** option from "blue" to "red."
   – Select one or more sample scripts from the **Samples** list and modify, if necessary.
5. (Optional) Click **Apply** to update the graph with your changes without closing the window. Unlike other property changes, script changes to not take effect until you click **Apply** or **OK.**
6. Click **OK** to save your changes.

   One of the following occurs:
   – The element that you created appears on the graph.
   – An error message appears if the script contains an error. Select **View > Log** to read about the error, and then correct the script.

To delete a script that you created, select the script and then select the **Delete selected script** button (**–**).

Change the Drawing Order of Graphical Elements

The graphical elements are drawn in the order in which they are listed. The first element on the list is drawn first, so it appears *behind* all other graphical elements. If one element hides another, you can rearrange the order of the elements.

To Rearrange Graphical Elements

1. Right-click the graph and select **Customize.**
2. Select the element that you want to move.
3. Click the **Move up in drawing order** button (⬆) or **Move down in drawing order** (⬇) button one or more times until the elements are in the order in which you want them drawn.
4. Click **OK.**

Save Your Customizations

The graph customizations apply to the current graph and are also used when you redo an analysis. To re-create your graph at a later time with its customizations, select **Script** from the red triangle menu, and then select one of the **Save** options. For example, you can save the script to the data table, which applies your customized properties each time you run the script. See "Saving Your Steps as a Script," p. 187 for details.

Copy Your Customizations

You can copy your customizations from one graph to another. All objects that you created or modified, such as colored text or lines, are pasted to the other graph.

To copy lines from one graph to another, see "Copying and Pasting Graph Contents," p. 207.

To Copy other Objects

1. Right-click in the graph with custom elements and select **Edit > Copy Customizations.**
2. Right-click in the destination graph and select **Edit > Paste Customizations.**

Note: The customizations are pasted into all similar graphs when you press CTRL, click, and then paste.

The objects appear on the current graph and are added to the list of customized elements.

Note: The copy customizations feature copies only the elements that you added or modified. It does not copy the other contents of the graph.

Chapter 7

Reshaping Data

Subset, Concatenate, Join, and More

You can perform a wide variety of data management tasks on JMP data:

- Create a new data table from a subset of rows and columns
- Sort by any number of columns
- Stack multiple columns into a single column
- Split a column into two or more columns
- Transpose rows and columns
- Concatenate multiple tables end to end
- Join two tables side by side
- Update columns in a table with values from another table

Contents

Creating a Subset Data Table

You can produce a new data table that is a subset of all rows and columns, only highlighted rows and columns, or randomly selected rows from the active data table.

To create a subset:

1. Select **Tables > Subset**.

Figure 7.1 The Subset Window

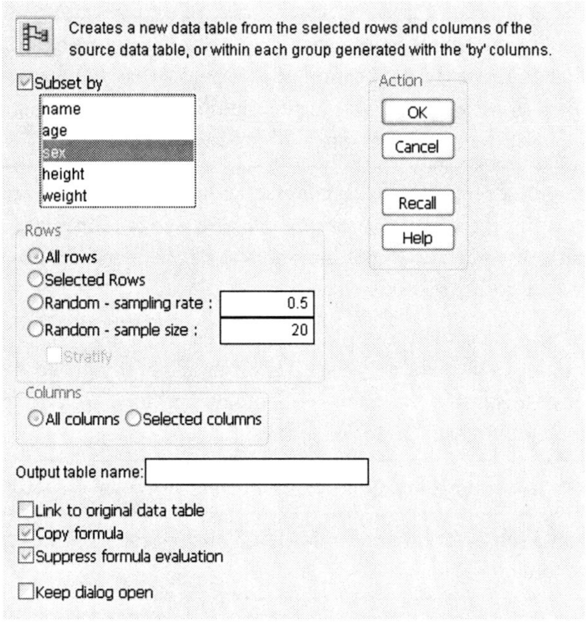

2. Specify the content that you want to subset. Select any combination of the following:
   – Subset by (the levels within selected columns)
   – Rows (all, selected, or random)
   – Columns (all or selected)

   For descriptions of these options, see Table 7.1.
3. Customize your subset table further using the additional options. See Table 7.1.
4. Click **OK** to create the subset table.

Table 7.1 Description of the Subset Window

Subset by	To subset by the levels of a column, select **Subset by** and select the columns that you want to categorize for the subset. Consider the fact that many new data tables might be created. A new data table appears for each level of the column that you specified in the **Subset** window.
All Rows	Creates a subset table that contains all rows from the active table.
Selected Rows	Creates a subset table that contains only the selected rows from the active table.
Random - sampling rate	Creates a subset table whose data is a random proportion of the active data table. Enter the proportion of the sample that you want in the text box. For example, if you want a random 50% of the data to be included in the new table, enter 0.5 in the text box.
Random - sample size	Creates a subset table whose data is a random sample of the active data table. Enter the size of the sample that you want in the text box. For example, if you want 16 random rows to be included in the new table, enter 16 into the text box. If you select a random sample that is the entire source table, the result is a random shuffle of the rows of the data table. If you specify columns to stratify, the result is a random shuffle of each of the rows for each group. See "Stratified Subsets," p. 233.
All columns	Creates a subset table that contains all columns from the active table.
Selected columns	Creates a subset table that contains only the selected columns from the active table.
Output table name	To give a name to the subset table, type a name in the box beside **Output table name**.
Link to original data table	To keep the subset table linked to the original table, click the box beside **Link to original data table**. When you change values in one table, the other table is updated.
Copy formula	To include formulas from the original table in the output columns, click the box beside **Copy formula**. Include all columns needed for the calculation of the formula. This option is not present when there are no formulas in the data table.
Suppress formula evaluation	To prevent JMP from evaluating columns' formulas when the new table is created, click the box beside **Suppress formula evaluation**. This option is not present when there are no formulas in the data table.

Table 7.1 Description of the Subset Window *(Continued)*

Keep dialog open	To keep this window open after clicking **OK**, check the box beside **Keep dialog open**.

Stratified Subsets

If you specify a sample size and add stratification columns, the sample size represents the size per stratum, rather than the size of the whole subset.

Figure 7.2 Stratified Subsets

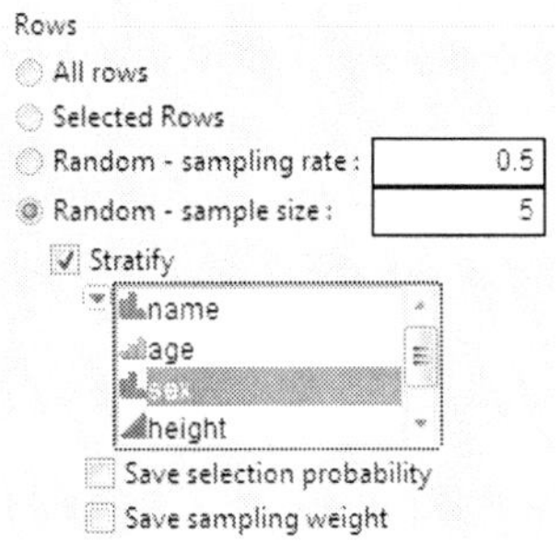

For stratified random samples with a specified sample size, two columns can be saved: **Selection Probability** and **Sampling Weight.** Check the corresponding check box to save these columns.

Creating a Subset Data Table from a Report

These two methods produce linked subsets of a data table.

Using a Histogram

Once you have produced output that contains a histogram (by selecting **Analyze > Distribution**), you can use the histogram to create a new data table. The new data table contains the data in the histogram's highlighted bars.

To create a subset, double-click a highlighted bar. Or, right-click (Ctrl-click on the Macintosh) anywhere in the histogram and select **Subset** from the menu. The subset table appears, as shown in Figure 7.3.

Figure 7.3 Subset Created from a Histogram

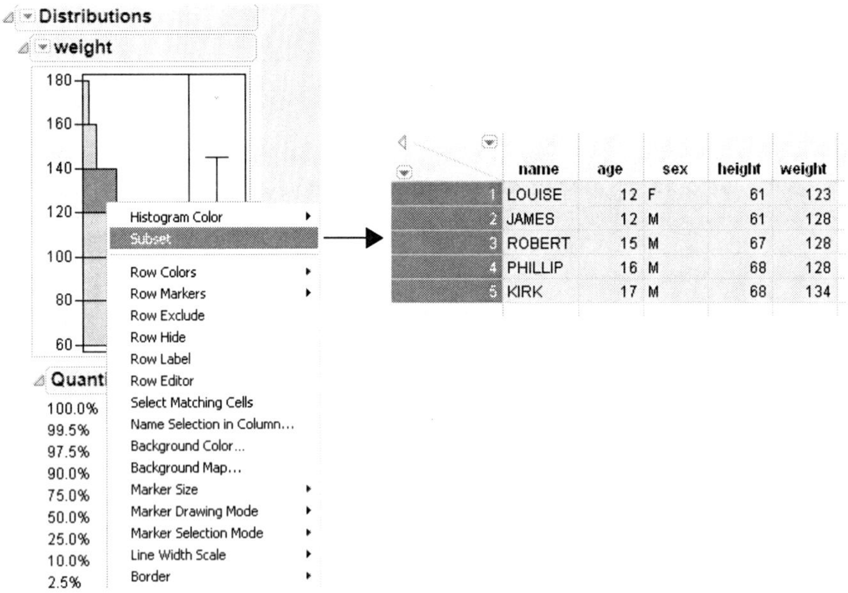

Using a Pareto Plot

Once you have produced output that contains a Pareto Plot (by selecting **Graph > Pareto Plot**), you can use the Pareto Plot to create a new data table. The new data table contains the data in the Pareto Plot's highlighted bars. To create a subset, double-click a highlighted bar.

Sorting Data Tables

You can sort a JMP data table by columns in either ascending or descending order. By default, columns sort in ascending order. You can either create a new table that contains the sorted values, or you can replace the original table with the sorted table.

If columns contain value labels, sorting is based on the actual data values, not the value labels. (See "Using Value Labels," p. 151.) However, the value labels are displayed in the sorted data table.

If your sorted column uses either the Value Ordering property or the List Check property, the column is sorted according to that order.

Example of Sorting Data Tables

1. Open the Popcorn.jmp sample data table.
2. Select **Tables > Sort**.

Figure 7.4 The Sort Window

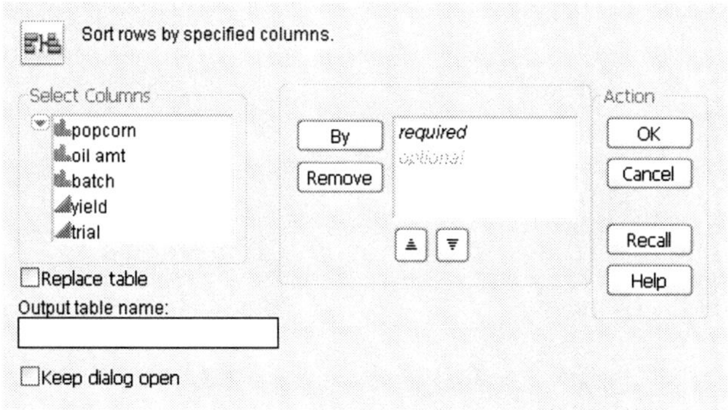

3. Highlight the names of the columns that you want to sort by. For this example, select popcorn and yield.
4. Click **By** to add the columns to the sort list.

   The columns that you add to the list establish the order of precedence for sorting. The first column in the list is the major sort field. Each variable thereafter is sorted within the previous variable in the sort list. You can drag and drop within the By list to change the sort order.
5. Customize your sort further using the additional options. For this example, highlight yield and click the descending button.

   For a complete list of options, see Table 7.2.
6. Type a name for the new sorted table in the box beside **Output table name**. For this example, type sorted popcorn.

Figure 7.5 Completed Sort Window

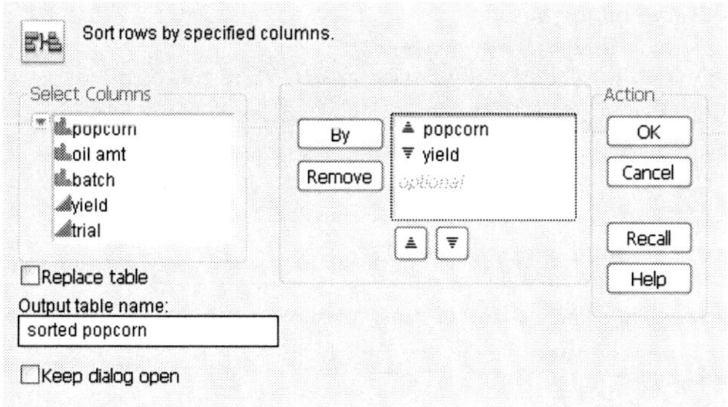

7. Click **OK**.

Figure 7.6 Sorted in Ascending and Descending Order

	popcorn	oil amt	batch	yield	trial
1	gourmet	lots	small	18.0	1
2	gourmet	lots	small	16.0	2
3	gourmet	little	small	15.9	2
4	gourmet	little	small	12.1	1
5	gourmet	lots	large	9.8	2
6	gourmet	lots	large	9.2	1
7	gourmet	little	large	8.6	1
8	gourmet	little	large	8.2	2
9	plain	lots	small	10.6	1
10	plain	lots	large	10.4	1
11	plain	little	small	10.1	2
12	plain	little	small	9.9	1
13	plain	little	large	8.8	2
14	plain	lots	large	8.8	2
15	plain	little	large	8.2	1
16	plain	lots	small	7.4	2

Table 7.2 Description of the Sort Window

Select Columns Filter Menu	Contains options to search and filter through columns. See "Columns Filter Menu," p. 10 in the "Preliminaries" chapter.
Replace Table	To replace the original data table with the sorted table instead of creating a new table with the sorted values, click the box beside **Replace Table**. This option is not available if there are any open report windows generated from the original table.
Output table name	(Optional) To give a name to the sorted table, type a name in the box beside **Output table name**.
Keep dialog open	To keep this window open after clicking **OK**, check the box beside **Keep dialog open**.
By	Adds the columns that you want to sort by. The columns that you add to the list establish the order of precedence for sorting. The first column in the list is the major sort field. Each variable thereafter is sorted within the previous variable in the sort list.
Remove	Removes any highlighted columns.
ascending and descending buttons ( )	To change the list order (ascending or descending) of the values for the grouping variables, in the By variable list, select a variable and click the appropriate ascending or descending button ( ). The icon beside the variable changes to indicate the sorting order.

Stacking Columns

You can rearrange your data table by stacking two or more columns into a single new column, preserving the values from the other columns. Or, you can stack a set of columns into multiple groups. The various ways you can stack columns are explained in Table 7.3.

To stack columns:

1. Select **Tables > Stack.**

Figure 7.7 Stack Window

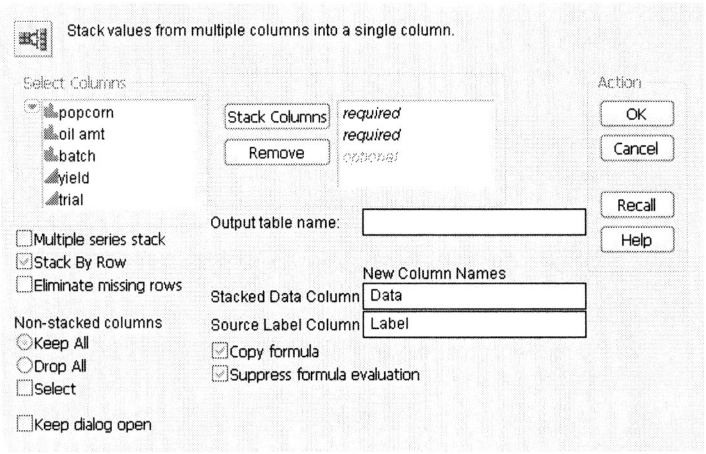

2. Highlight the names of the columns that you want to stack and click **Stack Columns.**
3. Customize your stacking further using the additional options. See Table 7.3.
4. Click **OK.**

Table 7.3 Description of the Stack Window

Select Columns Filter Menu	Contains options to search and filter through columns. See "Columns Filter Menu," p. 10 in the "Preliminaries" chapter.

Table 7.3 Description of the Stack Window *(Continued)*

Multiple series stack	To stack selected columns into two or more columns, check the box beside **Multiple series stack**. Specify the number of columns into which you want the selected columns to be stacked by entering the number into the **Number of Series** box. This box appears when you check the box beside **Multiple series stack**. Select the **Contiguous** option if the series consists of adjacent columns. See "Example of Stacking Into More Than One Column (Using the Multiple Series Stack Option)," p. 240. **Note**: The order in which you add columns to the box on the right determines the group to which they belong.
Stack by Row	Leaving **Stack by Row** unchecked stacks one column underneath another. Checking it stacks columns by rows.
Eliminate missing rows	To eliminate missing data from the new table, check the box beside **Eliminate missing rows.** If **Stack by Rows** is checked also, only rows with all data missing are eliminated.
Non-stacked columns	Includes or drops non-stacked columns from the new data table. Select one of these options: **Keep All** select this option if you want the new table to contain all of the non-stacked columns from the original table. **Drop All** select this option if you want to include only the stacked columns in the new table, and you do not want to include any non-stacked columns. **Select** enables you to choose the non-stacked columns that you want to include or drop in the new table.
Keep dialog open	To keep this window open after clicking **OK**, check the box beside **Keep dialog open.**
Stack Columns	Adds the columns that you want to stack.
Remove	Removes any highlighted columns.
Output table name	(Optional) To name the new table, type a name in the box beside **Output table name.**
Stacked Data Column	To assign a name to the column that will contain the data for the stacked columns, enter a name in the **Stacked Data Column** box. Leave the box empty if you do not want this column to appear in the new table.

Table 7.3 Description of the Stack Window *(Continued)*

Source Label Column	To assign a name to the column that will contain the original table's column names, enter the name in the **Source Label Column** box. The default name is Label. Leave the box empty if you do not want this column to appear in the new table.
Copy formula	To include formulas from the original table in the output columns, click the box beside **Copy formula**. This option is not present when there are no formulas in the data table.
Suppress formula evaluation	To prevent JMP from evaluating columns' formulas when the new table is created, check the box beside **Suppress formula evaluation.** This option is not present when there are no formulas in the data table.

Example of Stacking into One Column

A researcher has two columns in their data table representing yield, and they want to stack the two columns into a single column. (This new single column would be called Data by default.)

1. Open the Popcorn Trials.jmp sample data table.
2. Select **Tables > Stack.**
3. Select yield1 and yield2 and click **Stack Columns.**

Figure 7.8 Completed Stack Window

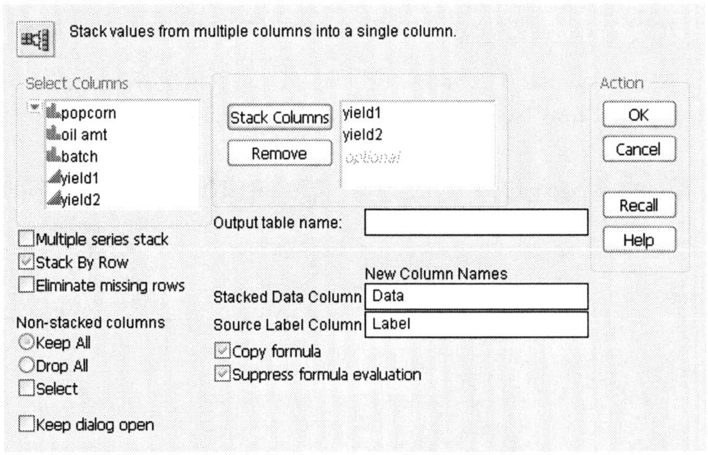

4. Click **OK.**

Figure 7.9 Stacked and Original Data Tables

stacked data table

	popcorn	oil amt	batch	Label	Data
1	plain	little	large	yield1	8.2
2	plain	little	large	yield2	8.8
3	gourmet	little	large	yield1	8.6
4	gourmet	little	large	yield2	8.2
5	plain	lots	large	yield1	10.4
6	plain	lots	large	yield2	8.8
7	gourmet	lots	large	yield1	9.2
8	gourmet	lots	large	yield2	9.8
9	plain	little	small	yield1	9.9
10	plain	little	small	yield2	10.1
11	gourmet	little	small	yield1	12.1
12	gourmet	little	small	yield2	15.9
13	plain	lots	small	yield1	10.6
14	plain	lots	small	yield2	7.4
15	gourmet	lots	small	yield1	18.0
16	gourmet	lots	small	yield2	16.0

original data table

	popcorn	oil amt	batch	yield1	yield2
1	plain	little	large	8.2	8.8
2	gourmet	little	large	8.6	8.2
3	plain	lots	large	10.4	8.8
4	gourmet	lots	large	9.2	9.8
5	plain	little	small	9.9	10.1
6	gourmet	little	small	12.1	15.9
7	plain	lots	small	10.6	7.4
8	gourmet	lots	small	18.0	16.0

The Label column represents the **Source Label Column** that identifies the source of the data. Its values are the column names in the original table from which the stacked values originated.

Example of Stacking Into More Than One Column (Using the Multiple Series Stack Option)

Suppose that a researcher has data on blood pressure readings. The readings were taken over three days: Monday, Wednesday, and Friday. Three readings were taken each day, at 8am, 12pm, and 6pm.

1. Open the Blood Pressure.jmp sample data table.

Figure 7.10 Blood Pressure.jmp

	Subject	Dose	BP 8M	BP 12M	BP 6M	BP 8W	BP 12W	BP 6W	BP 8F	BP 12F	BP 6F
1	1	A	183	174	180	174	178	181	171	178	171
2	2	A	173	181	181	170	179	176	175	185	188
3	3	A	181	189	177	188	175	182	183	183	180
4	4	A	181	177	182	176	173	184	183	187	183
5	5	A	184	180	176	172	175	176	170	190	183
6	1	B	171	181	180	183	176	174	185	182	184
7	2	B	175	180	180	183	180	173	184	185	180
8	3	B	179	173	183	188	193	170	179	185	170
9	4	B	178	192	177	185	172	179	187	184	177
10	5	B	181	187	182	182	177	177	184	183	184
11	1	Control	182	186	178	177	177	177	183	191	175
12	2	Control	178	184	187	174	175	176	180	181	184
13	3	Control	186	180	181	171	173	181	179	181	189
14	4	Control	175	180	177	185	171	182	176	185	170
15	5	Control	181	175	181	179	185	184	184	173	185
16	1	Placebo	181	178	179	178	182	179	188	183	173
17	2	Placebo	179	175	189	191	178	177	182	188	171
18	3	Placebo	177	187	172	180	183	178	183	190	178
19	4	Placebo	176	178	182	177	178	177	176	184	184
20	5	Placebo	184	171	182	181	186	183	184	172	179

Each BP (blood pressure) column is delineated according to the date and time. The BP 8M column corresponds to readings that were taken at 8am on Monday; the BP 12W column corresponds to readings that were taken on 12pm on Wednesday; and so on. The researcher wants to stack all of the blood pressure columns into three columns that correspond to each day: Monday, Wednesday, and Friday.

2. Select **Tables > Stack.**
3. Select all of the BP readings and click **Stack Columns**.

   The order of the columns reflects the way that the columns in the series should be grouped.
4. Select **Multiple series stack.**
5. Next to **Number of Series**, type 3.
6. Since you want the columns to be stacked vertically, select **Contiguous**.
7. Rename the **Stacked Data Column** from Data to BP (for blood pressure).
8. Rename the **Source Label Column** from Label to Day.

Figure 7.11 Completed Stack Window

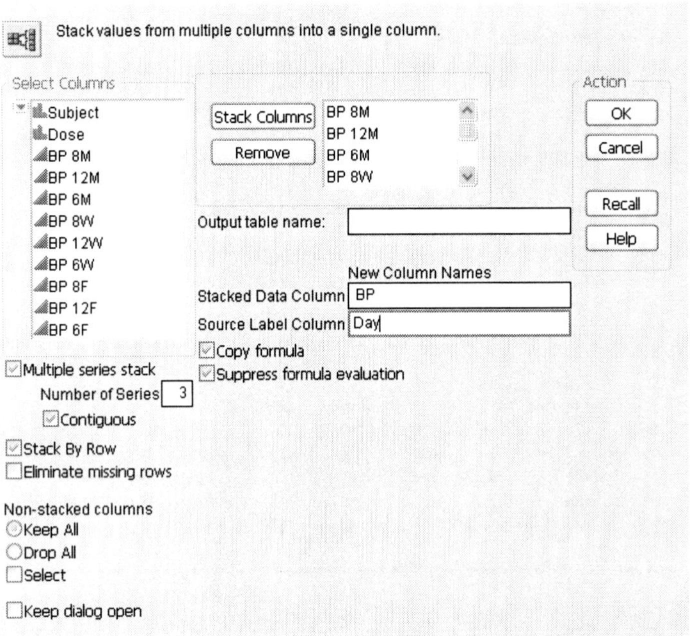

9. Click **OK**.

Figure 7.12 Stacked Data Table

	Subject	Dose	Day	BP	Day 2	BP 2	Day 3	BP 3
1	1	A	BP 8M	183	BP 8W	174	BP 8F	171
2	1	A	BP 12M	174	BP 12W	178	BP 12F	178
3	1	A	BP 6M	180	BP 6W	181	BP 6F	171
4	2	A	BP 8M	173	BP 8W	170	BP 8F	175
5	2	A	BP 12M	181	BP 12W	179	BP 12F	185
6	2	A	BP 6M	181	BP 6W	176	BP 6F	188
7	3	A	BP 8M	181	BP 8W	188	BP 8F	183
8	3	A	BP 12M	189	BP 12W	175	BP 12F	183
9	3	A	BP 6M	177	BP 6W	182	BP 6F	180
10	4	A	BP 8M	181	BP 8W	176	BP 8F	183
11	4	A	BP 12M	177	BP 12W	173	BP 12F	187
12	4	A	BP 6M	182	BP 6W	184	BP 6F	183
13	5	A	BP 8M	184	BP 8W	172	BP 8F	170
14	5	A	BP 12M	180	BP 12W	175	BP 12F	190
15	5	A	BP 6M	176	BP 6W	176	BP 6F	183
16	1	B	BP 8M	171	BP 8W	183	BP 8F	185
17	1	B	BP 12M	181	BP 12W	176	BP 12F	182
18	1	B	BP 6M	180	BP 6W	174	BP 6F	184
19	2	B	BP 8M	175	BP 8W	183	BP 8F	184
20	2	B	BP 12M	180	BP 12W	180	BP 12F	185
21	2	B	BP 6M	180	BP 6W	173	BP 6F	180

In the stacked data table, note the following:

- The first Day column represents Monday.
- The Day 2 column represents Wednesday.
- The Day 3 column represents Friday.

Splitting Columns

You can create a new data table from the active table by splitting one column into several new columns. This column is split according to the values found in another column, referred to as the **Split By** column. You can also split columns according to the values of one or more grouping variables.

To split columns:

1. Select **Tables > Split**.

Figure 7.13 Split Window

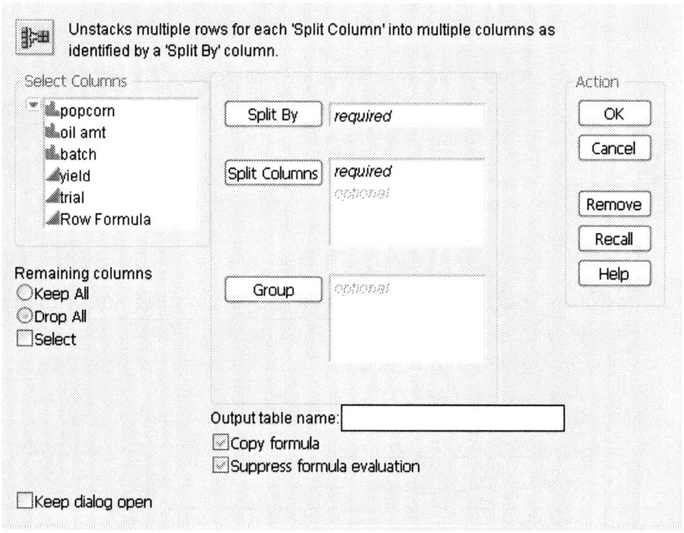

2. Highlight the names of the column or columns that you want to split and click **Split Columns**.
3. Highlight a column whose values you want to use as the basis for splitting the column.
4. Click **Split By**.
5. Customize your splitting further using the additional options. See Table 7.4.
6. Click **Split**.

Table 7.4 Description of the Split Window

Select Columns Filter Menu	Contains options to search and filter through columns. See "Columns Filter Menu," p. 10 in the "Preliminaries" chapter.
Keep All	To include all columns in the new table, click the **Keep All** option in the **Remaining Columns** area.
Drop All	To include only columns used in the split in the new table, click the **Drop All** option in the **Remaining Columns** area.
Select	To select which columns to keep in the new table, click the **Select** option in the **Remaining Columns** area. Then select which columns to keep in the new table.
Keep dialog open	To keep this window open after clicking **OK**, check the box beside **Keep dialog open**.
Split By	Adds the column whose values you want to use as the new column names, and as the basis for splitting the column.
Split Columns	Adds the column or columns that you want to split.
Group	Specify a **Group** variable when you want your data to be split within each group of the selected variable. Each group results in a row in the output table. **Note**: You must also specify the required variables, **Split By** and **Split Columns**. If your grouping variable contains unequal groups, or if your grouping variable is not grouped in order (is random) then you must specify a **Group** variable. The **Group** variable ensures that your data is restructured properly.
Output table name	(Optional) To assign a name to the new table, enter the name in the box beside **Output table name**.
Copy formula	To include formulas from the original table in the output columns, click the box beside **Copy formula**. This option is not present when there are no formulas in the data table.
Suppress formula evaluation	To prevent JMP from evaluating columns' formulas when the new table is created, click the box beside **Suppress formula evaluation**. This option is not present when there are no formulas in the data table.

Examples of Splitting Columns

This section contains two examples using **Split**. In the first example, one column is split by a second column. See "Splitting a Column: Basic Example," p. 245. The second example uses a **Group** variable. See "Splitting a Column: Grouping Rows Example," p. 247.

Splitting a Column: Basic Example

In the Popcorn.jmp data table shown in Figure 7.14, the data in the trial column shows that there are two trials, 1 and 2.

Figure 7.14 Original Data Table Showing Two Trials

	popcorn	oil amt	batch	yield	trial
1	plain	little	large	8.2	1
2	gourmet	little	large	8.6	1
3	plain	lots	large	10.4	1
4	gourmet	lots	large	9.2	1
5	plain	little	small	9.9	1
6	gourmet	little	small	12.1	1
7	plain	lots	small	10.6	1
8	gourmet	lots	small	18.0	1
9	plain	little	large	8.8	2
10	gourmet	little	large	8.2	2
11	plain	lots	large	8.8	2
12	gourmet	lots	large	9.8	2
13	plain	little	small	10.1	2
14	gourmet	little	small	15.9	2
15	plain	lots	small	7.4	2
16	gourmet	lots	small	16.0	2

For this example, split the yield column into two new columns, one for trial 1 and one for trial 2. Proceed as follows:

1. Open the Popcorn.jmp sample data table.
2. Select **Tables > Split**.
3. Select the yield column and click **Split Columns**.
4. Select the trial column and click **Split By**.
5. Under Remaining columns, select **Keep All**.

   The default is **Drop All**, which omits any columns that are not in the **Split By**, **Split Columns**, or **Group** fields. Selecting **Keep All** includes these columns in the new table.
6. (Optional) Type Yield column split by Trial column in the Output table name field.

Figure 7.15 Settings for Splitting yield Column by trial Column

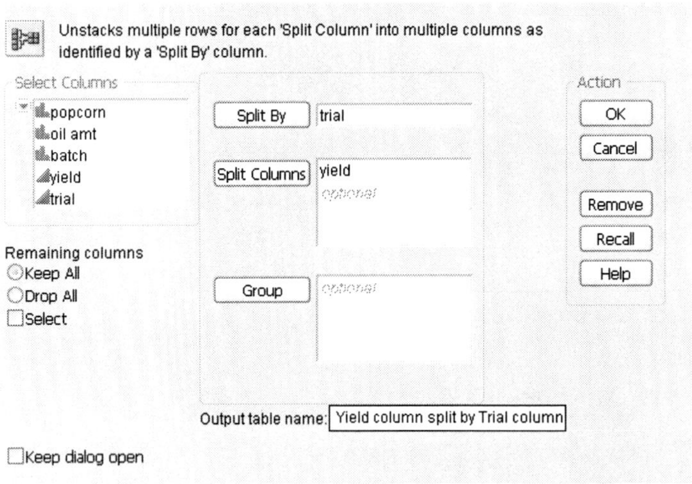

7. Click **OK**.

   A new data table is created that looks like Figure 7.16. Notice the following:

   – The yield and trial columns are gone.

   – The data table has two new columns, named after the unique values (1 and 2) from the original trial column.

   – The values from the original yield column are now split into the new columns named 1 and 2.

   – The columns other than trial and yield are exactly the same as they were in the original table.

Figure 7.16 New Table Created by Splitting yield Column by trial Column

	popcorn	oil amt	batch	1	2
1	plain	little	large	8.2	8.8
2	gourmet	little	large	8.6	8.2
3	plain	lots	large	10.4	8.8
4	gourmet	lots	large	9.2	9.8
5	plain	little	small	9.9	10.1
6	gourmet	little	small	12.1	15.9
7	plain	lots	small	10.6	7.4
8	gourmet	lots	small	18.0	16.0

8. (Optional) Rename the new columns to give them meaningful names. For example, rename 1 to yield (trial 1) and rename 2 to yield (trial 2).

Figure 7.17 Rename the New Columns for Clarity

	popcorn	oil amt	batch	yield (trial 1)	yield (trial 2)
1	plain	little	large	8.2	8.8
2	gourmet	little	large	8.6	8.2
3	plain	lots	large	10.4	8.8
4	gourmet	lots	large	9.2	9.8
5	plain	little	small	9.9	10.1
6	gourmet	little	small	12.1	15.9
7	plain	lots	small	10.6	7.4
8	gourmet	lots	small	18.0	16.0

Splitting a Column: Grouping Rows Example

The Drug Measurements.jmp sample data table contains measurements of three different drugs (a, b, and c) administered to 12 different subjects. You want to split the measurement into different columns, one for each drug type. You also want to group the measurements by subject.

1. Open the Drug Measurements.jmp sample data table.

Figure 7.18 Drug Measurements.jmp Sample Data Table

	Subject	Drug Type	Measurement
1	1	a	91
2	1	b	99
3	1	c	91
4	2	b	96
5	3	a	95
6	3	c	93
7	3	b	99
8	4	a	97
9	4	c	96
10	4	b	93
11	5	a	95
12	5	b	100
13	5	c	94
14	6	a	92
15	6	b	96
16	6	c	95
17	7	a	93
18	7	b	96
19	8	a	100
20	8	b	94
21	8	c	92
22	9	a	98
23	9	b	99
24	9	c	97
25	10	a	91
26	10	b	96
27	10	c	99
28	11	a	98
29	11	b	100
30	11	c	93
31	12	b	95
32	12	c	95

2. Select **Tables > Split**.

3. Select Drug Type and click **Split By.**
4. Select Measurement and click **Split Columns.**

Notice that the Subject variable contains unequal groups. Most of the subjects were given all three drugs, but subject 2 was only given one drug, and subjects 7 and 12 were given only two drugs. In this situation, to ensure that the correct measurements are associated with the correct subject, specify Subject as the **Group** variable.

5. Select Subject and click **Group.**
6. Click **OK.**

Figure 7.19 Drug Measurements.jmp Split by a Grouping Variable

	Subject	a	b	c
1	1	91	99	91
2	2	•	96	•
3	3	95	99	93
4	4	97	93	96
5	5	95	100	94
6	6	92	96	95
7	7	93	96	•
8	8	100	94	92
9	9	98	99	97
10	10	91	96	99
11	11	98	100	93
12	12	•	95	95

You can see that the appropriate missing values appear for subjects 2, 7, and 12.

Transposing Rows and Columns

You can create a new JMP table that is a transposed version of the active data table. The columns of the active table are the rows of the new table, and its rows are the new table's columns.

When you transpose columns, you do the following:

- Select the columns to be transposed.
- Specify a "label" column, from which the new columns get their names (optional).
- Specify "by" columns, which tells JMP to transpose data within groups (optional).

Note: Columns that you want to transpose must have the same data type. Also, if columns contain value labels, transposing uses the actual data values, not the value labels. (See "Using Value Labels," p. 151.)

To transpose rows and columns

1. Open a data table that contains the rows and columns that you want to transpose.

2. Select **Tables > Transpose.**

Figure 7.20 Transpose Window

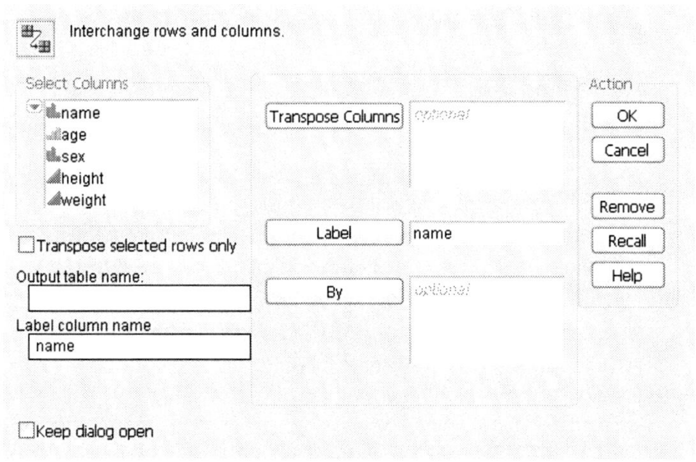

3. Highlight the column name(s) you want to transpose in the **Select Columns** box on the left.
4. Click **Transpose Columns.**
5. Customize your transposed table further using the additional options. See Table 7.5.
6. Click **OK.**

Table 7.5 Description of the Transpose Window

Select Columns Filter Menu	Contains options to search and filter through columns. See "Columns Filter Menu," p. 10 in the "Preliminaries" chapter.
Transpose selected rows only	To transpose only rows that are currently highlighted in the active table, click the box beside **Transpose selected rows only.**
Output table name	(Optional) To name the subset table, type a name in the box beside **Output table name.**
Label column name	(Applicable only if you have specified a **Label** column.) Specify an alternative name for the Label column. Otherwise, the default column name is **name.**
Keep dialog open	To keep this window open after clicking **OK**, check the box beside **Keep dialog open.**
Transpose Columns	Adds the columns that you want to transpose.

Table 7.5 Description of the Transpose Window *(Continued)*

Label	To use the data from a column in the original table as the column names in the new table: 1. Highlight a column from the **Select Columns** box on the left. 2. Click **Label**. The column name appears in the **Label** box. The default column name is Row followed by the row number from the original data table. You can specify an alternative name for the column using the **Label column name** option. Only one column is created for each distinct value in the label column. Therefore, if there are duplicate values in the label column, JMP creates only one column for the duplicated value using the value from the last duplicated row.
By	To organize the transposed columns into groups based on the columns that you put into the **By** box: 1. Highlight the column name(s) in the **Select Columns** box whose values you want to see as a group. 2. Click **By**.

Rules for Transposing

Table 7.6 describes the rules that apply to transposing.

Table 7.6 Rules for Transposing

If	Then
The original table has columns but no rows	The new table contains one column that lists those column names.
The original table has one column and it is assigned to **Label**	Its values become the column names in the transposed table.
The original table has multiple columns and contains a label column	JMP automatically inserts the label column into the **Label** box when the window appears. You can remove this column if you do not want it to appear.
There is no label column in the original table	The column names in the transposed table are Row 1, Row 2, ..., Row n where *n* is the number of rows in the original table.

Examples of Transposing

Simple Example of Transposing

1. Open the Materials1.jmp sample data table.
2. Select **Tables > Transpose**.
3. Select plastic, tin, and gold and click **Transpose Columns**.
4. Click **OK**.

Figure 7.21 Simple Transpose

Original table

	plastic	tin	gold
1	1	2	3
2	4	5	6

Transposed table

	Label	Row 1	Row 2
1	plastic	1	4
2	tin	2	5
3	gold	3	6

The original table in Figure 7.21 has two rows and three continuous columns called plastic, tin, and gold. The transposed table has a row for each of the three columns in the original table, and columns named Row 1 and Row 2 for the original table's rows. The additional column called Label has the column names (plastic, tin, and gold) from the original table as values.

Example Using the Label Button

1. Open the Materials2.jmp sample data table.
2. Select **Tables > Transpose**.
3. Select plastic, tin, and gold and click **Transpose Columns**.
4. Select item and click **Label**.

Figure 7.22 Completed Transpose Window

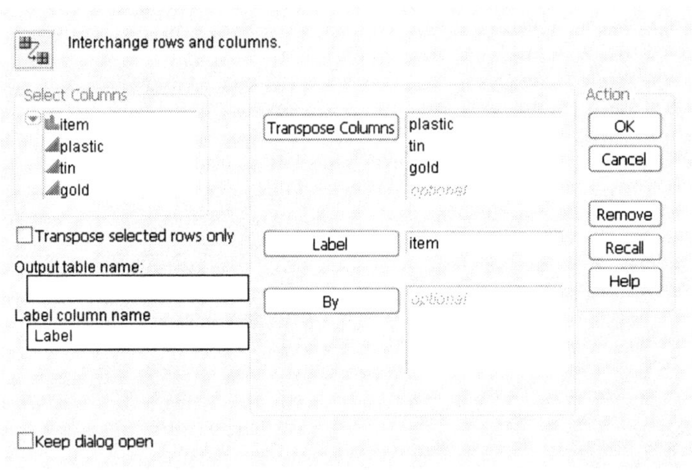

5. Click **OK**.

Figure 7.23 Transpose with a Label

original table

	item	plastic	tin	gold
1	nails	1	2	3
2	hooks	4	5	6

transposed table

	Label	nails	hooks
1	plastic	1	4
2	tin	2	5
3	gold	3	6

The values from the item column in the original table are used as column labels in the transposed table.

Example Using a By Group

1. Open the Animals Subset.jmp sample data table.
2. Select **Tables > Transpose**.
3. Select subject and miles and click **Transpose Columns**.
4. Select season and click **Label**.
5. Select species and click **By**.

Figure 7.24 Completed Transpose Window

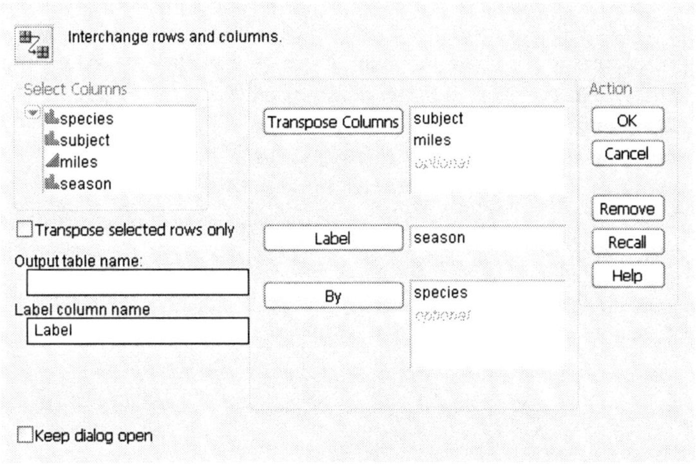

6. Click **OK**.

Figure 7.25 Transpose Using a By Group

original table

	species	subject	miles	season
1	FOX	3	4	fall
2	FOX	3	3	winter
3	FOX	3	6	spring
4	FOX	3	2	summer
5	COYOTE	1	4	fall
6	COYOTE	1	2	winter
7	COYOTE	1	7	spring
8	COYOTE	1	8	summer

transposed table

	species	Label	fall	winter	spring	summer
1	COYOTE	subject	1	1	1	1
2	COYOTE	miles	4	2	7	8
3	FOX	subject	3	3	3	3
4	FOX	miles	4	3	6	2

The transposed table contains values that have been transposed in groups.

Concatenating Data Tables

When you concatenate data tables in JMP, you combine rows from two or more data tables. You can create a new data table *or* you can append rows to the first data table. If a column name is the same in the data tables that you want to concatenate, then the column in the new data table lists the values from all of the data tables in the order of concatenation. If the two original data tables have columns with different names, those columns are included in the new data table showing missing values.

To concatenate two data tables with the same column names:

1. Select **Tables > Concatenate.**

Figure 7.26 Concatenate Window

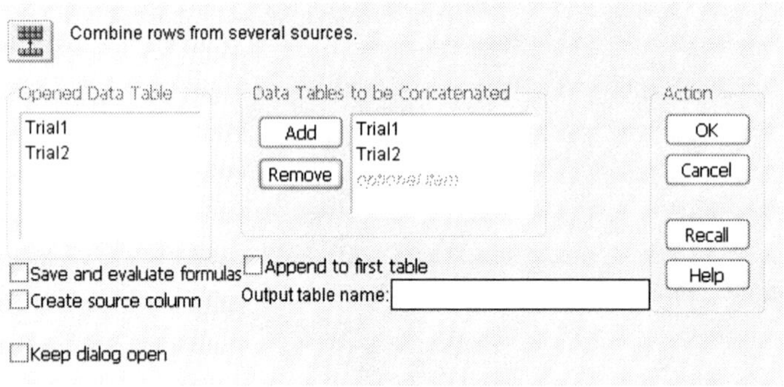

2. Highlight the names of the data tables you would like to combine, and click **Add**.

   You can concatenate as many data tables as you choose, and you can also add the same data tables multiple times. The number of rows in the new data tables is the sum of the number of rows in all the data tables.

3. (Optional) Click the **Save and evaluate formulas** choice to request that JMP include all formulas.

   If you do not select this option, no formulas are included in the new data table.

Note: If columns with the same name have different formulas, then the formula from the first data table is saved in the concatenated data table.

4. (Optional) Click the **Create source column** choice to add a column called Source Table to the new data table.

   This column identifies the name of the source data table in the corresponding rows.

5. (Optional) Select the **Append to first table** choice to append rows to the data table listed first in the **Data Tables to be Concatenated** field instead of creating a new data table.

6. (Optional) Type a name for the new data table in the **Output table name** field.

   If you do not type a name, JMP names the data table Untitled*x* (for example, Untitled1). The **Output table name** field is not available if you selected the **Append to first table** choice.

7. Click **OK**.

Example of Concatenating Data Tables

Suppose you want to concatenate two data tables—Trial1 and Trial2—into a new data table.

1. Open the Trial1.jmp and Trial2.jmp sample data tables.
2. From the Trial1.jmp table, select **Tables > Concatenate**.
3. In the **Opened Data Table** list, select Trial2 and click **Add**.

Figure 7.27 Concatenate Settings

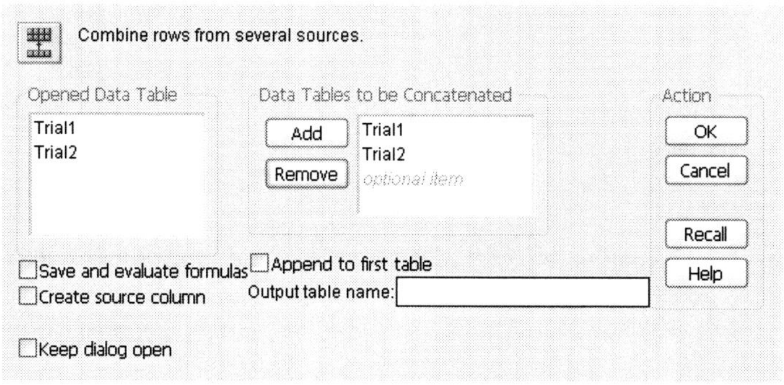

4. Click **OK**.

The data tables combine into a new concatenated table with all of the rows from the first data table followed by all of the rows from the second data table. See Figure 7.28.

Figure 7.28 Result of Concatenating Two Data Tables

Trial1.jmp

	popcorn	oil amt	batch	yield
1	plain	little	large	8.2
2	gourmet	little	large	8.6
3	plain	lots	large	10.4
4	gourmet	lots	large	9.2
5	plain	little	small	9.9
6	gourmet	little	small	12.1
7	plain	lots	small	10.6
8	gourmet	lots	small	18.0

Trial2.jmp

	popcorn	oil amt	batch	yield
1	plain	little	large	8.8
2	gourmet	little	large	8.2
3	plain	lots	large	8.8
4	gourmet	lots	large	9.8
5	plain	little	small	10.1
6	gourmet	little	small	15.9
7	plain	lots	small	7.4
8	gourmet	lots	small	16.0

new concatenated data table

	popcorn	oil amt	batch	yield
1	plain	little	large	8.2
2	gourmet	little	large	8.6
3	plain	lots	large	10.4
4	gourmet	lots	large	9.2
5	plain	little	small	9.9
6	gourmet	little	small	12.1
7	plain	lots	small	10.6
8	gourmet	lots	small	18.0
9	plain	little	large	8.8
10	gourmet	little	large	8.2
11	plain	lots	large	8.8
12	gourmet	lots	large	9.8
13	plain	little	small	10.1
14	gourmet	little	small	15.9
15	plain	lots	small	7.4
16	gourmet	lots	small	16.0

Concatenated data tables always have a column for every column name found in the original data tables. However, if the column names do not match exactly, they are not merged. For example, in Figure 7.28, if the yield column was instead named yield1 and yield2, a separate column would be created for each in the concatenated data table.

Joining Data Tables

You can combine two data tables into one new table by selecting **Tables > Join**. Tables can be joined in three different ways:

- By combining them according to row number. See "How to Join Data Tables," p. 257, and "Example of Joining by Row Number," p. 260.
- In a Cartesian fashion, where you form a new table consisting of all possible combinations of the rows from two original tables. See "How to Join Data Tables," p. 257, and "Examples of a Cartesian Join," p. 264.
- By matching the values in one or more columns that exist in both data tables. See "How to Join Data Tables," p. 257, and "Example of Joining by Matching Columns," p. 266.

How to Join Data Tables

To join two data tables into a new data table:

1. Open the two data tables that you want to join.
2. Select **Tables > Join**.

   In the window that appears, the names of all open tables appear below **Join...with**, as shown in Figure 7.29.

Figure 7.29 The Join Window

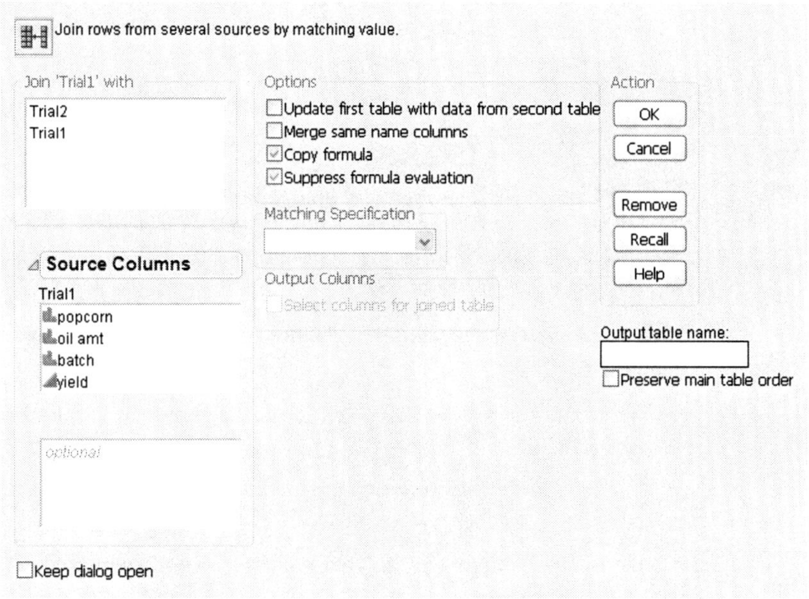

3. In the **Join...with** box, select the table to join with the active table.
4. From the **Matching Specification** area, select the option that specifies how to join the tables.
5. Type the name of the new table in the text box beside **Output table name**.
6. Customize your join further using the additional options. See Table 7.7.
7. Click **OK** to create the joined data table.

Table 7.7 Description of the Join Window

Keep dialog open	To keep this window open after clicking **OK**, check the box beside **Keep dialog open**.
Update first table with data from second table	When **Update first table with data from second table** is checked, column data from the second table change the data of the same name columns in the original table. Note the following: • JMP does not replace data with missing values. • The output table uses the same columns as the original table. Thus, when you use **Update first table with data from second table**, **Select Columns for joined table** is not applicable. • The **Update first table with data from second table** option is available only when joining by row number or by matching columns.
Merge same name columns	Click the box beside **Merge same name columns** if you want the data from the second table to replace the data of the same name columns in the original table. Note that missing values in the first table are replaced by nonmissing values in the second. If you join by matching columns, the new table contains a column named Match Flag: • If a one (1) appears in this column, the data originated from the first (active) table. • If a two (2) appears in this column, the data originated from the second table. • If a three (3) appears in this column, the data was found in both the first and second tables.
Copy formula	To include formulas from the original table in the output columns, click the box beside **Copy formula**.
Suppress formula evaluation	To prevent JMP from evaluating columns' formulas during the creation of the new table, click the box beside **Suppress formula evaluation**.

Table 7.7 Description of the Join Window *(Continued)*

By Matching Columns	When you select this option, the window expands, prompting you to select columns whose values and data types must match in both tables for rows to be joined. You should: 1. Highlight a column name from each list in the **Source Columns** area. The first highlighted column in the top list pairs with the first highlighted column in the bottom list, the second columns are paired, and so on. Rows join only if values and data types match for all the column pairs. 2. Click **Match**. The selected pair of columns appears in the **Match columns** box. Matching columns do not have to have the same names and do not have to be in the same relative column position in both tables. 3. (Optional) To only include the first match found, check the boxes associated with **Drop multiples** in both tables. Only the first match found is written to the new table. If you specify this option for one table, the first match value is joined with all matches in the other table. If you do not check the boxes associated with **Drop multiples** in either table, a Cartesian join is performed within each group of matching column values. 4. (Optional) To include all rows from the data table, even when there is no matching value, check the boxes associated with **Include non-matches**. You can specify this option for either or both data tables being joined.
By Row Number	Joins the two tables side by side.
Cartesian Join	Joins two tables using a Cartesian fashion, where it forms a new table consisting of all possible combinations of the rows from two original tables. JMP crosses the data in the first table with the data in the second to display all combinations of the values in each set.
Select Columns for joined table	Select the **Select Columns for joined table** option if you want to select a subset of columns from either table for inclusion in the output table. Follow these steps: 1. In the **Source Columns** area, highlight the columns from each table that you want to include in the new table. 2. Click **Select** in the **Output Columns** area.
Output table name	To give a name to the joined table, type a name in the box beside **Output table name**.
Preserve main table order	(Available only if the By Matching Columns option is specified) Maintains the order of the original data table in the joined table, instead of sorting by the matching columns.

Examples of Joining Data Tables

The following sections give examples of how to use the **Join** command.

Example of Joining by Row Number

Joining tables by row number joins the two tables side by side. The new table has all of the columns from both tables, unless you specify to include only certain columns.

Joining Tables with an Unequal Number of Rows

If the two tables that you want to join have an unequal number of rows, the new table contains values for the rows found in both tables.

1. Open the Species1.jmp and Species2.jmp sample data tables.

   Notice that the Species1.jmp table has two rows, and the Species2.jmp table has four rows.
2. From the Species1.jmp table, select **Tables < Join.**
3. In the **Join...with** box, select Species2.
4. From the **Matching Specification** area, select **By Row Number.**
5. Click **OK.**

Figure 7.30 Joining Tables by Row Number

Species1.jmp

	species
1	FOX
2	COYOTE

Species2.jmp

	season
1	spring
2	summer
3	fall
4	winter

joined table

	species	season
1	FOX	spring
2	COYOTE	summer
3		fall
4		winter

When one table with two rows is joined with a table with four rows, then the new table contains four rows.

Joining Columns with the Same Name

If the two tables have column names that are the same, the names of these columns in the new table appear as "column name of table name." For example, suppose that you want to combine the eight rows from the Trial1.jmp and Trial2.jmp data tables shown in Figure 7.32 into a single table. You want to combine them so that the new table contains all of the columns from both tables.

1. Open the Trial1.jmp and Trial2.jmp sample data tables.
2. From the Trial1.jmp data table, select **Tables > Join.**
3. In the **Join...with** box, select Trial2.
4. From the **Matching Specification** menu, select **By Row Number.**

Figure 7.31 The Completed Join Window

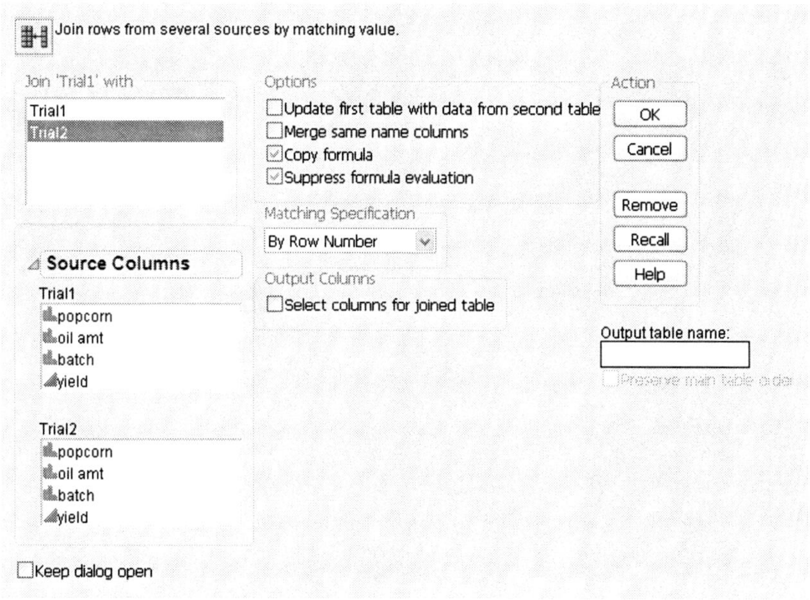

5. Click **OK**.

Figure 7.32 Original Tables and the Joined Table

Trial1.jmp

	popcorn	oil amt	batch	yield
1	plain	little	large	8.2
2	gourmet	little	large	8.6
3	plain	lots	large	10.4
4	gourmet	lots	large	9.2
5	plain	little	small	9.9
6	gourmet	little	small	12.1
7	plain	lots	small	10.6
8	gourmet	lots	small	18.0

Trial2.jmp

	popcorn	oil amt	batch	yield
1	plain	little	large	8.8
2	gourmet	little	large	8.2
3	plain	lots	large	8.8
4	gourmet	lots	large	9.8
5	plain	little	small	10.1
6	gourmet	little	small	15.9
7	plain	lots	small	7.4
8	gourmet	lots	small	16.0

joined table

	popcorn of Trial1	oil amt of Trial1	batch of Trial1	yield of Trial1	popcorn of Trial2	oil amt of Trial2	batch of Trial2	yield of Trial2
1	plain	little	large	8.2	plain	little	large	8.8
2	gourmet	little	large	8.6	gourmet	little	large	8.2
3	plain	lots	large	10.4	plain	lots	large	8.8
4	gourmet	lots	large	9.2	gourmet	lots	large	9.8
5	plain	little	small	9.9	plain	little	small	10.1
6	gourmet	little	small	12.1	gourmet	little	small	15.9
7	plain	lots	small	10.6	plain	lots	small	7.4
8	gourmet	lots	small	18.0	gourmet	lots	small	16.0

Note that if a column name is the same in the two original tables, the output column name is qualified by the source table name. For example, the column names in the new table appear as <variable name> of table name.

Joining Only Specified Columns

Suppose that you do not want all of the columns from the original data tables to be in the joined table. Proceed as follows:

1. Open the Trial1.jmp and Trial2.jmp sample data tables.
2. From the Trial1.jmp data table, select **Tables > Join**.
3. In the **Join...with** box, select Trial2.
4. From the **Matching Specification** menu, select **By Row Number**.

5. Click **Select columns for joined table** to specify the subset of columns that you want to include.
6. In the Source Columns list, select popcorn and yield from the Trial1 list and select yield from the Trial2 list.

   Because identical data exists in the popcorn column of both tables, you only need to select one column.
7. Click **Select**.

Figure 7.33 Completed Join Window

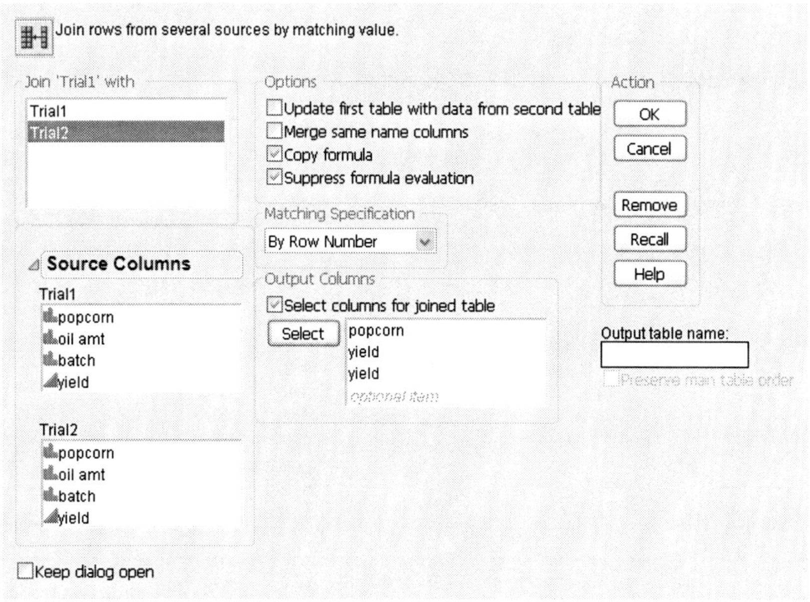

8. Click **OK**.

Figure 7.34 Joining Only Specified Columns

	popcorn	yield of Trial1	yield of Trial2
1	plain	8.2	8.8
2	gourmet	8.6	8.2
3	plain	10.4	8.8
4	gourmet	9.2	9.8
5	plain	9.9	10.1
6	gourmet	12.1	15.9
7	plain	10.6	7.4
8	gourmet	18.0	16.0

Examples of a Cartesian Join

When doing a Cartesian join, JMP joins two tables in a Cartesian fashion, where a new table is created that consists of all possible combinations of the rows from two original tables. This creates cases in the output table where there are one case for each combination of column values.

Example 1

1. Open the Species1.jmp and Species2.jmp sample data tables.
2. From the Species1.jmp table, select **Tables > Join**.
3. In the **Join...with** box, select Species2.
4. From the **Matching Specification** menu, select **Cartesian Join**.
5. Click **OK**.

Figure 7.35 Joining Tables Using Cartesian Join

Species1.jmp

	species
1	FOX
2	COYOTE

Species2.jmp

	season
1	spring
2	summer
3	fall
4	winter

joined table

	species	season
1	FOX	spring
2	FOX	summer
3	FOX	fall
4	FOX	winter
5	COYOTE	spring
6	COYOTE	summer
7	COYOTE	fall
8	COYOTE	winter

The data in Species1.jmp is crossed with the data in Species2.jmp to produce the joined table, which shows all combinations of the values in each set.

Example 2

In this example, you must use the **Tables > Join** command twice. The first join combines Oil Amount.jmp table with the Batch.jmp table using the **Cartesian** option. The second join combines the resulting table (Cartesian oil amount + batch) with the Popcorn Type.jmp table and produces a final table with all tables joined.

1. Open the Oil Amount.jmp, Batch.jmp, and Popcorn Type.jmp sample data tables.
2. From the Oil Amount.jmp table, select **Tables > Join**.
3. In the **Join...with** box, select Batch.
4. From the **Matching Specification** menu, select **Cartesian Join**.
5. Under **Output table name**, type Oil Amount and Batch.

Figure 7.36 Completed Join Window

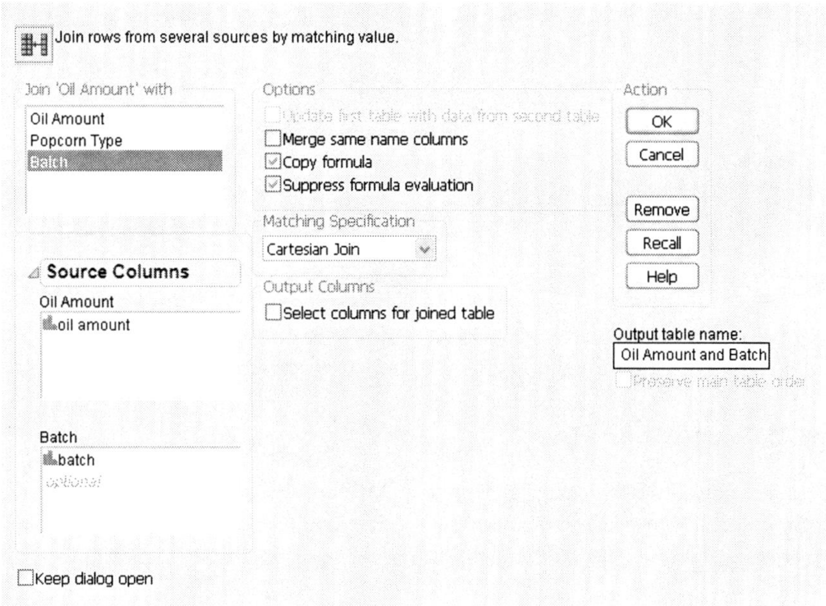

6. Click **OK**.

Figure 7.37 Oil Amount and Batch Joined Table

	oil amount	batch
1	little	small
2	little	large
3	lots	small
4	lots	large

The joined table contains all of the columns from the Oil Amount.jmp and Batch.jmp tables. Add the Popcorn Type.jmp columns, as follows:

7. From the Oil Amount and Batch table that you just created, select **Tables > Join.**
8. In the **Join...with** box, select Popcorn Type.
9. From the **Matching Specification** menu, select **Cartesian Join.**
10. Click **OK.**

Figure 7.38 Oil Amount and Batch Joined with Popcorn Type

	oil amount	batch	popcorn
1	little	small	gourmet
2	little	small	plain
3	little	large	gourmet
4	little	large	plain
5	lots	small	gourmet
6	lots	small	plain
7	lots	large	gourmet
8	lots	large	plain

The final table contains all of the columns from all three original tables. Keep in mind that the number of rows produced by a Cartesian join is the product of the number of rows in the original tables.

Example of Joining by Matching Columns

When you select to join data tables by matching columns, JMP finds specified column(s) values that exist in both tables and combines all values associated with the specified column(s) values into a new data table. In order to join by matching columns, the columns must have the same data type (numeric, character, or row state).

Joining Tables with the Same Rows in Different Order

1. Open the Students1.jmp and Students2.jmp sample data tables.

Figure 7.39 Students1 and Students2 Data Tables

Students1.jmp

	name	age	sex
1	KATIE	11	F
2	TIM	11	F
3	LOUISE	11	F
4	JEFFREY	11	M
5	JANE	11	F
6	JACLYN	11	F
7	ALICE	11	F
8	JAMES	11	F
9	ROBERT	11	F
10	BARBARA	11	F
11	CAROL	11	M
12	SUSAN	11	F

Students2.jmp

	name	height	weight
1	KATIE	56	85
2	LOUISE	57	69
3	JACLYN	62	104
4	JUDY	61	85
5	LILLIE	51	51
6	TIM	62	85
7	JAMES	54	81
8	ROBERT	58	96
9	ALICE	56	84
10	BARBARA	53	64
11	PATTY	57	84
12	SUSAN	60	77

The Students1.jmp data table contains names, ages, and sexes of the students. The Students2.jmp data table contains names, height, and weight of the students. Instead of working with two separate tables, you would like to combine the tables into one. Notice that the students' names are not in the same order in both tables. For example, Alice is in row 7 and row 9.

2. From the Students1.jmp data table, select **Tables > Join.**
3. In the **Join...with** box, select Students2.

Because both tables have one column (name) that contains the same values, you need to tell JMP that they are matches. JMP then examines each of the values in the name column of the first table to determine whether there was a corresponding value in the second table's name column. For example, it detects that Alice is located in both tables. It creates a name column in the new table with Alice as a value. It then takes the age and sex of Alice from table one and puts it in the new table. Then it takes the height and weight of Alice and puts them in the new table.

4. Select **By Matching Columns** in the Matching Specification area.
5. From the Students1 and Students2 lists, select name.
6. Click **Match**.
7. You want the new table to contain only one row for each name, so check the **Drop multiples** boxes for both tables. See Figure 7.40.

Figure 7.40 Completed Join Window

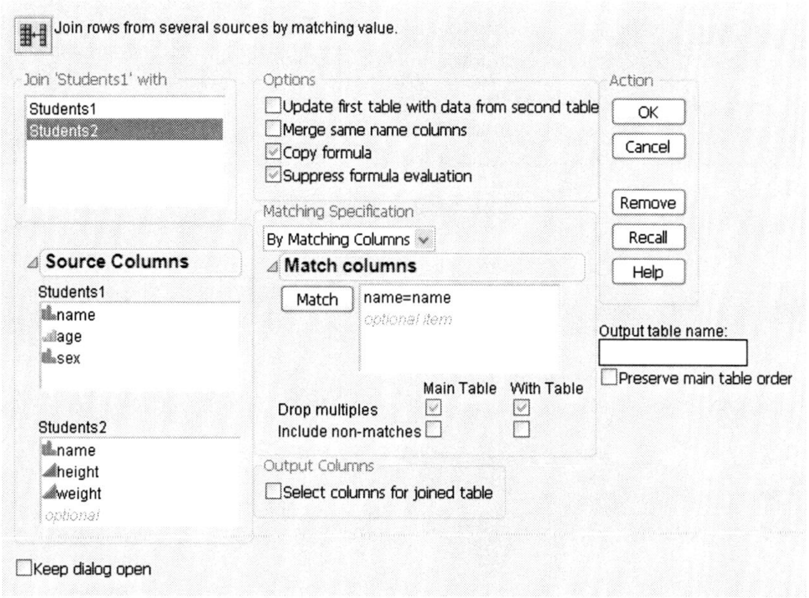

8. Click **OK**.

Figure 7.41 Joined Table

	name of Students1	age	sex	name of Students2	height	weight
1	ALFRED	11	M	ALFRED	59	87
2	ALICE	11	F	ALICE	56	84
3	AMY	11	M	AMY	55	74
4	BARBARA	11	F	BARBARA	53	64
5	CAROL	11	M	CAROL	60	95
6	CHRIS	11	M	CHRIS	58	85
7	CLAY	12	F	CLAY	63	108
8	DANNY	12	F	DANNY	59	89
9	DAVID	11	F	DAVID	56	73
10	EDWARD	11	M	EDWARD	58	101
11	ELIZABETH	11	M	ELIZABETH	61	87
12	FREDERICK	11	M	FREDERICK	58	75

Joining Tables with Different Numbers of Rows and Different Column Names

Suppose that Sarah and Joe are performing a popcorn experiment. They are popping different types of popcorn (gourmet and plain) in different amounts of oil. They are recording the amount (yield) of popcorn that is produced. Sarah gave you the first trial data in a file named Trial1.jmp. Joe gave you the second trial data in a file named Little.jmp. You want to combine the two tables into one table.

1. Open the Trial1.jmp and Little.jmp sample data tables.

Figure 7.42 Two Tables

Trial1.jmp

	popcorn	oil amt	batch	yield
1	plain	little	large	8.2
2	gourmet	little	large	8.6
3	plain	lots	large	10.4
4	gourmet	lots	large	9.2
5	plain	little	small	9.9
6	gourmet	little	small	12.1
7	plain	lots	small	10.6
8	gourmet	lots	small	18.0

Little.jmp

	popcorn	oil	batch	yield
1	plain	little	large	8.8
2	gourmet	little	large	8.2
3	plain	little	small	10.1
4	gourmet	little	small	15.9

2. From the Trial1.jmp table, select **Tables > Join.**
3. In the **Join...with** box, select Little.

You can see that three of the columns (**popcorn**, **oil amt/oil**, and **batch**) contain the same values in both tables. Identify these columns as matches. Also, because Sarah and Joe gave the **oil** and **oil amt** columns different names, you can tell JMP that **oil amt** and **oil** match.

4. Select **By Matching Cols** in the Matching Specification area.
5. From the Trial1 list, select **popcorn**, **oil amt**, and **batch**.
6. From the Little list, select **popcorn**, **oil**, and **batch**.
7. Click **Match**.

Looking at the two data tables, you can see that they have different numbers of rows. Trial1.jmp has values for eight experimental conditions, and Little.jmp has values for only four of those conditions. Sarah completed her experiment, but Joe only partially completed his experiment. You want the joined table to contain all of the rows in Trial1.jmp, even if that row in the Little.jmp table contains a missing value.

8. Select the **Include non-matches** boxes for both tables.

In the joined table, you only want one column for **popcorn**, one column for **oil**, and one column for **batch**. However, you want two columns for yield: one representing the yield from Trial1.jmp, and another representing the yield from Little.jmp.

9. Select the box beside **Select columns for joined table**.
10. From the Trial1 list, select all of the columns.
11. Click **Select**.
12. From the Little list, select **yield**.
13. Click **Select**.

Figure 7.43 Completed Join Window

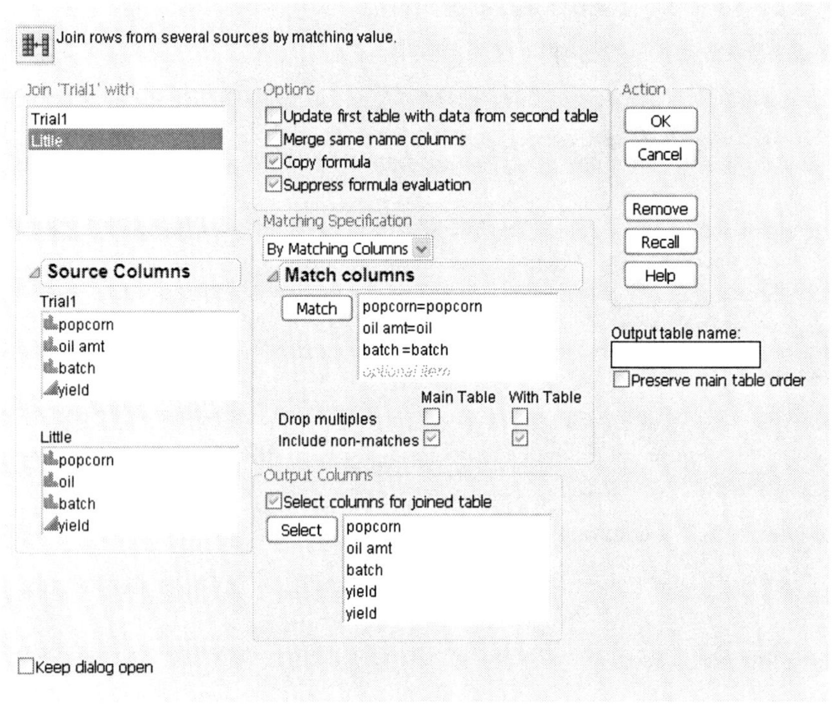

14. Click **OK**.

Figure 7.44 Trial1.jmp and Little.jmp Joined

	popcorn	oil amt	batch	yield of Trial1	yield of Little
1	gourmet	little	large	8.6	8.2
2	gourmet	little	small	12.1	15.9
3	gourmet	lots	large	9.2	•
4	gourmet	lots	small	18.0	•
5	plain	little	large	8.2	8.8
6	plain	little	small	9.9	10.1
7	plain	lots	large	10.4	•
8	plain	lots	small	10.6	•

The joined table is sorted by the matching columns. Note that the yield column from the Little.jmp table (Yield of Little) has missing values indicating no matching values with the Trial1.jmp table.

Updating a Data Table

If you have two data tables and would like to update your original table with data from a new table, select **Tables > Update**. The **Update** command is a special case of **Join** in place. It is a Join with the **Update** option checked, and it does not result in a new table.

Before you update a table, make sure that the name of the column containing the values that you want to replace is the same as the name of the column containing the data that you want to replace it with.

To replace values in the active table with those found in another open table

1. Click on the original table that you want to update (this is the table whose values you want to replace) to make it the active table.
2. Select **Tables > Update.**

Figure 7.45 Updating a Table

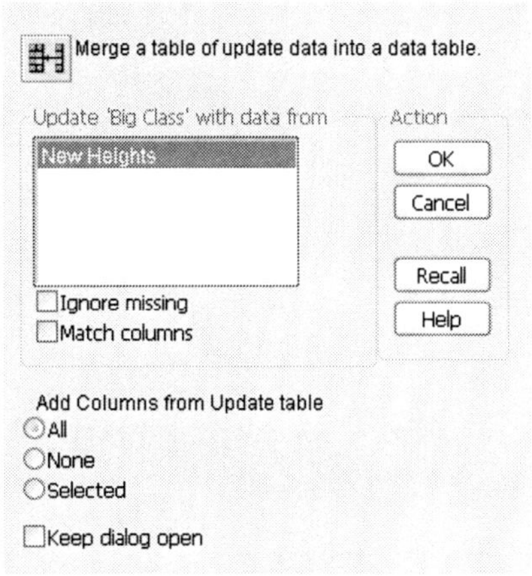

3. Highlight the new table containing the data that you want to transfer to the original table.
4. If you do not want JMP to replace the values in the original table with any missing values found in the new table, select the box next to **Ignore missing.** The original table retains its original values if they correspond to missing values in the new table.
5. If the two tables have one or more columns whose values uniquely describe each row, JMP uses those columns as the *match column values*. That is, JMP updates the rows whose match column values coincide. JMP uses these columns to preserve the sorted order of the data. If your tables do not have matching column values, you can incorporate the updated values according to their row order by continuing here. To proceed with tables containing matching column values, see “To update a table using matching columns,” p. 272.
6. From the Add Columns from Update table area, select an option. Using these options, you can add columns (that do not exist) from the new table into the original table.

- Choose **All** to add all columns from the new table into the original table.
- Choose **Selected** to add only columns that you have selected from the new table into the original table.
- Choose **None** if you do not want to add any non-existent columns from the new table into the original table.

7. Click **OK.**

To update a table using matching columns

1. Follow the first four steps outlined in the previous section, "To replace values in the active table with those found in another open table," p. 271.
2. Select **Match columns.**

Figure 7.46 This Window Appears When You Click Match Columns

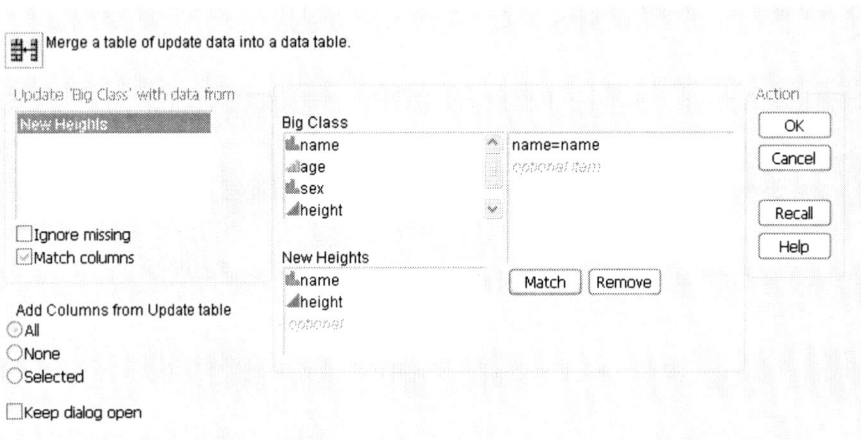

3. Highlight the two column names (in the respective tables) that you want to match.
4. Click **Match.**
5. (Optional) Repeat to match more columns.
6. Click **OK.**

Note: Unless the **None** option is selected in the Add Columns from Update table area, any columns that have different names from the columns in the table that you are updating (and that have not been assigned matches) are appended as separate columns.

Example of Updating a Data Table

Suppose a researcher has a data table containing height measurements for students. The researcher receives an updated table that contains more recent measurements of the students' heights. The researcher wants to avoid scrolling through the data tables to find the students whose height has changed, and copying and pasting the new values. Using the **Update** command, the researcher can quickly update the original data table with the new height values.

1. Open the Big Class.jmp and New Heights.jmp sample data tables.

   The Big Class.jmp table contains the original data, and the New Heights.jmp table contains the updated data.

Figure 7.47 Big Class.jmp and New Heights.jmp

Big Class.jmp (original data table)

	name	age	sex	height	weight
1	KATIE	12	F	59	95
2	LOUISE	12	F	61	123
3	JANE	12	F	55	74
4	JACLYN	12	F	66	145
5	LILLIE	12	F	52	64
6	TIM	12	M	60	84
7	JAMES	12	M	61	128
8	ROBERT	12	M	51	79
9	BARBARA	13	F	60	112
10	ALICE	13	F	61	107
11	SUSAN	13	F	56	67
12	JOHN	13	M	65	98
13	JOE	13	M	63	105
14	MICHAEL	13	M	58	95
15	DAVID	13	M	59	79

New Heights.jmp (updated data table)

	name	height
1	KATIE	62
2	LOUISE	61
3	JANE	58
4	JACLYN	68
5	LILLIE	52
6	TIM	64
7	JAMES	63
8	ROBERT	56
9	BARBARA	62
10	ALICE	61
11	SUSAN	58
12	JOHN	68
13	JOE	64
14	MICHAEL	60
15	DAVID	61

2. From the Big Class.jmp table, select **Tables > Update**.
3. In the **Update...with data from** box, select New Heights.
4. Select **Match columns**.
5. In the Big Class and New Heights lists, select name.
6. Click **Match**.

   This tells JMP to use name as the match column value, since it is the column whose values uniquely describe each row.

Figure 7.48 Updating Big Class.jmp with New Heights.jmp

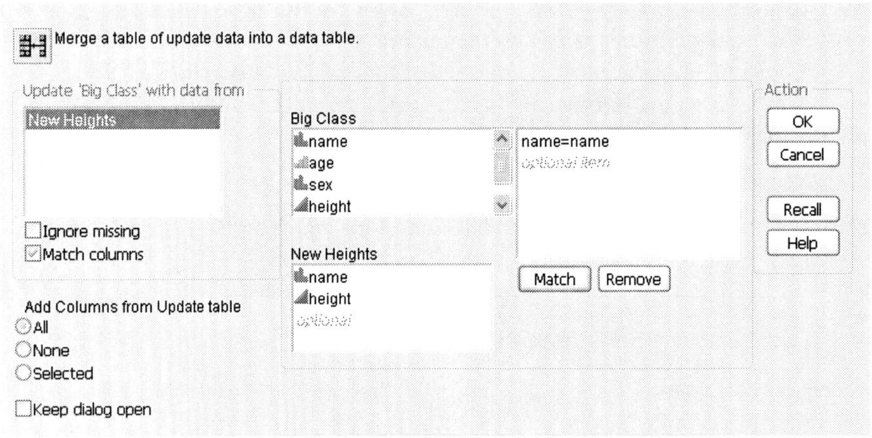

7. Click **OK**.

Figure 7.49 The Updated and the Original Big Class.jmp Tables

Updated table

	name	age	sex	height	weight
1	KATIE	12	F	62	95
2	LOUISE	12	F	61	123
3	JANE	12	F	58	74
4	JACLYN	12	F	68	145
5	LILLIE	12	F	52	64
6	TIM	12	M	64	84
7	JAMES	12	M	63	128
8	ROBERT	12	M	70	79
9	BARBARA	13	F	62	112
10	ALICE	13	F	61	107
11	SUSAN	13	F	58	67
12	JOHN	13	M	68	98
13	JOE	13	M	64	105
14	MICHAEL	13	M	60	95
15	DAVID	13	M	61	79

Original table

	name	age	sex	height	weight
1	KATIE	12	F	59	95
2	LOUISE	12	F	61	123
3	JANE	12	F	55	74
4	JACLYN	12	F	66	145
5	LILLIE	12	F	52	64
6	TIM	12	M	60	84
7	JAMES	12	M	61	128
8	ROBERT	12	M	51	79
9	BARBARA	13	F	60	112
10	ALICE	13	F	61	107
11	SUSAN	13	F	56	67
12	JOHN	13	M	65	98
13	JOE	13	M	63	105
14	MICHAEL	13	M	58	95
15	DAVID	13	M	59	79

You can see that the height values in the updated table no longer match the values in the original table. The values have been updated to use the newer values from the New Heights.jmp table.

Chapter 8

Summarizing Data

The Summarize and Tabulate Commands

You can perform a wide variety of data management tasks on JMP data:

- Create a table that contains columns of summary statistics
- Tabulate data so it is displayed in a tabular format

Contents

Summarizing Columns

The **Tables > Summary** command calculates various summary statistics, including the mean and median, standard deviation, minimum and maximum value, and so on.

In a summary table:

- A single row exists for each level of a grouping variable that you specify. If no grouping variable is specified, a single row exists for the full data table.
- When there are several grouping variables, the table contains rows for each combination of levels of all the grouping variables.
- In addition to one column for each grouping variable, the table contains frequency counts in a column named N Rows with counts for each grouping level.
- The summary table can be linked to its source table. When you select rows in the summary table, the corresponding rows are highlighted in its source table.
- If the source table's column(s) contain value labels, the value labels are displayed in the new table.
- A summary table is not saved when you close it unless you select **File > Save As** to give it a name and location.

Creating a Summary Table

To create a summary table

1. Open a data table.
2. Select **Tables > Summary.** The window in Figure 8.1 appears.
3. Highlight the columns you want to summarize.

Note: For details about the options in the red triangle menu, see "Columns Filter Menu," p. 10 in the "Preliminaries" chapter.

Figure 8.1 The Summary Window

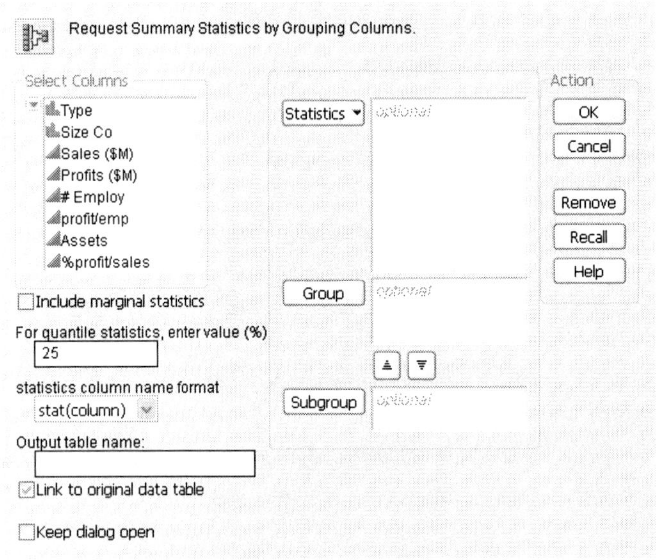

4. Add summary statistics, groups, subgroups, and select any options needed:
   - "Adding Summary Statistics," p. 278
   - "Using One or More Grouping Columns," p. 279
   - "Using Quantile Statistics," p. 280
   - "Changing the Format of the Statistics Column Name," p. 280
   - "Linking to the Original Data Table," p. 280
   - "Keeping the Summary Window Open," p. 280
   - "Creating a Two-Way Table of Summary Statistics by Adding a Subgroup Variable," p. 280
5. Name the summary table by typing a name in the box beside **Output table name**.
6. Click **OK**.

Adding Summary Statistics

You can add columns that display summary statistics (such as mean, standard deviation, median, and so on) for any numeric column in the source table.

1. In the Summary window, highlight the column that you want to use in calculating the statistics.
2. Click the **Statistics** button.
3. Select one of the standard univariate descriptive statistics from the **Statistics** drop-down menu. The statistics are described in "Explanation of Statistics," p. 282.

Using One or More Grouping Columns

If you want the statistics summarized by group, highlight the column(s) that you want to be your grouping variables and click **Group** to move the variable into the grouping variables list. See "Example of Creating a Summary Table," p. 283, for an example. If you add only grouping variables, the summary table shows a count for each group.

Changing the Order of the Grouping Variables

To change the order of the grouping variables (ascending or descending order), select a variable in the grouping variable list and click the ascending or descending button ( ). The icon beside the variable changes to indicate the sorting order.

You can also change the order of the grouping variables using the Value Ordering column property. See "Ordering Values in Columns," p. 153.

Including Marginal Statistics

To add marginal statistics (for the grouping variables) to the output columns, click the box beside **Include marginal statistics.** In addition to adding marginal statistics for each grouping variable, JMP adds rows at the end of the table that summarize each level of the first grouping variable. For example, proceed as follows:

1. Open the Companies.jmp sample data table.
2. Select **Tables > Summary.**
3. Select Profits ($M) and click **Statistics.**
4. Select **Mean.**
5. Select Type and Size Co and click **Group.**
6. Select **Include marginal statistics.**
7. Click **OK** (or **Create**). See Figure 8.2 at left.

Figure 8.2 Summary Table with and without Marginal Statistics

	Type	Size Co	N Rows	Mean(Profits ($M))
1	Computer	big	4	1,089.93
2	Computer	medium	2	-85.75
3	Computer	small	14	44.94
4	Computer		20	240.87
5	Pharmaceutical	big	5	894.42
6	Pharmaceutical	medium	5	698.98
7	Pharmaceutical	small	2	156.95
8	Pharmaceutical		12	690.08
9			32	409.32

	Type	Size Co	N Rows	Mean(Profits ($M))
1	Computer	big	4	1,089.93
2	Computer	medium	2	-85.75
3	Computer	small	14	44.94
4	Pharmaceutical	big	5	894.42
5	Pharmaceutical	medium	5	698.98
6	Pharmaceutical	small	2	156.95

Compare the summary table with marginal statistics (at left) to the summary table without marginal statistics (at right). You can see that the marginal statistics are added, and a row showing that there are 32 total Computer and Pharmaceutical companies.

Using Quantile Statistics

To add specific quantile statistics, follow these steps:

1. In the box under **For quantile statistics, enter value (%)** type the desired quantile value (%) for the first quantile (for example, 25).
2. Select the applicable column and click **Statistics.**
3. Select **Quantiles.**
4. (Optional) Repeat this process for any additional quantiles.

Changing the Format of the Statistics Column Name

To change the format of the statistics column name in the summary table, select from one of the formats in the **statistics column name format** menu. Table 8.1 illustrates the available options. Assume that you are creating a summary table of the mean profits for a company. Your original column name is Profits ($M).

Table 8.1 Statistics Column Name Format Options and Examples

Option	Example
stat (column)	Mean (Profits ($M))
column	Profits ($M)
stat of column	Mean of Profits ($M)
column stat	Profits ($M) Mean

Linking to the Original Data Table

You can select whether to link the summary table to the original data table. By default, the **Link to original data table** option is selected. If you want to edit the data in the summary table, deselect the **Link to original data table** option. When the summary table is linked to the original data table, you cannot edit the data in the summary table, since that would modify and compromise the original data.

Within linked tables, if you drag and drop columns from the summary table into the column heading of a new column in the original data table, the values are expanded as if they were matched by grouping columns.

Keeping the Summary Window Open

If you select the **Keep dialog open** option, the Summary window remains open after you click **Create.** Notice that once you select this option, the **OK** button is replaced by a **Create** button.

Creating a Two-Way Table of Summary Statistics by Adding a Subgroup Variable

1. Highlight the column(s) that you want to be the nested variable(s). These are your "subgroup variable(s)."

2. Click **Subgroup** to move the variable(s) into the subgroup list.
3. Highlight the column for which you want statistics summarized by subgroup.
4. In the **Statistics** list, select the specific statistic that you want.
5. Click **OK**.

For details on the types of statistics, see “Explanation of Statistics,” p. 282.

Adding a Statistics Column to an Existing Summary Table

After you have created a summary table, you can add columns of descriptive summary statistics for any numeric column in the source table. To do so, from an existing summary table, click on the upper red triangle in the data grid and select **Add Statistics Column**.

Example of Adding a Statistics Column to an Existing Table

Suppose that you have already created a summary table, and you want to add more statistics to the existing summary table.

1. Open the Companies.jmp sample data table.
2. Select **Tables > Summary**.
3. Select Type and Size Co and click **Group**.
4. Click **OK** (or **Create**).
5. From the red triangle menu in the upper left corner of the data table grid, select **Add Statistics Column**.

Figure 8.3 Creating a Summary Statistics Column from Within a Data Table

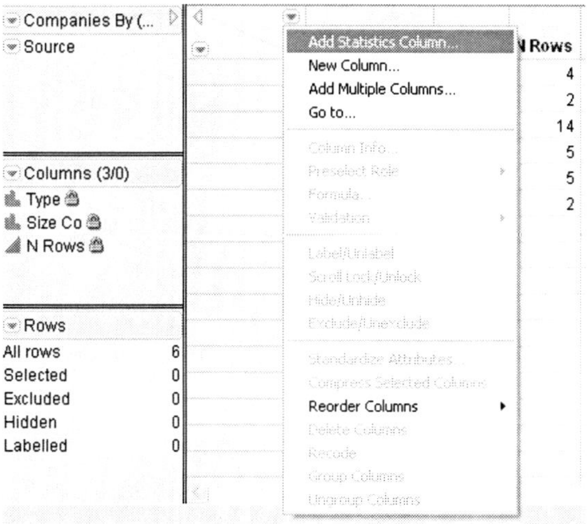

A modified version of the Summary window appears.

6. Select the column that you want, click **Statistics**, and select the specific statistic that you want. For this example, select profit/emp and click **Statistics**, then select **Mean**.
7. Click **OK**.

Figure 8.4 Example of a Summary Table with a Summary Statistics Column

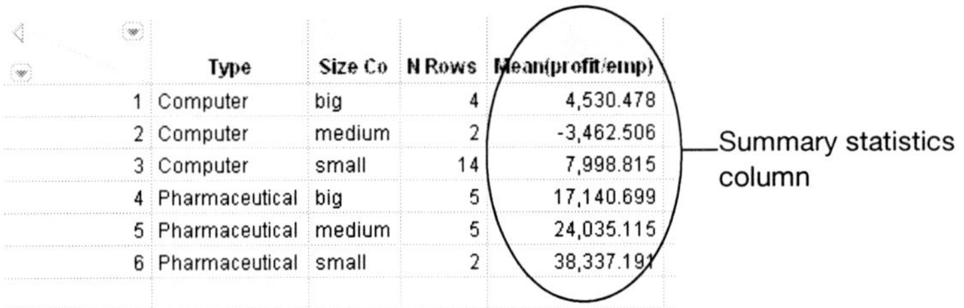

	Type	Size Co	N Rows	Mean(profit/emp)
1	Computer	big	4	4,530.478
2	Computer	medium	2	-3,462.506
3	Computer	small	14	7,998.815
4	Pharmaceutical	big	5	17,140.699
5	Pharmaceutical	medium	5	24,035.115
6	Pharmaceutical	small	2	38,337.191

The Mean(profit/emp) column is added to the existing summary table.

Explanation of Statistics

You can add columns of descriptive summary statistics for any numeric column in the source table by clicking the **Statistics** button (Figure 8.5) and making a selection from the menu.

Figure 8.5 Adding Summary Statistics

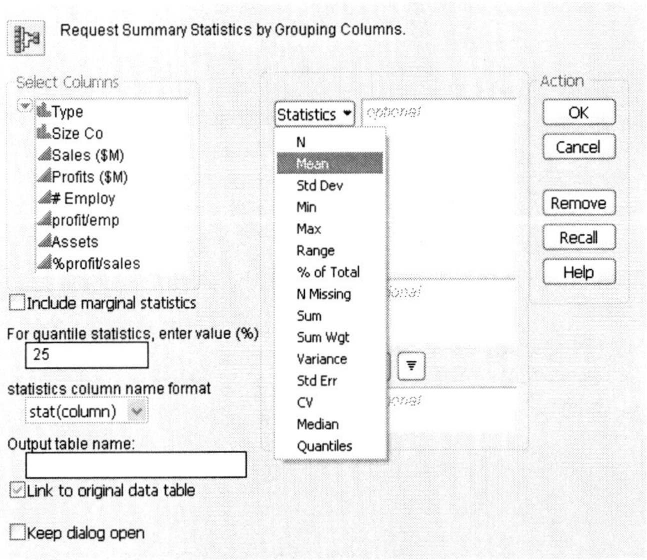

The **Statistics** menu gives these summary statistics for numeric columns:

- **N** Is the number of nonmissing values.
- **Mean** Is the arithmetic mean of a column's values. It is the sum of nonmissing values (and if defined, multiplied by the **weight** variable) divided by the **Sum Wgt**.
- **Std Dev** Is the sample standard deviation, computed for the nonmissing values. It is the square root of the sample variance.
- **Min** Is the smallest nonmissing value in a column.
- **Max** Is the largest nonmissing value in a column.
- **Range** Is the difference between **Max** and **Min**.
- **% of Total** Is the percent of the total count for each group. Or, if you have so specified, the percent of nonmissing values of the column to the total count for each group.
- **N Missing** Is the number of missing values.
- **Sum** Is the sum of all values in a column.
- **Sum Wgt** Is the sum of all weight values in a column. (See "Giving Columns a Preselected Analysis Role," p. 143.) Or, if no column is assigned the **weight** role, **Sum Wgt** is the total number of nonmissing values.
- **Variance** Is the sample variance, computed for the nonmissing values. It is the sum of squared deviations from the mean, divided by the number of nonmissing values minus one.
- **Std Err** Is the standard error of the mean. It is the standard deviation divided by the square root of **N**. If a column is assigned the role of weight, then the denominator is the square root of the sum of the weights.
- **CV** (Coefficient of Variation) Is the measure of dispersion, which is the standard deviation divided by the mean multiplied by one hundred.
- **Median** Is the 50th percentile, which is the value where half the data are below and half are above or equal to the 50th quantile (median).
- **Quantiles** Gives the value at which the specific percentage of the argument is less than or equal to. For example, 75% of the data is less than the 75th quantile. The summary window has an edit box for entering the quantile percentage that you want.

Example of Creating a Summary Table

Suppose a researcher is working with Companies.jmp, which groups companies by Type and Size. Follow along with this next example by opening Companies.jmp from the Sample Data folder that was installed when you installed JMP.

Suppose the researcher wants to:

- Create a table that shows the average profit per employee for small, medium, and big computer and pharmaceutical companies. In other words, create a table that contains a row for each size company and a column for the mean profit per employee of each type of company.

- Create it so the cells hold the mean for the subgroup (defined by the intersection of the row and column).

1. Open the Companies.jmp sample data table.
2. Select **Tables > Summary**.
3. Select Size Co and click **Group**.

   The researcher selects Size Co as the grouping variable because he wants the values in that column to become rows in the new table.
4. Select profit/emp and click **Statistics**.
5. Select **Mean**.
6. Select Type and click **Subgroup**.

   This tells JMP to create a column for the average profit per employee (Mean(profit/emp)) for each level (computer, pharmaceutical) of subgroup variable (type).

Figure 8.6 shows the completed Summary window and the resulting summary table.

Figure 8.6 Summary Statistics for a Subgroup

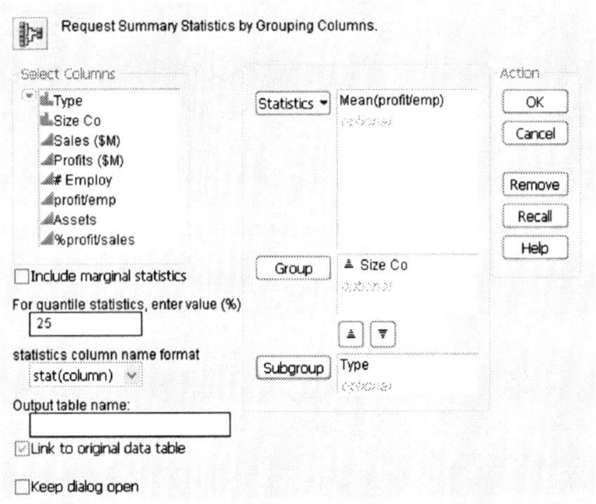

	Size Co	N Rows	Mean(profit/emp, Computer)	Mean(profit/emp, Pharmaceutical)
1	big	9	4,530.478	17,140.699
2	medium	7	-3,462.506	24,035.115
3	small	16	7,998.815	38,337.191

Tabulating Data

Use the **Tables > Tabulate** command for constructing tables of descriptive statistics. The tables are built from grouping columns, analysis columns, and statistics keywords. Through its interactive interface for defining and modifying tables, the Tabulate command provides a powerful and flexible way to present summary data in tabular form, as shown in Figure 8.7.

Figure 8.7 Examples of Tables

	sex											
	F						M					
	age						age					
	12	13	14	15	16	17	12	13	14	15	16	17
	weight	weight	weight	weight	weight	weight	weight	weight	weight	weight	weight	weight
Min	64	67	81	92	112	116	79	79	92	104	128	134
Mean	100.2	95.3	96.6	102.0	113.5	116.0	97.0	94.3	103.9	110.8	128.0	153.0
Max	145	112	142	112	115	116	128	105	119	128	128	172

		sex							
		Female				Male			
		marital status				marital status			
		Married		Single		Married		Single	
		Mean	Std Dev	Mean	Std Dev	Mean	Std Dev	Mean	Std Dev
country	size	age	age	age	age	age	age	age	age
American	Large	33.6	8.107	41.0	.	34.7	3.931	32.0	6.265
	Medium	31.4	5.827	29.0	9.258	31.3	5.413	32.1	11.05
	Small	31.0	5.657	29.0	9.539	31.8	4.813	26.5	6.455
European	Large	34.0	7.071	28.0	.	.	.	26.0	.
	Medium	31.0	5.06	28.7	5.508	32.3	5.62	31.0	10.13
	Small	29.8	6.611	28.0	1.414	33.8	4.381	25.7	2.517
Japanese	Large	25.0	.	.	.	32.0	.	.	.
	Medium	30.5	4.993	28.0	3.071	32.3	3.878	27.4	5.016
	Small	29.6	4.251	31.1	9.562	29.8	5.357	28.7	4.739
country									
American		31.9	6.452	30.0	9.115	32.6	4.919	31.0	8.179
European		31.0	5.612	28.3	3.559	33.3	4.608	28.4	7.328
Japanese		29.8	4.54	30.1	8.113	30.9	4.822	28.3	4.781

		Mean		
Type	Size Co	Profits ($M)	Sales ($M)	profit/emp
Computer	big	1089.9	20597.48	4530.478
	medium	-85.75	3018.85	-3462.51
	small	44.94	1758.06	7998.815
	All	240.87	5652.02	6159.015
Pharmaceutical	big	894.42	7474.04	17140.70
	medium	698.98	4261.06	24035.11
	small	156.95	1083.75	38337.19
	All	690.08	5070.25	23546.12
All	All	409.32	5433.86	12679.18
Type				
Computer		240.87	5652.02	6159.015
Pharmaceutical		690.08	5070.25	23546.12
All		409.32	5433.86	12679.18

How to Create a Table in Tabulate

A report in Tabulate consists of one or more column tables concatenated side by side, and one or more row tables concatenated top to bottom. A report might have only a column table or a row table.

Creating a table using the interactive table is an iterative process:

1. Click and drag the items (column name from the column list or statistics from the keywords list) from the appropriate list. Refer to the description of the elements in the interactive table in "Elements of a Table in Tabulate," p. 286.

Note: For details about the options in the red triangle menu, see "Columns Filter Menu," p. 10 in the "Preliminaries" chapter.

2. Drop the items on the dimension (row table or column table) where you want to place the items' labels. (See "Clicking and Dragging Items," p. 290, and "Elements of a Table in Tabulate," p. 286, for details.)
3. After creating a table, add to it by repeating the above process. The table updates to reflect the latest addition. If there are already column headings or row labels, you decide where the addition goes relative to the existing items.

Elements of a Table in Tabulate

In Tabulate, a table is defined by its column headings and row labels. They are referred to as the column table and the row table. For a description of column tables and row tables, see "Column Tables and Row Tables," p. 292.

Grouping Columns

Grouping columns are columns that you want to use to classify your data into categories of information. They can have character, integer, or even decimal values, but the number of unique values should be limited.

Note the following:

- If there is more than one grouping column, Tabulate constructs distinct categories from the hierarchical nesting of the values of the columns. For example, from the grouping columns Sex with values F and M, and the grouping column Marital Status with values Married and Single, Tabulate constructs four distinct categories: F and Married, F and Single, M and Married, M and Single.
- You can specify grouping columns for column tables as well as row tables. Together they generate the categories that define each table cell.
- Tabulate does not include observations with a missing value for one or more grouping columns by default. You can include them by checking the **Include missing for grouping columns** option.
- To specify codes or values that should be treated as missing, use the Missing Value Codes column property. You can include these by checking the **Include missing for grouping columns** option. See "Specifying Missing Value Codes," p. 151.

Analysis Columns

Analysis columns are any numeric columns for which you want to compute statistics. They are usually continuous columns. Tabulate computes statistics on the analysis columns for each category formed from the grouping columns.

Note that all the analysis columns have to reside in the same dimension, either in the row table or in the column table.

Statistics

Tabulate supports a list of standard statistics. The list is displayed in the control panel. You can drag any keyword from that list to the table, just like you do with the columns. Note the following:

- The statistics associated with each cell are calculated on values of the analysis columns from all observations in that category, as defined by the grouping columns.

- All of the requested statistics have to reside in the same dimension, either in the row table or in the column table.
- If you drag a continuous column into a data area, it is treated as an analysis column.

Some of the keywords used in Tabulate are defined below. A comprehensive description is listed in "Explanation of Statistics," p. 282.

- N provides the number of nonmissing values in the column. This is the default statistic when there is no analysis column.
- Sum is the sum of all values in the column. This is the default statistic for analysis columns when there are no other statistics for the table.
- Quantiles gives the value at which the specific percentage of the argument is less than or equal to. For example, 75% of the data is less than the 75th quantile. You can request different quantiles by clicking and dragging the **Quantiles** keyword into the table, and then entering the quantile into the box that appears.
- % of Total computes the percentage of total of the whole population. The denominator used in the computation is the total of all the included observations, and the numerator is the total for the category. If there is no analysis column, the % of Total is the percentage of total of counts. If there is an analysis column, the % of Total is the percentage of the total of the sum of the analysis column. Thus, the denominator is the sum of the analysis column over all the included observations, and the numerator is the sum of the analysis column for that category. You can request different percentages by dragging the keyword into the table.
  - Dropping one or more grouping columns from the table to the % of Total heading changes the denominator definition. For this, Tabulate uses the sum of these grouping columns for the denominator.
  - To get the percentage of the column total, drag all the grouping columns on the row table and drop them on the % of Total heading (same as Column %). Similarly, to get the percentage of the row total, drag all grouping columns on the column table and drop them on the % of Total heading (same as Row %).
- Column % is the percent of each cell count to its column total if there is no analysis column. If there is an analysis column, the Column % is the percent of the column total of the sum of the analysis column.
- Row % is the percent of each cell count to its row total if there is no analysis column. If there is an analysis column, the Row % is the percent of the row total of the sum of the analysis column.
- All is a special keyword for grouping columns. It is used when you want to aggregate summary information for categories of a grouping column.

Example Using the All Keyword

Suppose one of the grouping columns in a table is Sex with two categories, F and M. Add the keyword All to create a third category that aggregates the statistics for both F and M.

1. Open the Big Class.jmp sample data table.
2. Select **Tables > Tabulate**.
3. Click sex and drag and drop it into the Drop zone for columns.

4. Click **Mean** and drag and drop it into the blank cell next to the number 18.
5. Click **height** and drag and drop it just below Mean.
6. Select **Add Analysis Columns.**
7. Click All and drag and drop it in the column name sex.

Figure 8.8 Using the All Keyword

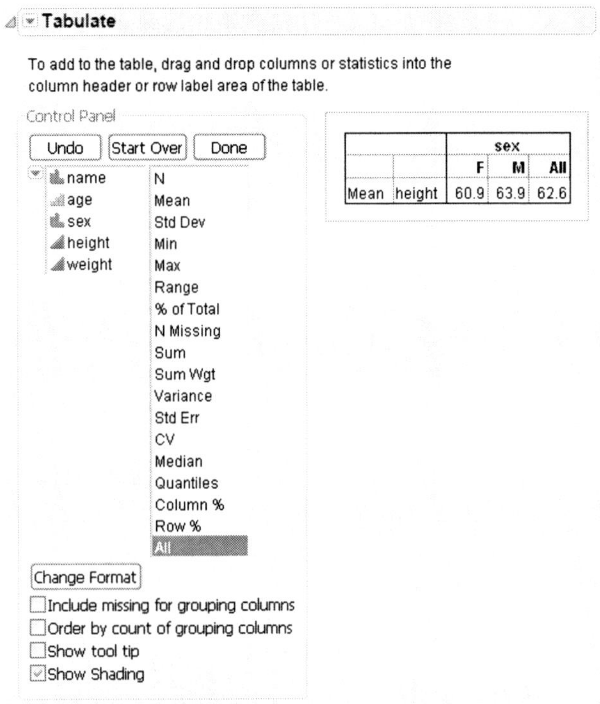

Columns by Categories

The **Columns by Categories** option is a variant of grouping columns that appears when you drag multiple columns to the table. They are independent grouping columns sharing a common set of values. When a set of grouping columns is used collectively as **Columns by Categories**, a crosstabulation of the column names and the categories gathered from these columns is generated. Each cell is defined by one of the columns and one of the categories. If **Columns by Categories** is defined on the Column table, then the corresponding categories are automatically used to define the row table.

Example of Columns by Categories

1. Open the Children's Popularity.jmp sample data table.

   This data table contains data on the importance of self-reported factors in children's popularity.

2. Select **Tables > Tabulate**.

3. Select Grades, Sports, Looks, and Money and drag and drop them into the Drop zone for rows.
4. Select **Add Columns by Categories**.

Figure 8.9 Columns by Categories

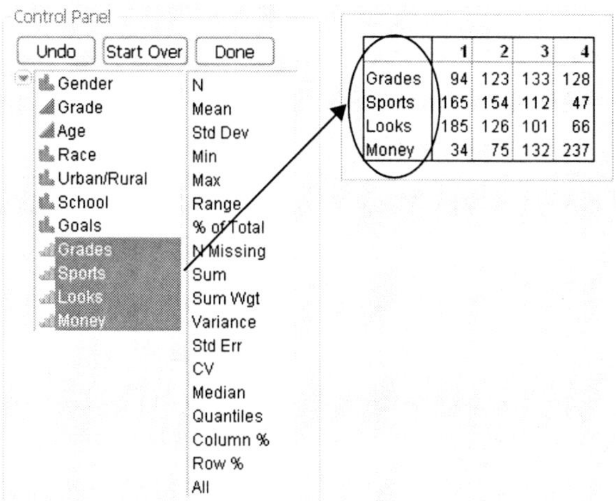

Tabulate the percentage of the one to four ratings of each category.

5. Drag and drop Gender into the empty heading at left.
6. Drag and drop % of Total above one of the numbered headings.
7. Drag and drop All above the number 4.

Figure 8.10 Tabulating the Percentages

		% of Total				
Gender		1	2	3	4	All
boy	Grades	8.16%	12.76%	14.02%	12.55%	47.49%
	Sports	26.57%	12.55%	6.49%	1.88%	47.49%
	Looks	9.21%	15.48%	12.34%	10.46%	47.49%
	Money	3.56%	6.69%	14.64%	22.59%	47.49%
girl	Grades	11.51%	12.97%	13.81%	14.23%	52.51%
	Sports	7.95%	19.67%	16.95%	7.95%	52.51%
	Looks	29.50%	10.88%	8.79%	3.35%	52.51%
	Money	3.56%	9.00%	12.97%	26.99%	52.51%

Break down the tabulation further by adding demographic data.

8. Drag and drop Urban/Rural below the % of Total heading.

Figure 8.11 Adding Demographic Data

		% of Total														
		Urban/Rural														
		Rural					Suburban					Urban				
Gender		1	2	3	4	All	1	2	3	4	All	1	2	3	4	All
boy	Grades	2.30%	2.93%	3.56%	5.02%	13.81%	2.72%	5.86%	5.02%	5.02%	18.62%	3.14%	3.97%	5.44%	2.51%	15.06%
	Sports	6.49%	3.97%	2.72%	0.63%	13.81%	11.30%	4.60%	2.30%	0.42%	18.62%	8.79%	3.97%	1.46%	0.84%	15.06%
	Looks	3.14%	4.60%	2.72%	3.35%	13.81%	3.77%	5.86%	5.44%	3.56%	18.62%	2.30%	5.02%	4.18%	3.56%	15.06%
	Money	1.88%	2.30%	4.81%	4.81%	13.81%	0.84%	2.30%	5.86%	9.62%	18.62%	0.84%	2.09%	3.97%	8.16%	15.06%
girl	Grades	4.39%	4.39%	4.39%	4.18%	17.36%	2.51%	3.35%	2.93%	4.18%	12.97%	4.60%	5.23%	6.49%	5.86%	22.18%
	Sports	2.30%	6.69%	5.44%	2.93%	17.36%	1.67%	3.56%	5.65%	2.09%	12.97%	3.97%	9.41%	5.86%	2.93%	22.18%
	Looks	9.62%	3.35%	3.35%	1.05%	17.36%	7.95%	2.93%	1.26%	0.84%	12.97%	11.92%	4.60%	4.18%	1.46%	22.18%
	Money	1.05%	2.93%	4.18%	9.21%	17.36%	0.84%	3.14%	3.14%	5.86%	12.97%	1.67%	2.93%	5.65%	11.92%	22.18%

Clicking and Dragging Items

Each column heading has two zones: the upper and the lower. As you drag each column heading into a zone, the cursor transforms into a rectangle to indicate that you can drop the column heading at that position.

- Dropping in the upper zone places the new items above (before) the items on which the addition is dropped.
- Dropping in the lower zone places the addition below (after) the items on which the addition is dropped.

Each row label has two zones: the left and the right:

- Dropping in the left zone puts the new items to the left (before) of the items dropped on.
- Dropping in the right zone puts them to the right (after) of the items dropped on.

Note: You can drag columns from the Tables panel in the data table onto a Tabulate table instead of using the Tabulate Control Panel.

In a properly created table, all grouping columns are together, all analysis columns are together, and all statistics are together. Therefore, JMP does not intersperse a statistics keyword within a list of analysis columns. JMP also does not insert an analysis column within a list of grouping columns.

If the items' role is obvious, such as keywords or character columns, when you drag and drop, JMP populates the table automatically with the given items. Otherwise, a popup menu lets you choose the role for the items. Roles included on the popup menu are:

Add Grouping Columns Choose **Add Grouping Columns** if you want to use the variables to categorize the data. For multiple grouping columns, Tabulate creates a hierarchical nesting of the variable.

Add Analysis Columns Choose **Add Analysis Columns** if you want to compute the statistics of these columns.

Columns by Categories Choose **Columns by Categories** if the columns are independent grouping columns (in other words, no hierarchical nesting) sharing a similar set of distinct data values, and if you want a cross tabulation of the column by the categories layout.

Grouping Columns for Separate Tables Choose **Grouping Columns for Separate Tables** if you have multiple independent grouping columns and you want to generate separate tables for each grouping column.

Inserting a Grouping Column

To insert a grouping column, click and drag, and then release a column name or statistics keyword into the table. Select **Add Grouping Columns** from the menu that appears, as shown in Figure 8.12. If adding it as a grouping column is the only logical choice, JMP automatically inserts it as a grouping column; the popup menu does not appear.

Figure 8.12 Example of Adding a Grouping Column

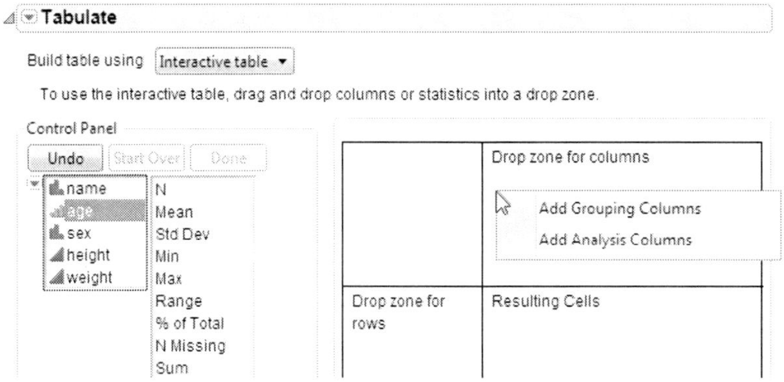

Inserting an Analysis Column

To insert an analysis column, click and drag, and then release a column name or statistics keyword into the table. Select **Add Analysis Columns** from the menu that appears, as shown in Figure 8.12.

Using the Dialog

If you prefer not to click and drag and build the table interactively, you can create a simple table using the Dialog interface. After selecting **Tables > Tabulate**, select **Dialog** from the drop-down menu beside Build table using, as shown in Figure 8.13. The window that appears is very similar to the Summary window, and the resultant table is like the layout of the summary table. (See "Summarizing Columns," p. 277.) You can change the table generated by the window in the same way that you would with one generated through drag and drop.

Figure 8.13 Using the Window

Column Tables and Row Tables

The Tabulate platform enables you to build sub-tables within a larger table. These sub-tables are called *row tables* and *column tables*, as illustrated in Figure 8.14 using Car poll.jmp.

Example of Row and Column Tables

1. Open the Car Poll.jmp sample data table.
2. Select **Tables > Tabulate**.
3. Drag size into the Drop zone for rows.
4. Drag and drop country to the left of the size heading.
5. Drag and drop Mean over the N heading.
6. Drag and drop Std Dev below the Mean heading.
7. Drag and drop age above the Mean heading.
8. Select **Add Analysis Columns**.
9. Drag and drop type to the far right of the table.
10. Drag and drop sex under the table.

Figure 8.14 Row and Column Tables

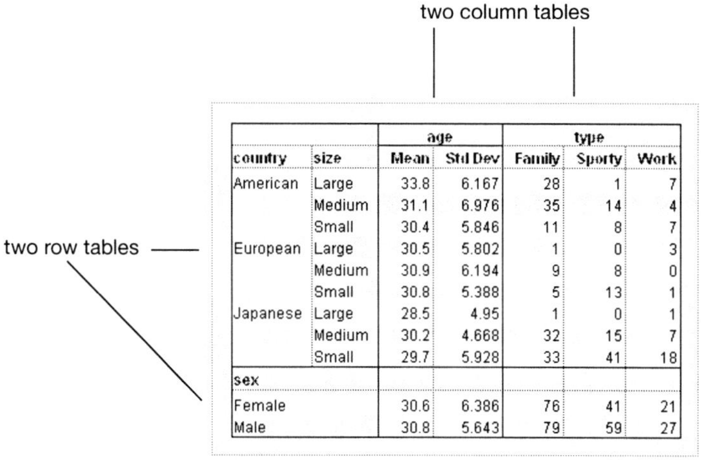

country	size	age Mean	age Std Dev	type Family	type Sporty	type Work
American	Large	33.8	6.167	28	1	7
	Medium	31.1	6.976	35	14	4
	Small	30.4	5.846	11	8	7
European	Large	30.5	5.802	1	0	3
	Medium	30.9	6.194	9	8	0
	Small	30.8	5.388	5	13	1
Japanese	Large	28.5	4.95	1	0	1
	Medium	30.2	4.668	32	15	7
	Small	29.7	5.928	33	41	18
sex						
Female		30.6	6.386	76	41	21
Male		30.8	5.643	79	59	27

Editing Tables in Tabulate

There are several ways to edit the items that you add to a table.

Changing Numeric Formats

The formats of each cell depend on the analysis column and the statistics. For counts, the default format has no decimal digits. For each cell defined by some statistics, JMP tries to determine a reasonable format using the format of the analysis column and the statistics requested. To override the default format:

1. Click the **Change Format** button at the bottom of the Tabulate window.
2. In the panel that appears (Figure 8.15), enter the field width, a comma, and then the number of decimal places that you want displayed in the table.
3. (Optional) If you would like JMP to determine the best format for you to use, type the word **Best** in the text box.

   JMP now considers the precision of each cell value and selects the best way to show it.
4. Click **OK** to implement the changes and close the Format section, or click **Set Format** to see the changes implemented without closing the Format section.

Figure 8.15 Changing Numeric Formats

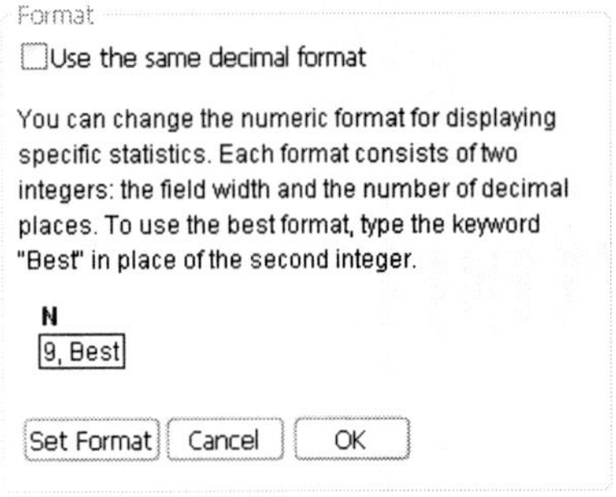

Deleting Items

After you add items to the table, you can remove them in any one of the following ways:

- Drag and drop the item away from the table.
- To remove the last item, click **Undo**.
- Right-click on an item and select **Delete**.

Removing Column Labels

Grouping columns display the column name atop the categories associated with that column. For some columns, the column name might seem redundant. Remove the column name from the column table by right-clicking on the column name and selecting **Remove Column Label**. To re-insert the column label, right-click on one of its associated categories and select **Restore Column Label**.

Editing Statistical Key Words and Labels

You can edit a statistical key word or a statistical label. For example, instead of Mean, you might want to use the word Average. To make edits, right-click on the word that you want to edit and select **Change Item Label**. In the box that appears, type the new label. Alternatively, you can type directly into the edit box.

If you change one statistics keyword to another statistics keyword, JMP assumes that you actually want to change the statistics, not just the label. It would be as if you have deleted the statistics from the table and added the latter.

Additional Tabulate Options

At the bottom of the Tabulate window, check boxes enable you to control plot elements, tooltips, and shading.

Figure 8.16 Tabulate Options

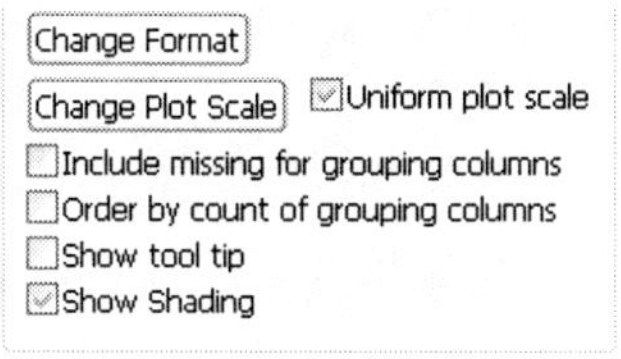

Change Format Enables you to change the numeric format for displaying specific statistics. See "Changing Numeric Formats," p. 293.

Change Plot Scale (Only appears if **Show Chart** is selected from the red triangle menu.) Enables you to specify a uniform custom scale.

Uniform plot scale (Only appears if **Show Chart** is selected from the red triangle menu.) Deselect this box to have each column of bars use the scale determined separately from the data in each displayed column.

Include missing for grouping columns Creates a separate group for missing values in grouping columns. When unchecked, missing values are not included in the table. Note that any missing value codes that you have defined as column properties are taken into account.

Order by count of grouping columns Changes the order of the table to be in ascending order of the values of the grouping columns.

Show tool tip Displays tips that appear when you move the mouse over areas of the table.

Show Shading Displays gray shading boxes in the table when there are multiple rows.

Other options are available from the red triangle menu next to Tabulate:

Show Table Displays the summarized data in tabular form.

Show Chart Displays the summarized data in bar charts that mirrors the table of summary statistics. The simple bar chart enables visual comparison of the relative magnitude of the summary statistics. By default, all columns of bars share the same scale. You can have each column of bars use the scale determined separately from the data in each displayed column, by clearing the **Uniform plot scale** check box. You can specify a uniform custom scale using the **Change Plot Scale** button. The charts are either 0-based or centered on 0. If the data are all nonnegative, or all non-positive, the charts baseline is at 0. Otherwise, the charts are centered on 0.

Show Control Panel Displays the control panel for further interaction.

Show Test Build Panel Displays the control area that lets you create a test build using a random sample from the original table. See "Using Large Amounts of Data (the Test Build Feature)," p. 296, for details.

Make Into Data Table Makes a data table from the report. There is one data table for each row table, since labels of different row tables might not be mapped to the same structure.

Script Displays options for saving scripts, redoing analyses, and viewing the data table. See "Saving Your Steps as a Script," p. 187, for details.

Using Large Amounts of Data (the Test Build Feature)

If you have a very large data table, you might want to use a small subset of the data table to try out different table layouts to find one that best shows the summary information. In this case, JMP generates a random subset of the size as specified and uses that subset when it builds the table. To use the test build feature:

1. From the red triangle menu next to Tabulate, select **Show Test Build Panel**.
2. Enter the size of the sample that you want in the box under **Sample Size (>1) or Sampling Rate (<1)**, as shown in Figure 8.17. The size of the sample can be either the proportion of the active table that you enter or the number of rows from the active table.

Figure 8.17 The Test Build Panel

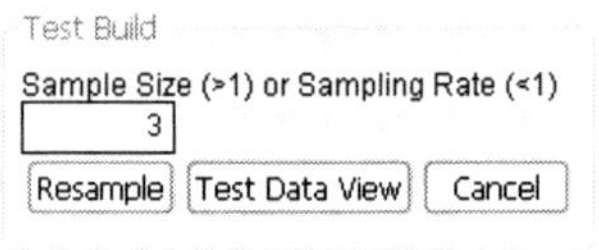

3. Click **Resample**.
4. To see the sampled data in a JMP data table, click the **Test Data View** button. When you dismiss the test build panel, Tabulate uses the full data table to regenerate the tables as designed.

Example of Tabulating Data

This example contains the following procedures:

1. "Creating a Table of Counts," p. 296
2. "Creating a Table Showing Statistics," p. 298
3. "Rearranging the Table Contents," p. 299

Creating a Table of Counts

Suppose you would like to view a table that contains counts for how many people in the survey own Japanese, European, and American cars.

1. Open the Car Poll.jmp sample data table.
2. Select **Tables > Tabulate**.
3. Click country and drag it into the Drop zone for rows.

Figure 8.18 Adding column to the Table

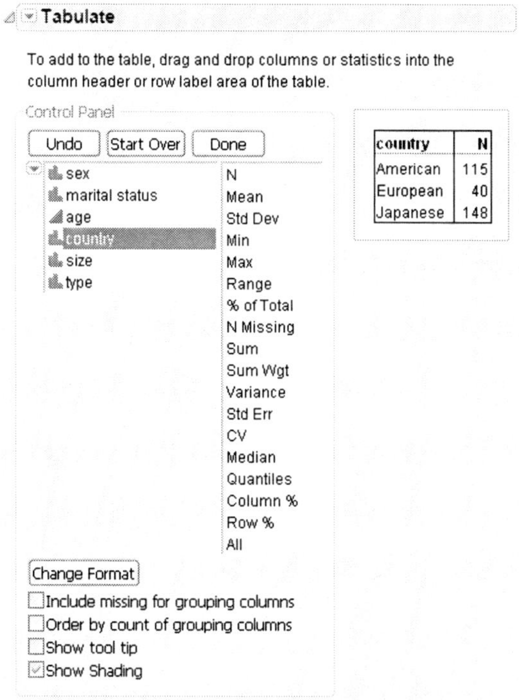

Now add further statistics and variables to the table. You would like to see a count of people who drive Japanese, European, and American cars broken down by the size of the car.

4. Click size and drag and drop it to the right of the country heading.

Figure 8.19 Adding Size to the Table

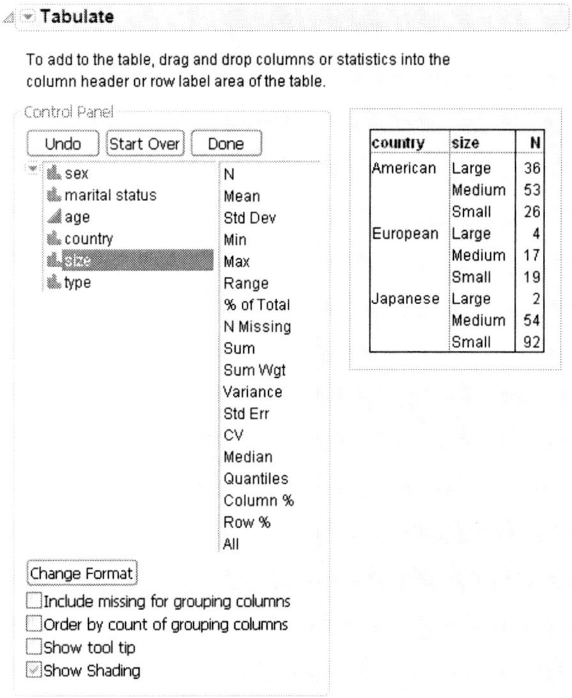

Creating a Table Showing Statistics

Now, suppose you would like to see the average and the standard deviation of the age of people who own each size car:

1. Start from Figure 8.19. Click age and drag and drop it to the right of the size heading.
2. Select **Add Analysis Columns**.
3. Click Mean and drag and drop it over Sum.
4. Click Std Dev and drag and drop it below Mean.

   This places Std Dev below Mean in the table. Dropping Std Dev above Mean places Std Dev above Mean in the table.

Figure 8.20 Table that Includes the Mean and Standard Deviation of Age

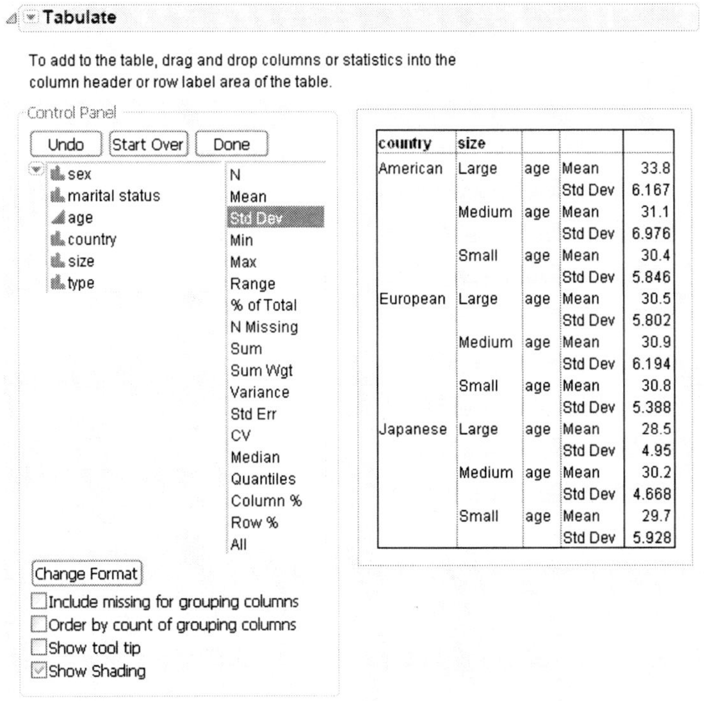

country	size			
American	Large	age	Mean	33.8
			Std Dev	6.167
	Medium	age	Mean	31.1
			Std Dev	6.976
	Small	age	Mean	30.4
			Std Dev	5.846
European	Large	age	Mean	30.5
			Std Dev	5.802
	Medium	age	Mean	30.9
			Std Dev	6.194
	Small	age	Mean	30.8
			Std Dev	5.388
Japanese	Large	age	Mean	28.5
			Std Dev	4.95
	Medium	age	Mean	30.2
			Std Dev	4.668
	Small	age	Mean	29.7
			Std Dev	5.928

Rearranging the Table Contents

To rearrange the table contents, proceed as follows:

1. Start from Figure 8.20. Click on the size heading and drag and drop it to the right of the table headings. See Figure 8.21.

Figure 8.21 Moving size

country	size			
American	Large	age	Mean	33.8
			Std Dev	6.167
	Medium	age	Mean	31.1
			Std Dev	6.976
	Small	age	Mean	30.4
			Std Dev	5.846
European	Large	age	Mean	30.5
			Std Dev	5.802
	Medium	age	Mean	30.9
			Std Dev	6.194
	Small	age	Mean	30.8
			Std Dev	5.388
Japanese	Large	age	Mean	28.5
			Std Dev	4.95
	Medium	age	Mean	30.2
			Std Dev	4.668
	Small	age	Mean	29.7
			Std Dev	5.928

→

			size		
country			Large	Medium	Small
American	age	Mean	33.8	31.1	30.4
		Std Dev	6.167	6.976	5.846
European	age	Mean	30.5	30.9	30.8
		Std Dev	5.802	6.194	5.388
Japanese	age	Mean	28.5	30.2	29.7
		Std Dev	4.95	4.668	5.928

2. Click on age and drag and drop it under the **Large** heading.
3. Select both Mean and Std Dev, then drag and drop them under the **Large** heading.

Figure 8.22 The Result of Moving Items

	size					
	Large		Medium		Small	
	Mean	Std Dev	Mean	Std Dev	Mean	Std Dev
country	age	age	age	age	age	age
American	33.8	6.167	31.1	6.976	30.4	5.846
European	30.5	5.802	30.9	6.194	30.8	5.388
Japanese	28.5	4.95	30.2	4.668	29.7	5.928

Now your table clearly presents the data. It is easy to see the mean and standard deviation of the car owner age broken down by car size and country.

Chapter 9

Formula Editor

Constructing a Formula

As described in "Giving Columns a Formula to Compute Values," p. 146, you can create a column whose values are computed by a formula and store that formula as part of a column's information. Formulas can be simple assignments of numeric, character, or row state constants, or they can contain complex evaluations based on conditional clauses. The JMP formula editor provides you these capabilities, and it also enables you to examine or change the formula at any time.

The formula editor window operates like a pocket calculator with buttons, displays, and a list of functions.

Figure 9.1 The Formula Editor

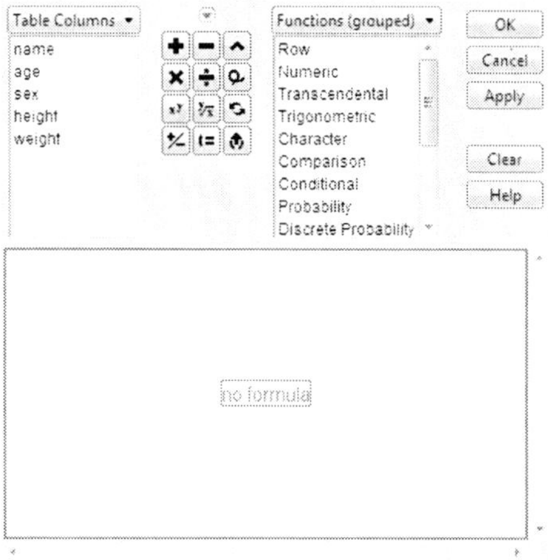

Formulas are an integral part of a data table for the following reasons:

- They are stored as part of a column's information when you save the data table.
- You can examine or change them at any time by opening the formula editor.
- Their values can be linked to, or dependent on, the values in other columns. Their values are automatically recomputed whenever you edit the values in the columns to which the formula is linked.
- Their values are locked in the data table so they cannot be manually edited.

This chapter describes the formula editor and shows how to build formulas. For details about each function, see "Formula Functions Reference," p. 473.

Contents

Creating a Formula

A formula is an expression stored in a column that performs operations in order to insert values into that column. Formulas can perform mathematical operations, such as addition and multiplication, or they can compare column values or join values by referring to other areas in the same data table. Formulas can consist of any JMP Scripting Language (JSL) command. Once you insert a formula into a column, the column is locked: its values can be edited only by changing or removing the formula.

There are three basic steps to building a formula:

1. Open the formula editor by right-clicking the column name to which you want to apply the formula and selecting **Formula**.

   or

   Double-click the column name to which you want to apply the formula, select **Formula** from the **Column Properties** menu, and then click **Edit Formula**.

2. Select an empty formula element in the formula editing area by clicking it. (See Figure 9.2.)

Note: An element is selected when there is a red outline around it. All terms within the smallest nesting box relative to the place that you clicked become selected. The subsequent actions apply to those combined elements.

---

3. Add expressions, functions, and terms from the formula work panel. They are applied to the highlighted red box. The following sections in this chapter provide detailed instructions on how to add constants, elements, operators, and functions.

Figure 9.2 Building a Formula

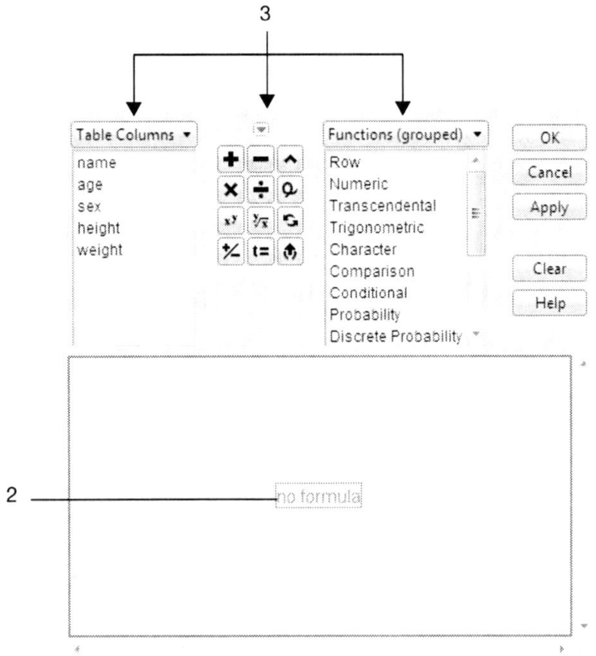

See "Using Basic Formula Editor Features," p. 323, for an example of how to use the formula editor.

Referencing Columns and Table Variables

You can create a formula that refers to values found in other parts of the data table, such as other columns and table variables.

When a formula uses values in other columns, the values in the column with that formula are dependent on the values in those other columns. Whenever a column that the formula refers to changes, the dependent column also changes. If you delete the referenced column, empty terms appear in the column containing the formula.

If you create a formula that refers to values found in table variables, those table variables must already exist in the data table. Table variables are character strings that are available to the entire table. Their names are displayed in the table panel at the left of the data table. (See "Adding Table Variables," p. 81, for details)

To build a formula that references values found in columns or table variables

1. Open the formula editor by right-clicking a column name in the data table and selecting **Formula**.
2. Select a box in the formula editing area by clicking it. (It is selected when there is a red outline around it.)

3. Click the menu on the formula element browser (see Figure 9.3) and select either **Table Columns** or **Table Variables**. Highlight an element from that category's list. The element is added to the selected box in the formula editing area.

Figure 9.3 Formula Element Browser with Table Columns Selected

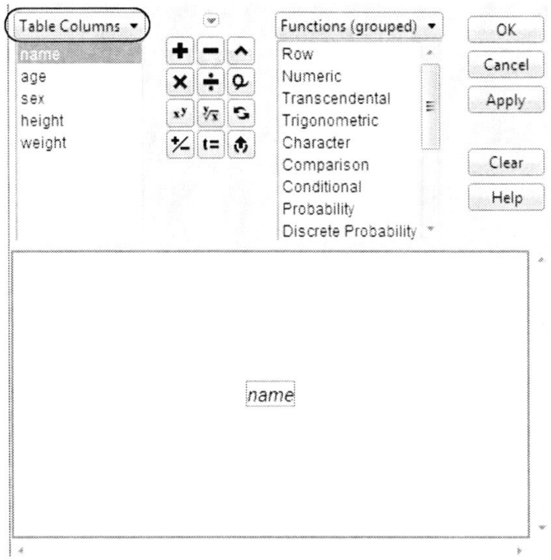

See "Using Basic Formula Editor Features," p. 323, for an example of referencing a column in a formula.

Note: In a formula, when you reference a column using value labels, hover your mouse over the value label to see the actual data value.

Using Local Variables

You can create and use temporary numeric variables in expressions. You can use ordinary local variables or you can use parameters, which are special types of local variables. Local variables exist only for the evaluation of the formula in which they are defined. They appear in formulas as bold italic terms.

Local variables are most often used with **Assignment** functions, which can assign expressions to local variables that are used in a complex equation. (See "Assignment Functions," p. 517 in the "Formula Functions Reference" chapter, for details.) This technique can sometimes simplify building an equation and improve the efficiency of its evaluation.

To build a formula that references values found in local variables

Step 1: Create the Local Variable

1. Open the formula editor by right-clicking a column name in the data table and selecting **Formula**.
2. Select a box in the formula editing area by clicking it. (It is selected when there is a red outline around it.)
3. Click the menu on the formula element browser and select **Local Variables** from the formula element browser menu.
4. Click **New Local Variable**.
5. Type a name for the local variable. By default, local variables have the names *t0*, *t1*, and so on, and have missing values.
6. Assign a starting value, and click **OK**.
7. (Optional) To copy, edit, or delete a local variable, right-click (Ctrl-click on the Macintosh) its name and select **Copy**, **Edit**, or **Delete**.

Step 2: Insert a Local Variable Into a Formula

1. Select a term in the formula editing area by clicking it. (It is selected when there is a red outline around it.)
2. Click the local variable name in the **Local Variables** list. It appears in the formula as a bold italic term.

Note: Another way to create local variables is to use the t= button on the formula editor keypad, which automatically creates and displays local variables and places a semicolon after it. See “Adding Operators,” p. 308, for details.

See “Using Local Variables in a Formula,” p. 325, for an example of referencing local variables in a formula.

Incorporating Parameters

Parameters are special types of local variables that are recognized as model parameters in some platforms, such as Nonlinear fitting. They can be used in formulas just as ordinary local variables can.

To view examples of parameters

1. In the Sample Data folder that was installed when you installed JMP, open the Nonlinear Examples folder and then US Population.jmp.
2. Right-click the column name x-formula and select **Formula**.
3. Click the menu on the formula element browser and select **Parameters**.

Notes:

- After completing a nonlinear fit or after using the **Reset** button in the nonlinear control panel, the parameter’s value is the most recent value computed by the nonlinear platform.

- Each time the fitting algorithm takes a step, the updated parameter values are shown in the Nonlinear report.
- When you paste a formula with parameters into a column, the parameters are automatically created for that column unless it has existing parameters with the same names.

To build a formula that references values found in parameters

Step 1: Create the Parameter

1. Open the formula editor by right-clicking a column name in the data table and selecting **Formula**.
2. Make sure a term is selected (a red highlight is around it in the formula editing area) in the formula editing area.
3. Click the menu on the formula element browser and select **Parameters**.
4. Click **New Parameter**.
5. Type a name for the parameter. By default, parameters have the names *b0*, *b1*, and so on, and have missing values.
6. Assign a starting value. It is important to enter this value when using a parameter in a model for the nonlinear platform. After completing a nonlinear fit, the parameter's value is the most recent value computed by the nonlinear platform.
7. (Optional) To add several parameters (one for each level of a categorical variable, for example) at once, select **Expand into categories, selecting column**. Then select the column for which you want to expand the parameter.
8. Click **OK**.
9. (Optional) To copy, edit, or delete a parameter, right-click (Ctrl-click on the Macintosh) its name and select **Copy**, **Edit**, or **Delete**.

Step 2: Insert a Parameter Into a Formula

1. Select a term in the formula editing area by clicking it. (It is selected when there is a red outline around it.)
2. Click the parameter name in the **Parameters** list. The parameter appears in the formula as bold type.

Inserting Constants

Formulas can be simple assignments of numeric, character, or row state constants, or they can contain complex evaluations based on conditional clauses. Constants include commonly used numeric terms, such as e, pi, -1, 0, 1, and 2. There are two ways to add a constant value to a formula:

- Type them in manually using the keyboard
- Select them from the formula element browser, as shown in Figure 9.4.

Figure 9.4 The Formula Element Browser with Constants Selected

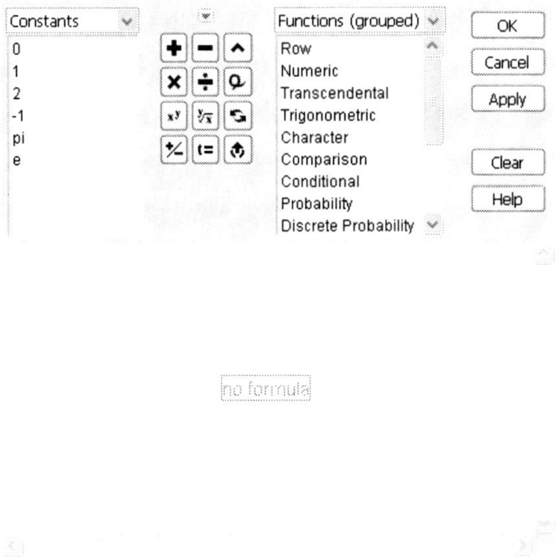

To add constants to a formula

1. Open the formula editor by right-clicking a column name in the data table and selecting **Formula**.
2. Select a box in the formula editing area by clicking it. (It is selected when there is a red outline around it.)
3. Either type in a number or click the menu at the top of the formula element browser and select **Constants**, as shown in Figure 9.4. Then click a value in the list that appears: **0**, **1**, **2**, **-1**, **pi**, **e**. The value appears in the outlined box.
4. Complete the remainder of the formula using the keypad and functions. (See "Referencing Columns and Table Variables," p. 304, "Adding Operators," p. 308, and "Using Functions," p. 312.)

Adding Operators

You can add operators to a formula by clicking the keypad (shown in Figure 9.5), which is a set of buttons used to help build formulas. It includes common operators, which are also referred to as functions.

Figure 9.5 The Keypad

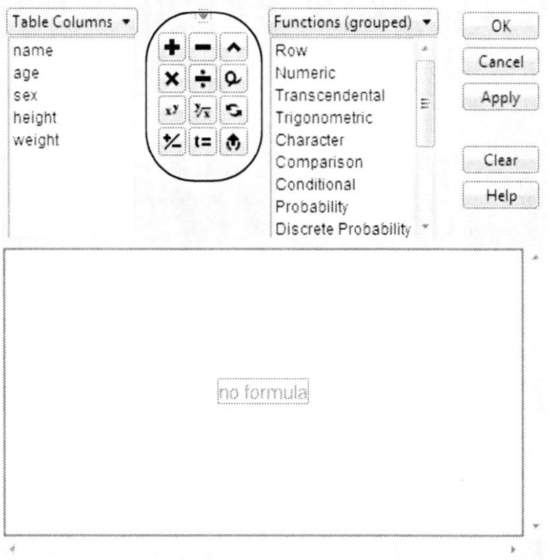

To build a formula using operators found on the on-screen keypad

1. Open the formula editor by right-clicking a column name in the data table and selecting **Formula**.
2. Select a box in the formula editing area by clicking it. (It is selected when there is a red outline around it.) The operator performs its action on the area that is highlighted.
3. Select the column or variable that you want to use in your formula.
4. Click the keypad button(s). Table 9.1 shows an example of using the keypad.

Table 9.1 Adding a Subtraction Operation to a Formula

Click the minus key on the keypad to add a subtraction operation.	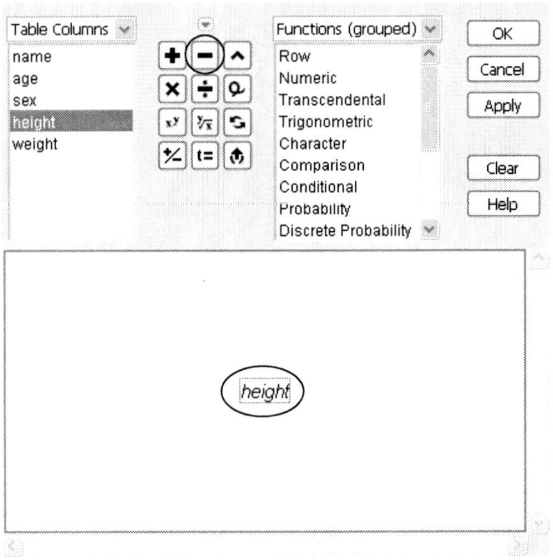
	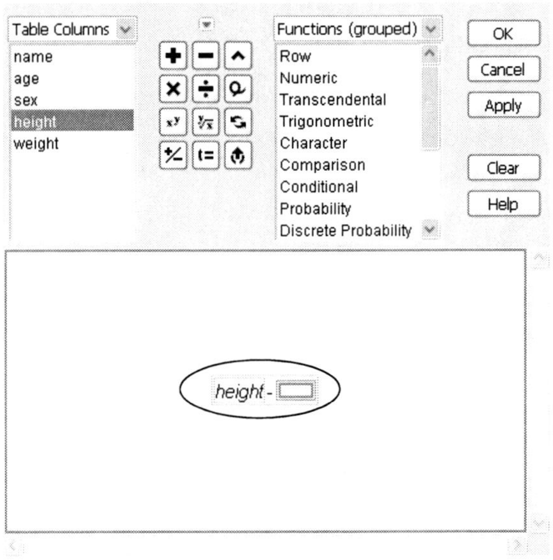

Keypad Reference

Table 9.2 describes the on-screen keypad buttons.

Table 9.2 Keypad Buttons in the Formula Editor

Key	Name	Description
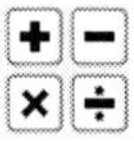	Arithmetic buttons	Work as they normally do on a pocket calculator, providing addition, multiplication, subtraction, and division operators.
	Insert	Inserts a new clause or function argument. First select the existing clause or argument that you want the new element to follow, and then click this button. The new clause appears and is selected. You can also insert a new clause or argument by typing a comma.
	Delete	Deletes an element's value, or deletes a clause. The delete button functions the same as the Delete key on the keyboard.
	Exponent	Raises a given value to a specified power. It has an exponent of two by default.
	Root	Calculates the specified root of the radicand. It has an implied index of two (a square root), which is not displayed.
	Switch terms	Looks at the operator that is central to the selected expression and switches the expressions on either side of that operator.
	Unary sign function	Inverts the sign of the argument. Apply the unary sign function to variable expressions or use it to enter negative constants.
	Local variable	Creates and displays a local variable and assigns it the value of the selected expression. The local variable has the default name *t0* in an expression and a semicolon after it. See "Using Local Variables," p. 305, for details about creating and inserting local variables. See "Using Local Variables in a Formula," p. 325, for an example.
	Delete expression (peel)	Replaces the outermost expression with the first argument. You can repeat this process to delete a formula term by term. See "Using the Delete Expression Key," p. 328, for an example.

Using Functions

You can add many types of functions to a formula. All of these functions are organized in the function browser. The browser groups collections of functions in lists organized both alphabetically (**Functions (all)**) and by topic (**Functions (grouped)** as shown in Figure 9.6. Use the function browser to specify the type of calculation that you want to perform on the elements in a formula.

Figure 9.6 The Functions Browser

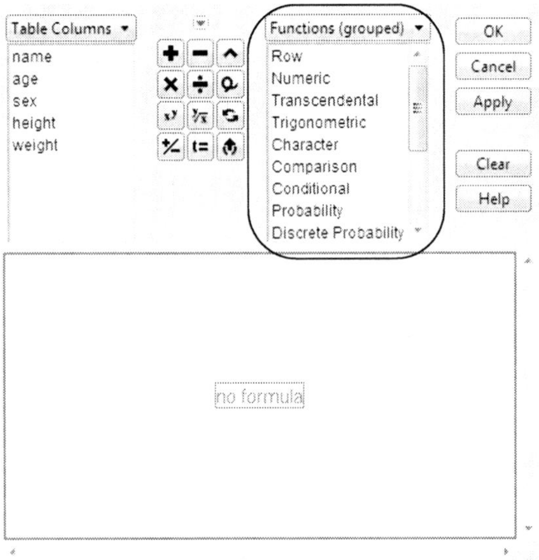

To create a formula that contains a function

1. Open the formula editor by right-clicking a column name in the data table and selecting **Formula**.
2. Select an expression in the formula editing area by clicking it. (It is selected when there is a red outline around it.) The function performs its action on the area that is highlighted.
3. Click the menu in the function browser to view the groups of functions.
4. Select a group of functions to view. See Table 9.3 for details.

   The functions that belong to that group are then displayed in the list below the menu. The function groups are briefly described in the following list.
5. Click any function in the **Functions** list to apply it to the selected item. When you click some items, you reveal a submenu from which you should make a selection.

Note: Most functions give hints about appropriate arguments through gray words inserted in the boxes in the formula editing area. Functions also show a small caret in the argument area if additional arguments can be added.

6. Continue to build the formula by highlighting terms and clicking items in the formula element browser, keypad, or function browser.

Table 9.3 Groups of Functions

Group	Functions Included
Functions (all)	Displays a list of all available functions in alphabetical order. For details about individual functions, see the "Formula Functions Reference" chapter.
Functions (grouped)	Displays a list of all available functions grouped according to topic.
Row	Displays a list of functions that contains miscellaneous functions such as **Lag**, **Dif**, **Subscript**, **Row**, and **NRow**. See "Row Functions," p. 475 in the "Formula Functions Reference" chapter.
Numeric	Displays a list of functions that are terms commonly used in formulas. See "Numeric Functions," p. 476 in the "Formula Functions Reference" chapter.
Transcendental	Displays a list of functions that are functions such as natural log, common log, exponential, root, factorial, combinatorial, beta, and gamma. See "Transcendental Functions," p. 477 in the "Formula Functions Reference" chapter.
Trigonometric	Displays a list of functions that are the standard trigonometric functions: sine, cosine, tangent, inverse functions, and hyperbolic functions. See "Trigonometric Functions," p. 479 in the "Formula Functions Reference" chapter.
Character	Displays a list of functions that operate on character arguments such as trimming, finding the length of a string, converting between numbers and characters. See "Character Functions," p. 480 in the "Formula Functions Reference" chapter.
Comparison	Displays a list of functions that are the standard logical comparisons such as less than, less than or equal to, not equal to, and so forth. See "Comparison Functions," p. 488 in the "Formula Functions Reference" chapter.
Conditional	Displays a list of functions that are programming-like functions, such as **If**, **Match**, and **Choose**. See "Conditional Functions," p. 489 in the "Formula Functions Reference" chapter.
Probability	Displays a list of functions that compute probabilities and quantiles for standard statistical distributions, such as normal, Student's *t*, Chi-squared, and *F*-distributions. See "Probability Functions," p. 494 in the "Formula Functions Reference" chapter.
Discrete Probability	Displays a list of functions that compute discrete probabilities, such as Poisson, Gamma Poisson, and Hypergeometric. See "Discrete Probability Functions," p. 501 in the "Formula Functions Reference" chapter.

Table 9.3 Groups of Functions *(Continued)*

Group	Functions Included
Statistical	Displays a list of functions that calculate standard statistical quantities such as the mean or standard deviation. See "Statistical Functions," p. 503 in the "Formula Functions Reference" chapter.
Random	Displays a list of functions that generate random numbers based on predefined distributions such as the uniform, normal, Cauchy, and so on. There is also a function to randomize the order of table rows. See "Random Functions," p. 507 in the "Formula Functions Reference" chapter.
Date Time	Displays a list of functions that require arguments with the *date* data type, which is interpreted as the number of seconds since January 1, 1904. **Date Time** functions return values such as day, week, or month of the year. They can also compute dates and can find data intervals. See "Date Time Functions," p. 510 in the "Formula Functions Reference" chapter.
Row State	Displays a list of functions that assign or detect row state status of color, marker, label, hidden, excluded, or selected. See "Row State Functions," p. 513 in the "Formula Functions Reference" chapter.
Assignment	Displays a list of functions that place the value on the right side of the assignment operator into the variable on the left side of the operator. See "Assignment Functions," p. 517 in the "Formula Functions Reference" chapter.
Parametric Model	Lets you construct a **Linear**, **Interactions**, or **Full Quadratic** model for columns that you choose. After choosing the type of model, a window appears that lets you select the variables from which to construct the model. Hold the CTRL key and click to select more than one. See "Parametric Model Functions," p. 518 in the "Formula Functions Reference" chapter.
Finance	Lets you create formulas to calculate principal payments, interest rate, rate of return, and so on. See "Finance Functions," p. 519 in the "Formula Functions Reference" chapter.

Ordering Expressions in Formulas

As you build a formula, keep in mind that all functions have an order of precedence shown in the following table, where level one is the highest order of precedence. Expressions with a high order of precedence are evaluated before those at lower levels. When an expression has operators of equal precedence, it is evaluated from left to right. You can use parentheses to override other precedence rules when necessary because any expression within parentheses is always evaluated first. Terms have no order of precedence because they cannot be evaluated further. Table 9.4 shows the first six levels of the order of precedence.

Table 9.4 Order of Precedence of Operators in Formulas

Level 1	Parentheses
Level 2	Functions in the function browser lists, **And**, **Or**, **Not**
Level 3	*, ÷, **Modulo**
Level 4	+, -
Level 5	Comparisons: <, ≤, =, ≠, ≥, >, ≤ x <, < x ≤, < x <
Level 6	Logical Operators **&**, **|**, **!**

Note: When a function has an expression as its argument, the argument has a higher order of precedence than it would if enclosed in parentheses outside the function.

Building a Formula in Order of Precedence

It is best to build a formula starting with any expression that serves as an argument. This is because functions have a high order of precedence and are always grouped with their corresponding arguments. It is also a good idea to create expressions working from highest to lowest order of precedence when possible. If you need parentheses, be sure to type the open parenthesis before entering the expression to be enclosed.

For example, given a data table with the columns A, B, and C, use the following steps to compose the expression $A(B + C)$. Note that this expression is not the same as A x $B + C$, which evaluates as (A x B) + C.

To enter the expression:

1. Select **Table Columns** from the formula element browser list.
2. Click column A in the **Table Columns** list. A
3. Click the multiplication button × in the formula editor keypad. A*□
4. Type an open parenthesis: A*(□)
5. Click column B in the **Table Columns** list. A*(B)
6. Click the addition button + in the formula editor keypad. A*(B+□)
7. Click column C in the **Table Columns** list. A*(B+C)

Because the order of precedence determines which arguments are affected by each functions, it also affects the grouping of expressions. Select functions in the formula to verify how the order-of-precedence rules have been applied.

Structuring Formulas for Efficient Evaluation

Usually, it is not necessary to structure formulas with efficient evaluation in mind. Most formulas evaluate almost instantaneously regardless of their structure. This is because statistical functions and constant expressions are evaluated only once when a column's values are calculated.

However, when you are creating conditional expressions, keep in mind that **Match** evaluates faster and uses less memory than an equivalent **Condition** function, **If**. (Note that **Match** ignores trailing spaces, while **If** does not.)

Consider two formulas for predicting a child's height from his age shown in Figure 9.7. In each case there is a base height of 58.125 inches to which a quantity is added depending on the value of the **age** variable.

Figure 9.7 The Match Conditional Evaluates Faster Than the If Function

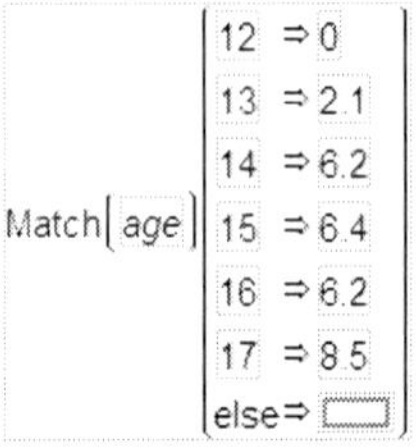

The **Match** conditional evaluates faster than the **If** function because the **age** variable is evaluated only once for each row in the data table. The **If** condition must evaluate the age variable at each **If** clause for each row until a clause evaluates as true.

Using Formula Editor Options

There are several options available to you as you create formulas. The following sections discuss each of these options.

Calculating Derivatives

The JMP formula editor can find and display the derivative of a function. The derivative is found with respect to the function argument (a single variable name) you highlight. Therefore, in order to differentiate with respect to *x*, *x* must be one of the arguments in the expression. The red triangle menu found above the keypad contains the **Derivative** command, as shown in Figure 9.8.

Figure 9.8 Select Derivative from the Keypad Menu Commands

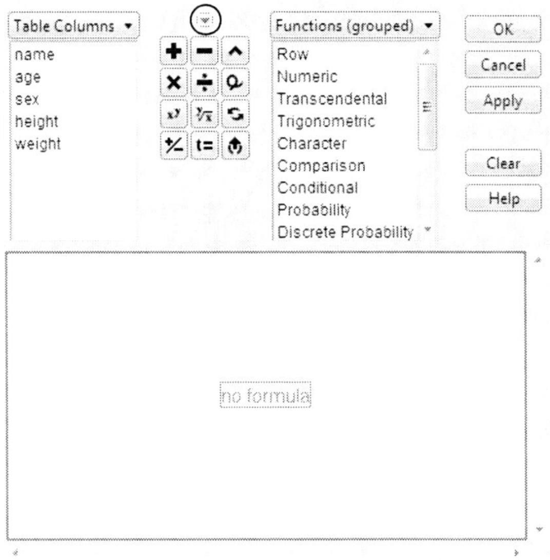

To calculate a derivative

1. Enter a function.
2. Highlight a variable.
3. Select **Derivative** from the menu. Figure 9.9 shows the completion of these steps.

Figure 9.9 Using the Derivative Option

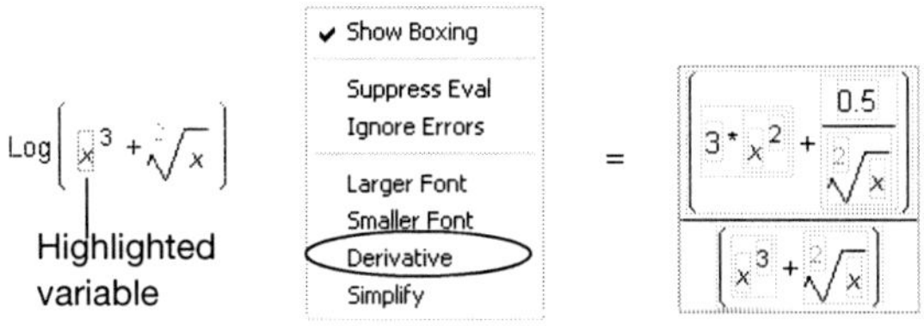

Simplifying Complex Formulas

When the formula editor contains a complex formula, JMP can simplify it using various algebraic rules. It can find constant expressions, distribute multiplication over addition, combine terms, and more.

To simplify complex formulas

1. Select a portion or all of the formula.
2. Click the red triangle icon above the keypad.
3. Select **Simplify** from the menu that appears, as shown in Figure 9.10.

Figure 9.10 Selecting Simplify (Left) Produces a Simplified Formula (Right)

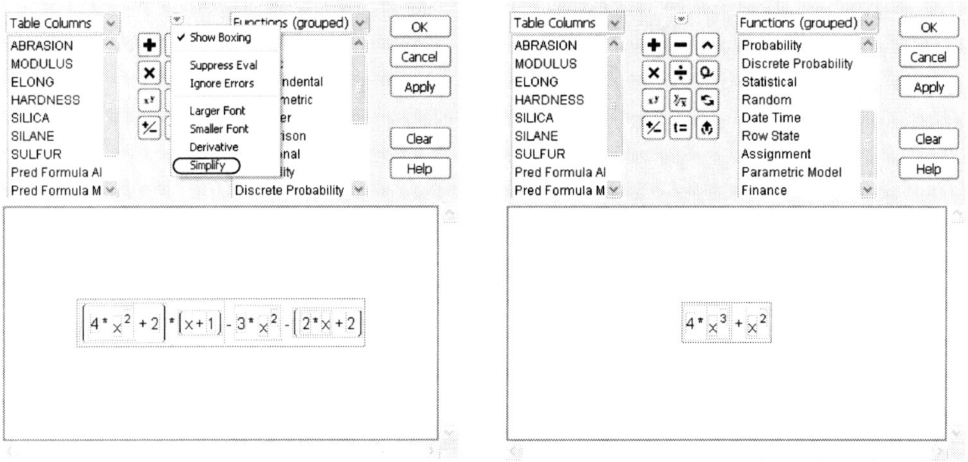

Evaluating Formulas

By default, JMP evaluates each formula that you create. You can turn this evaluation off, or you can use it before you have finished creating a formula.

Suppressing Evaluation

Turning off evaluation is a useful formula development mode for building complex formulas. You can turn off evaluation and build sections of a formula, and evaluate only to test it. In particular, you can close the formula editor and reopen it at a later time to continue building a formula without JMP evaluating it.

To suppress formula evaluation, click the red triangle icon above the keypad and select **Suppress Eval.** When evaluation is suppressed, the formula icon appears dimmed .

Note the following:

- If the icon appears to the right of the red triangle icon in the formula editor, it indicates that formula evaluation is suppressed for that formula.
- If the icon appears beside the column name in the column panel, it indicates that the values in the column result from a formula. When formula evaluation is suppressed, this plus icon becomes gray. (See "Icon Indicators," p. 144, for details.)

Note: The **Apply** button ignores the formula evaluation setting. Therefore, when formula evaluation is suppressed, clicking the **Apply** button overrides the suppression and evaluates the formula before it is applied to the column.

Ignoring Errors

Once you construct a formula and click **OK**, JMP checks behind the formula for error and alerts you of any errors that it finds. An error message appears for each error and asks whether you want to ignore further errors.

Sometimes you might want to suppress error messages while a formula is under development. For example, you might want the evaluated values for some rows without seeing an error message for each row that causes errors.

To have JMP ignore any errors that it finds

1. Create a formula.
2. Click the red triangle icon above the keypad and select **Ignore Errors**.

Viewing a Formula's Values from the Formula Editor

While in the Formula Editor, you can see the value of any expression within a formula with the **Evaluate** command. For column variables, you get the value at row 1. This is also true for parameters and expressions that evaluate to a constant value. To view values:

1. While in the formula editor, select the expression that you want to know about.
2. Right-click the selected expression.
3. Select **Evaluate**. The current value of the selected expression appears in a yellow box until you move the cursor.

Viewing a Formula in JSL

You have the option of entering or changing any part of a formula in *text mode*. Text mode displays the formula in JMP Scripting Language (JSL). The entire formula (or any of its terms) appears in text mode when you double-click the white space inside the boxed formula area. The elements of that box appear as plain text and you can then edit the formula as you would in any text editor.

Any element of a formula can be displayed as a scripting command and then edited. After editing formula scripting commands, click outside the formula to see its formatted form. For example, the text for the standardization of weight is `(:weight - Col Mean(:weight)) / Col Std Dev(:weight)`. To enter an **If** statement in text form, add pairs of arguments for each **If/Then** clause in the statement, and a single last argument for the else clause if needed. In text form, the **If** statement in Figure 9.11 looks like this: `If(:total!=0, (:count/:total)*100, 0)`.

Figure 9.11 An If Statement in Formula Mode

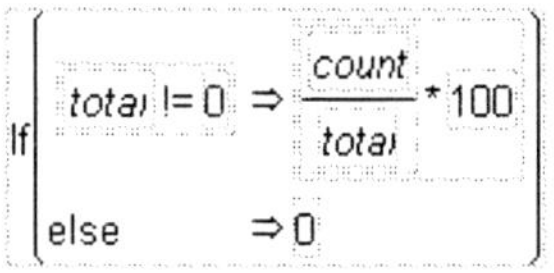

Editing Formulas

If you need to change a formula after you have exited the formula editor, right-click the column name in the data table and select **Formula**.

Correcting Mistakes

If you make a mistake while entering a formula, hold down the CTRL key and press Z. This reverses the effect of the last (undo-able) command.

Other commands to help modify formulas include the following:

- Click the delete button ( ) on the formula editor keypad to remove the selected expression.
- Use the cut, copy, and paste keyboard shortcut commands or right-click a highlighted part of the formula and select **Cut**, **Copy**, or **Paste** from the menu.
- To rearrange terms or expressions, click to see the hand grab the term. Drag to move formula pieces.

Selecting Expressions

Use the keyboard arrow keys to select expressions for editing. You can also use the arrow keys to view the formula's order of precedence when either parentheses or the boxing option are not present. (See "Hiding and Showing Boxing," p. 322.)

Clicking an operator (+, –, *, ÷) in an expression selects the operator and its operands. A red box appears around the items. Once an operator is selected:

- The left and right arrow keys move the selection across other associative operators having equal precedence within the expression.
- The up arrow extends the current selection by adding the operand and operator of higher precedence to the selection.
- The down arrow reduces the current selection by removing an operand and operator from the selection.

Deleting Functions

Deleting a function also deletes its arguments. Deleting a required argument or missing term from a function sometimes deletes the function as well. You can peel a function to delete it from its argument.

To peel a function from a single argument

1. Select the function.
2. Click the peel button in the formula editor keypad. Or, use the hand tool to drag the argument on top of its function.

Figure 9.12 Peeling an Argument

3. Complete formula changes.
4. Click **Apply**, and the new values fill the data table column automatically.
5. Once you have created a formula, you can change values in columns that are referenced by your formula. JMP automatically recalculates all affected values in the formula's column.

Cutting, Copying, and Pasting

You can cut or copy any expression or an entire formula and paste it into another formula display. Use the cut, copy, and paste shortcut commands or right-click a highlighted part of the formula and select **Cut**, **Copy**, or **Paste** from the menu. The following aspects apply when you cut, copy, and paste a formula:

- When you paste it into another formula display, the formula appears in formatted form.
- The formula is saved on the clipboard as a JSL statement. Thus, if you copy it into other applications, it appears as a JMP Scripting Language (JSL) statement.

Note: Press the Shift key on your keyboard, and then click the red triangle to reveal a command called **Copy As SAS Formula**.

Clicking and Dragging

You can drag any part of a formula that can be selected to any other location that can be selected.

To click and drag:

1. Place the arrow cursor inside an expression.
2. Click the expression. It is highlighted and the cursor changes to a hand cursor.
3. Drag across the formula. Destination expressions are highlighted.
4. Drag the selected expression to the new desired location. The selected expression is copied to the new location, where it replaces the existing expression.

Customizing Formulas

There are several ways that you can customize formulas in the formula editor. The following sections describe how to change font size, show and hide boxing, change the orientation of the formula, and close arguments.

Changing the Font Size

To incrementally increase or decrease the font used to display the formatted formula:

1. Click the red triangle icon above the keypad.
2. Select **Larger Font** or **Smaller Font**.
3. Repeat this process to further increase or decrease the font size.

Hiding and Showing Boxing

By default, JMP outlines specific terms within the formula. This is called *boxing*. Boxing is useful when you want to select and modify a specific portion of a formula, or need to determine the order of evaluation that takes place.

To turn boxing on or off:

1. Build a formula.
2. Click the red triangle icon above the keypad and select **Show Boxing**. When a check appears on the menu beside **Show Boxing**, the outline appears in the formula. When it does not, the outline does not appear.

Changing a Formula's Orientation

By default, JMP gauges the size of a formula and displays it in the formula editor in the best orientation (horizontally or vertically). However, if you create a long formula, you might want to display it in a different layout.

To change a formula's orientation:

1. Build a formula.
2. Highlight an argument or formula.
3. Right-click what you have highlighted.
4. Select **Orientation**.
5. Select from the **Orientation** options: **Best**, **Horizontal**, or **Vertical**.

Opening and Closing Arguments

When a formula is too large to fit on the screen, you can close formulas and arguments.

To close an argument:

1. Build a formula.
2. Highlight an argument or formula.
3. Right-click what you have highlighted.
4. Select **Close** or **Close Arguments** from the menu that appears.

Examples and Tutorials

To better familiarize yourself with building formulas, review the following examples and tutorials.

Using Basic Formula Editor Features

The following example uses the Big Class.jmp sample data table to walk you through using the basic features of the formula editor. You can find Big Class.jmp in the Sample Data folder, which was installed when you installed JMP.

Big Class.jmp has a column called weight. Suppose you want a new column that computes standardized weight values.

To create this column using a formula to obtain its values:

1. Open Big Class.jmp.
2. Select **Cols > New Column**.
3. Type the new name, Standard Weight, in the box beside **Column Name**.
4. Select **Formula** from the **Column Properties** menu, as shown in Figure 9.13.

Figure 9.13 Create a New Formula

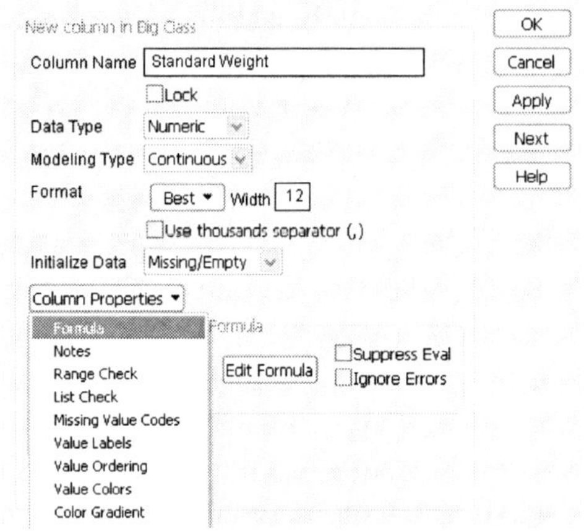

5. Click the empty formula element in the formula editing area to select it.

   When you create a formula and **Show Boxing** is checked, the selected portion of the formula is outlined with a thin red line. (See "Hiding and Showing Boxing," p. 322.) All terms within the smallest nesting box relative to the place that you clicked become selected, and the subsequent action applies to those combined elements.

Next, enter the formula that standardizes the weight values.

1. While the initial missing term is selected, click weight in the formula element browser column selector list.
2. Click the minus button [−] in the formula editor keypad.

   A new missing term appears after the minus sign as shown in Figure 9.14.
3. Click weight again.
4. Click the function browser menu, and select **Statistical.**
5. Select **Col Mean** from the **Statistical** list.
6. Select the entire expression.

   The red box should now enclose the whole formula.
7. Click the division button [÷] in the keypad.

   The result gives a selected missing denominator for the whole expression.
8. Click weight again from the column selector list.

   It becomes selected in the denominator.
9. Select **Col Std Dev** from the **Statistical** list.

   The completed formula should look like that in Figure 9.14.

Figure 9.14 Building a Formula

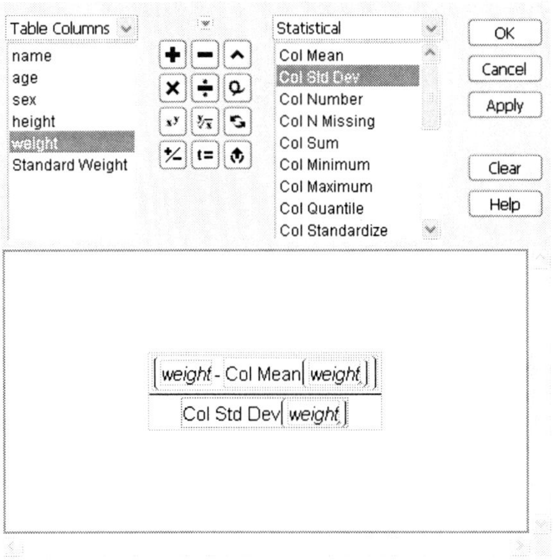

10. Close the formula editor by clicking **OK**.
11. Close the Column Info window by clicking **OK**.

In the data table, the new Standard Weight column fills with values. When a weight value changes, the calculated Standard Weight value automatically recalculates.

Using Local Variables in a Formula

Suppose you want to compute the slope in a simple linear regression of *y* on *x* using the standard formula shown in Figure 9.15. One way to do this is to create two local variables, and name them XY and Xsqrd. (See "Using Local Variables," p. 305.) Then assign them to the numerator and the denominator calculations of the slope formula. Delimit each assignment with a semicolon, as shown in Figure 9.15. (Statements in the formula editor are actually JSL programming statements. Multiple statements in a formula must be separated by a semicolons.) The slope computation is simplified to XY divided by Xsqrd.

Figure 9.15 Local Variables in a Formula

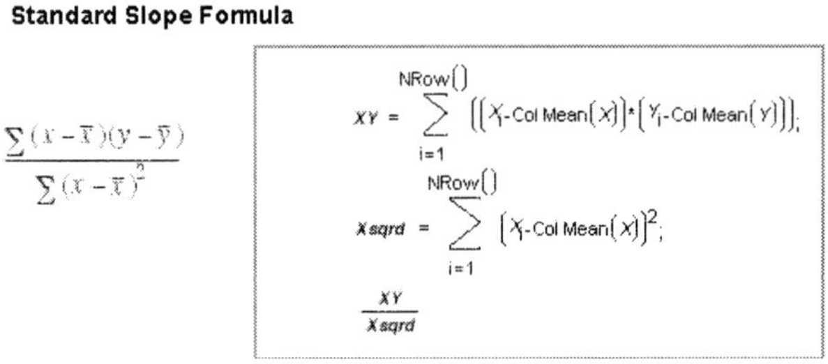

Note: You can also create local variables using the t= button on the on-screen keypad. Clicking this button creates a local variable with a default name in an expression and places a semicolon after it. The default name is *t1*, and additional local variables are named *t2*, *t3*, and so forth. You can change these default names by double-clicking and editing the name. However, you must be careful to rename each instance of the variable to avoid errors.

Using the Munger Function

The following examples show uses of the **Munger** function. In these examples, assume that there is a character column of names with "Veronica Layman" as one of its values. To simplify the examples, the literal name "Veronica Layman" is the search string instead of a column name.

See "Character Functions," p. 480 in the "Formula Functions Reference" chapter, for instructions on how to incorporate **Character** functions, such as **Munger**, into a formula.

Inserting Characters

This **Munger** example finds the blank between the first and last name, and inserts the middle initial "J." The formula **Munger("Veronica Layman", 1, " ", " J. ")** inserts the middle initial J., and evaluates as Veronica J. Layman.

Double quotes are required by the **Munger** function for literal strings, including strings that consist of a blank or when leading or trailing blanks are part of a string.

Deleting Characters

To delete one or more characters from a string:

1. Designate the characters to delete as the **Find** string in the **Munger** function.
2. Enter an empty **Replace** string: two quotation marks with nothing between them.

For example, the function, **Munger("Veronica Layman", 1, "onic", "")** removes the "onic" from Veronica and evaluates as Vera Layman.

Note: A **Replace** field with a null (no value) string enclosed in quotation marks is different from a **Replace** field with no value. If you delete the **Replace** string altogether, **Munger** shows the argument name ("**Replace**") in the formula editor window and behaves as if that optional argument does not exist. The resulting data type can also change from character to numeric, depending on the value of the **Find/Replace** argument.

Finding the Position (Index) of a Value

When the **Find/Length** field contains characters, **Munger** behaves like an index function and returns the numeric position of the first instance of the search string if it exists. For example, **Munger("Veronica Layman", 1, " ")** searches for a single blank and finds it in position nine. If the search string is not found, **Munger** returns a zero. This use of **Munger** produces the same result as the **Contains** function, as shown in "Character Functions," p. 480 in the "Formula Functions Reference" chapter.

Finding a Substring

Munger can extract substrings. For example, to extract only the first name, **Munger("Veronica Layman", 1, 8,)** starts at position one, reads through position eight. The remaining characters are ignored because the replace argument is not defined. This yields "Veronica" and produces the same result as the **Substring**, as shown in "Character Functions," p. 480 in the "Formula Functions Reference" chapter.

An alternative way to find a substring is with a start value, any negative find value, and a **no replace** argument. **Munger("Veronica Layman", 9, –1,)** returns "Layman".

Using the Match Conditional Function

This example walks you through using the Match conditional function.

Suppose that you want a **Match** conditional for the nominal variable Type from the Hot Dogs.jmp sample data table.

1. In the formula editor, select Type from the **Table Columns** list.
2. Select **Conditional** from the Functions list.
3. Select **Match.**
4. Select **Add Arguments from Data.**

   The values are automatically filled in (Figure 9.16).

Note: Rather than complete step 2 through step 4, hold down the SHIFT key, select **Conditional** from the Functions list, and then select **Match.**

Figure 9.16 Automatically Filling a Match Conditional Statement

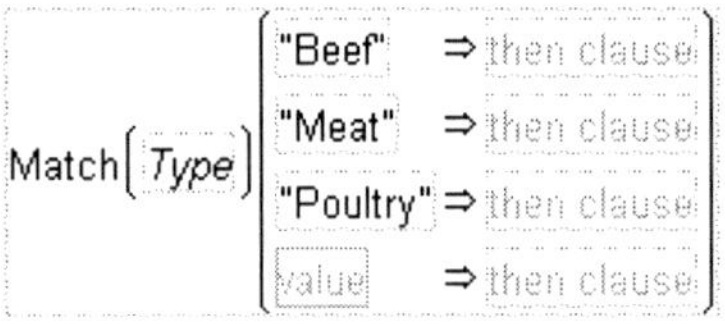

If you do not want the values filled in for you, select **Do Not Add** from the **Match** list instead of **Add Arguments from Data**.

Using the Delete Expression Key

Repeated clicks on the delete expression key produce the following sequence of steps.

Figure 9.17 Using the Delete Expression Key

1. Start with a formula.

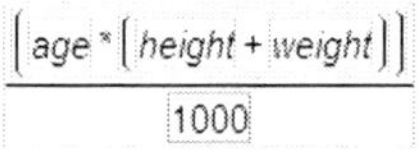

2. Select a formula element.

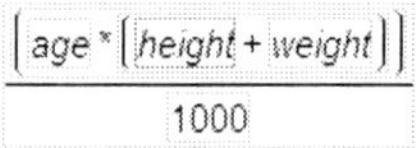

3. Click the delete expression button ( ).

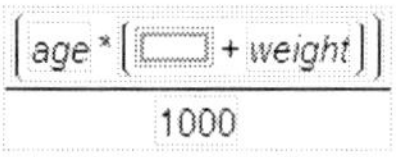

4. Click the delete expression again.

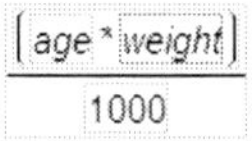

5. Click the delete expression again.

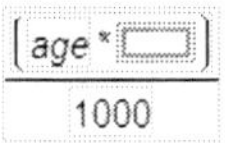

6. Click the delete expression again.

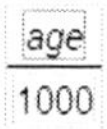

7. Click the delete expression again.

8. Click the delete expression again.

9. Click the delete expression again.

Using Keyboard Shortcuts

Table 9.5 describes the keyboard shortcuts that you can use in the formula editor.

Table 9.5 Keyboard Shortcuts in the Formula Editor

Insert This Item	**Using This Keyboard Shortcut**
A missing element (•)	DELETE
Subscript	[
() set of parentheses	(
* (multiplication symbol)	* on keypad or keyboard
+	+ on keypad or keyboard
–	– on keypad or keyboard
÷	/ on keypad or keyboard
+/-	Hold down the SHIFT key and press the - (MINUS) key
And	&
Not	!
Or	|
Power	^
New argument	,
x<y	<
x==y	=
x>y	>

Glossary of Terms

Remember that:

- Functions always *operate* on selected expressions.
- Arguments are always grouped with functions.
- To find which expressions serve as a function's arguments, select that function in the formula.

- The boxed groupings show how order-of-precedence rules apply and show which arguments are deleted when you delete a function. See "Ordering Expressions in Formulas," p. 314, for details.

Table 9.6 contains terms used for the parts of the formula itself.

Table 9.6 Terms Used in the Formula Editor

Element	The name of a constant, table variable, table column, local variable, or parameter that appears in the element browser list.	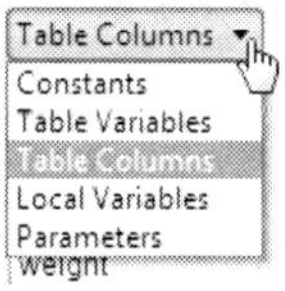
Argument	Any element or an entire expression (including mathematical operands) that is operated on by a function.	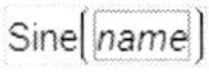
Term	Indivisible parts of an expression, such as constants and variables.	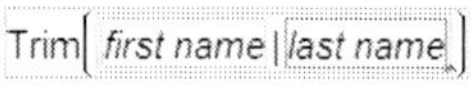
Expression	Any part of a formula that can be selected as a single unit, including terms, missing terms, and functions grouped with their arguments, as well as the entire formula.	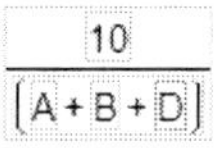
Clause	A complete segment in a conditional function.	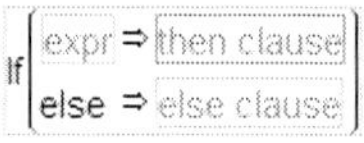
Function	A mathematical or logical operation that performs a specific action on one or more arguments. Functions include most items in the function browser list and all keypad operators. Functions always *operate* upon selected expressions, and arguments are always grouped with functions. To find which expressions serve as a function's arguments, select that function in the formula. The boxed groupings also show how order-of-precedence rules apply and show which arguments are deleted when you delete a function.	

Table 9.6 Terms Used in the Formula Editor *(Continued)*

Missing term	Any empty place holder for an expression, represented by an empty box.	
Missing value	Excluded or null data consisting of the missing value mark (•) for numeric data or null character strings for character data.	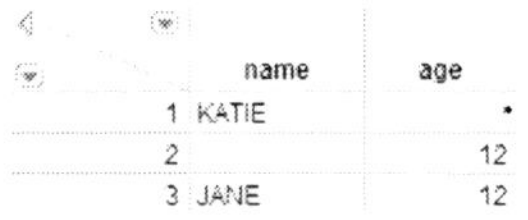

Chapter **10**

Personalizing JMP

Customizing the Interface and Adding New Features

This chapter describes how to customize JMP menus and toolbars to show only the commands that you need. For example, you might remove the **SAS** option in the **File** menu if you do not use SAS. Or you might assign a shortcut key to the **Run Script** command.

Customizing JMP also lets you set up JMP for groups with special interests. If one group does not design experiments, you could remove **DOE** from the JMP menu.

Another way to customize JMP is to write add-ins to extend JMP's functionality. An add-in can be any scriptable option. For example, you could add a **Graph** menu option that opens two data tables included in the add-in and runs a script to produce a bubble plot.

Contents

Personalize Toolbars and Menus (Windows Only)

Toolbars consist of buttons that execute commands and dividers that help you organize the buttons. In Windows, there are many ways to personalize toolbars. For example, you can create, rename, hide, and modify toolbars or buttons. You can also rearrange or delete toolbars and buttons that you create.

Menus consist of commands, submenus, and separators. As with toolbars and buttons, you can create, rename, hide, and modify menus. You can also rearrange or delete menus and menu items that you create.

You personalize toolbars and menus in the Menu Editor. The names of modified items are formatted to indicate which items were created, modified, or hidden, and which items are unsaved, as shown in Figure 10.1.

Figure 10.1 Menu Editor Options

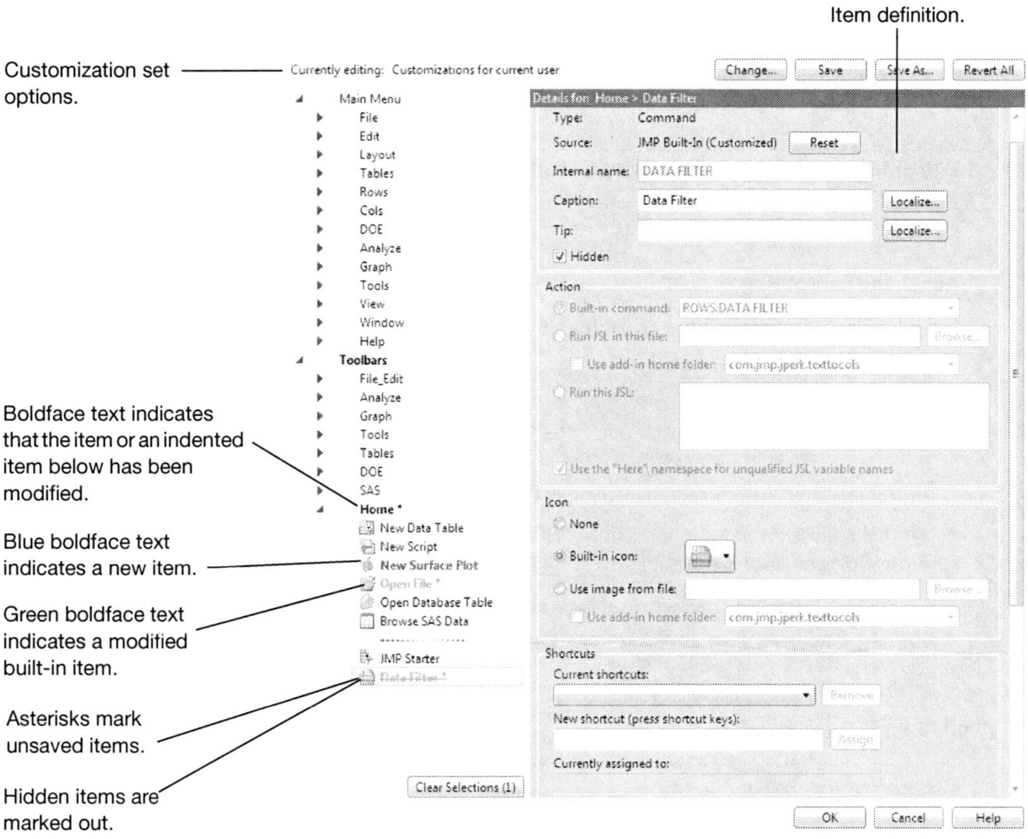

Change Customization Sets

JMP gives you the flexibility to specify which users see your customized toolbars and buttons. Your changes are saved in a text file with the .jmpcust extension called a *customization set*. By default, your customizations are shown only to the current user; other users who log on to the computer and open JMP do not see your personal toolbars and menus.

All customization sets include the built-in JMP buttons and menus. You can also base customization sets on other customization sets. For example, you might create a new customization set that shows the JMP add-in menus and toolbars along with your modifications.

When you modify the **Current user** customization set, JMP creates a backup of the set located in one of the following folders:

- C:\Users*user name*\AppData\Local\SAS\JMP\9\ (Windows 7 and Vista)
- C:\Documents and Settings*user name*\Local Settings\Application Data\SAS\JMP\9\ (Windows XP)

When you modify the **All users** customization set, the backup of the set is located in one of the following folders:

- C:\Users\All Users\SAS\JMP\9\ (Windows 7 and Vista)
- C:\Documents and Settings\All Users\Application Data\SAS\JMP\9\ (Windows XP)

The All Users folder does not appear when you try to browse for it, so you must type the path into the Windows Explorer address field.

Note: To find the customization set files in Windows Explorer, show all hidden files in the Windows Explorer Folder Options. See your operating system documentation for details.

Change the Customization Set

1. Select **View > Customize > Menus and Toolbars.**

   The Menu Editor appears.

2. Click **Change.**

   When you try to change the customization set, and changes to the selected set are not saved, you are prompted to save the changes. Click **Save Changes.**

3. Select the customization set in the **Customization Set to Modify** area. (By default, changes apply only to the **Current user** customization set.) Table 10.1 describes the options.

4. (Optional) To change the set on which the selected customization set is based, select the set in the **Starting Set** area. (Not available for the *All users* customization set.) Table 10.1 describes the options.

5. Click **OK.**

Table 10.1 Customization Set Options

Customization Set to Modify	
Current User	Only you see the changes. The customization file is called usercust.jmpcust.
All users	All users who open JMP on the computer see the changes. The customization file is called admincust.jmpcust.

Table 10.1 Customization Set Options *(Continued)*

JMP Add-In	Those who use the selected JMP Add-in, and select **JMP Add-In customizations** as the **Starting Set,** see the changes. When you modify the customization set for a disabled add-in, those changes do not appear until you enable the add-in.
Other file	Changes appear when the selected .jmpcust file is chosen as the customization set. You create this file in a text editor.
New	Changes appear when this new customization is selected. (After you make your changes, click **Save** to name the new file in the selected folder.)
Starting Set	
JMP built-in items	The selected customization set is based on the default JMP toolbars and menus. (Applies to all customization sets and cannot be deselected.)
All Users customizations	The selected customization set is based on the **All users** customization set. (Available for the **Other file** or **New** customization sets. Cannot be deselected from the **Current user** customization set.)
JMP Add-In customizations	The selected customization set is based on the JMP Add-In customizations. The add-in customizations are typically installed with the add-in. (Available for all customization sets except for **All users**.)
Current User customizations	The selected customization set is based on the **Current user** customization set. (Available only for **Other file** and **New** customization sets.)

Create Toolbars

JMP includes toolbars that you can add new buttons to, or you can create your own toolbars. Table 10.2 describes the toolbars.

Table 10.2 JMP Toolbars

File_Edit 	Shows icons for commands found in the **File** and **Edit** menus.
Analyze 	Shows icons for commands found in the **Analyze** menu.
Graph 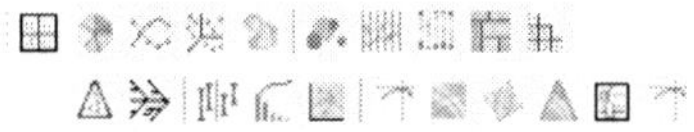	Shows icons for commands found in the **Graph** menu.

Table 10.2 JMP Toolbars *(Continued)*

Tools 	Shows icons of tools that you can click and use as your cursor. In some windows, the toolbar is hidden. A blue line appears instead. Hover your cursor over the blue line to show the toolbar.
Data_Tables_List 	Shows a list of open data tables. You select a data table in this list to make it the current table. Note that the current table is not necessarily the front window. To bring a table (or any window) to the front, select its name from the list in the **Window** menu.
Tables 	Shows icons for commands in the **Tables** menu.
DOE 	Shows icons for commands in the **DOE** menu.
SAS 	Shows icons for accessing and browsing SAS data and folders.
Home 	Shows frequently used icons from the **File** and **Edit** menus along with icons for opening the JMP Starter window and filtering data.

Create your own toolbars for frequently used commands that are not included in the default toolbars. You assign a command to the button and can add other properties such as shortcut keys, icons, and JMP Scripting Language (JSL) scripts.

When you want to create a toolbar based on an existing toolbar, make a copy of the existing toolbar and then change settings as necessary. See "Copy and Paste Menus, Menu Items, Toolbars, and Buttons," p. 346 for details.

Step 1: Create the Toolbar

1 Select **View > Customize > Menus and Toolbars**.

The Toolbars list appears on the left.

2 (Optional) Change the customization set to control which users see your customizations. See "Change Customization Sets," p. 335 for details.

3. With the Menu Editor still opened, right-click one of the toolbars, such as **File_Edit**.

Select any toolbar, because all new toolbars appear at the end of the Toolbars list.

4. Select **New Toolbar**.

The toolbar is inserted at the end of the list. The toolbar includes an untitled button, because all toolbars must have at least one button.

Figure 10.2 New Toolbar and Button

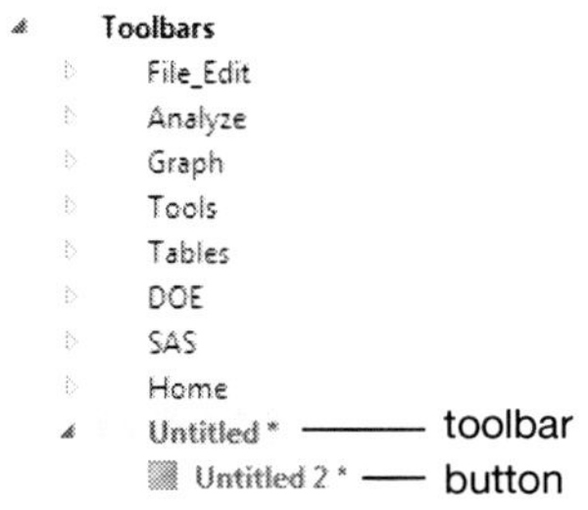

Step 2: Specify the Caption and Internal Name

Toolbars have several basic properties:

- The caption appears in the **View > Toolbars** list, which lets you show or hide the toolbar. Give each toolbar a unique name. JMP merges toolbars with the same name after you close and reopen the Menu Editor.
- The internal name, which is case insensitive. JMP identifies the location of an item by its internal name rather than its caption.

To specify these toolbar properties, follow these steps:

1. With the Menu Editor still opened, select the toolbar.

   The General properties appear on the right.

Figure 10.3 General Properties

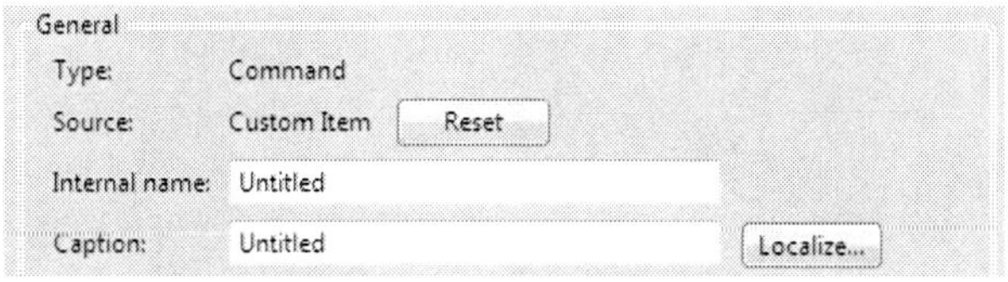

2. Enter the internal name next to **Internal name**.
3. Enter a unique name for the toolbar or button in the **Caption** box.

4. (Optional) To specify the translation for the caption, do the following:
   - Click the **Localize** button for the item that you want to customize.
   - Select the language.
   - Enter the translation next to **Text.**
   - Click **OK.**
5. Customize the untitled button as described in “Create Menu Items and Toolbar Buttons,” p. 341.
6 Click **Save** to save your changes.

Create Main Menus

On Windows, you can add your own menus to the main menu bar in JMP. These menus appear before or after menus such as **File**, **Edit**, and **Tables.** An untitled command is inserted automatically in the menu, because all menus must have at least one command.

When you want to create a main menu based on an existing menu, make a copy of the existing menu and then change settings as necessary. See “Copy and Paste Menus, Menu Items, Toolbars, and Buttons,” p. 346 for details.

Create a Main Menu

1 Select **View > Customize > Menus and Toolbars.**

The Main Menu list appears on the left.

Figure 10.4 Main Menus

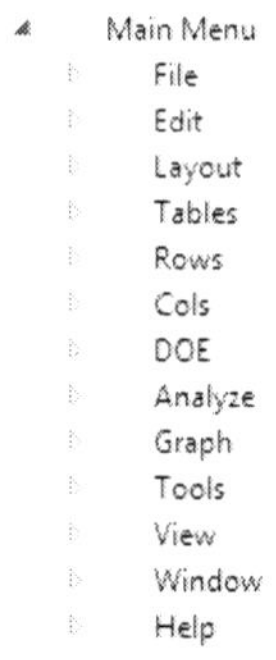

2. (Optional) Change the customization set to control which users see your customizations. See “Change Customization Sets,” p. 335 for details.
3. Right-click the menu next to which you want to add a new menu.
4. Select **Insert Before** or **Insert After.**

An untitled menu and a menu item are added to the list.

Figure 10.5 New Main Menu and Untitled Menu Item

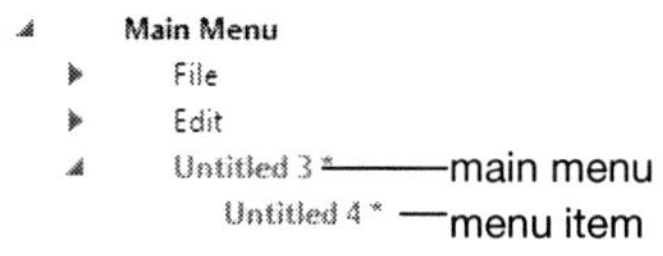

5 Complete "Step 2: Specify the Caption and Internal Name," p. 339 to rename the main menu.

6. Customize the menu item as described in"Create Menu Items and Toolbar Buttons," p. 341.

7 Click **Save** to save your changes.

Create Menu Items and Toolbar Buttons

Menus consist of the following menu items:

- Commands are the items that you click to execute a command (such as **Data Table** and **Script**).
- Submenus are menu items that you click to reveal more menu options. An example of submenu is **New** inside the top-level **File** menu. **New** is also a submenu because it contains commands (such as **Data Table** and **Script**).
- Separators ( ) are lines that divide or group commands and submenus.

The button on a toolbar is considered a type of command, so you also complete the following steps to create new toolbar buttons.

When you want to create a menu item or button based on an existing item, make a copy of the existing item and then change settings as necessary. See "Copy and Paste Menus, Menu Items, Toolbars, and Buttons," p. 346 for details.

Step 1: Create a New Menu Item or Button

1. Select **View > Customize > Menus and Toolbars.**

   The Menu Editor appears.

2 (Optional) Change the customization set to control which users see your customizations. See "Change Customization Sets," p. 335 for details.

3. Right-click where you want to add the menu item or button.

   A list of possible locations appears.

4. Select **Insert Before** or **Insert After.**

   The Specify Type window appears.

Figure 10.6 Specify Type Window

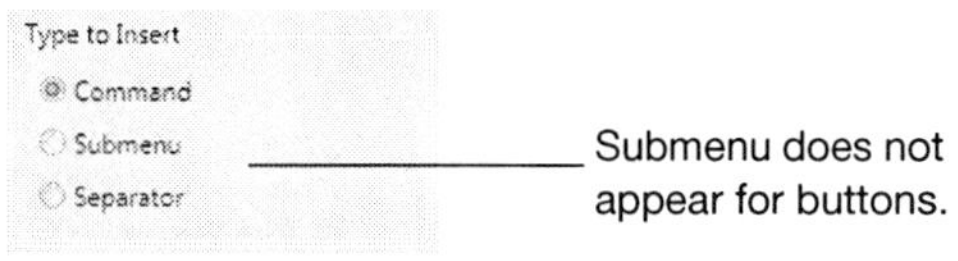

5. Do one of the following:
   - To create a button, select **Command**.
   - To create a menu within a menu, select **Submenu**. (Not available for buttons.)
   - To create divider between menu items or toolbar buttons, select **Separator**.
6. Click **OK**.

   The new untitled button, submenu, or separator is added.

Step 2: Specify the Caption, Tooltip, and Internal Name

Buttons have several basic properties:

- The caption appears in the **View > Toolbars** list, which lets you show or hide the toolbar. Give each toolbar a unique name. JMP merges toolbars with the same name after you close and reopen the Menu Editor.
- The tooltip appears when you place the cursor over a menu item or button.
- The internal name, which is case insensitive. JMP identifies the location of an item by its internal name rather than its caption.

Menu items include the same properties, but the tip does not appear in JMP.

To specify these properties, follow these steps:

1. With the Menu Editor still opened, select the menu item or button.

   The General properties appear on the right.

Figure 10.7 General Properties

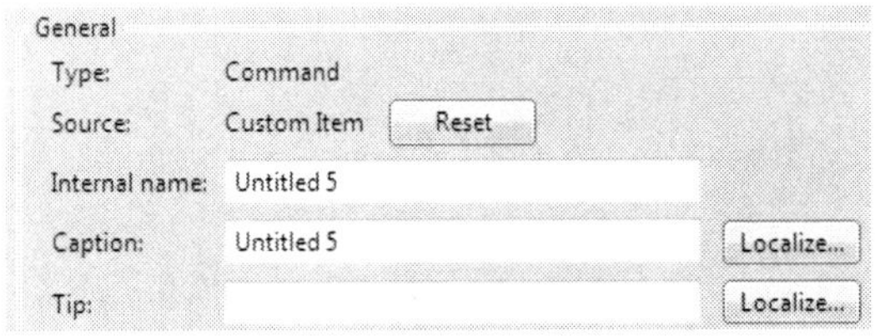

2. Enter the internal name next to **Internal name**.
3. Enter a name for the menu item or button in the **Caption** box.
4. Enter a description for the menu item or button in the **Tip** box.
5. (Optional) To specify the translation for the caption or tip, do the following:
   - Click the **Localize** button for the item that you want to customize.
   - Select the language.
   - Enter the translation next to **Text**.
   - Click **OK**.

6. Customize the untitled button or menu item as described in the following procedure.

Step 3: Assign Functionality

Clicking a menu item or button either executes a predefined command or runs a JSL script. The script can be stored in a separate file, or you can enter the JSL in the **Run this JSL** area of the Menu Editor. If users have access to a central location, such as a network, you typically want to run the script from that location. This also simplifies giving users access to updated scripts. Otherwise, store the JSL in the menu item or button definition.

To associate a menu item or button with an add-in JSL script, you have two options:

- Select **Run JSL in this file**, and then browse to find the external file.
- Type the relative path to the JSL script, and select the add-in from the **Use add-in home folder** list. For example, the following command runs the mds_application.jsl script from the selected add-in:

  ```
  $ADDIN_HOME(com.jmp.mdswithr)\mds_application.jsl
  ```

In addition, you can select **Use the "Here" namespace for unqualified JSL variable names** if you include namespaces in the internal or external JSL script. See the *Scripting Guide* book for details about namespaces.

To assign this functionality, follow these steps:

1. With the Menu Editor still opened, select the menu item or button that you want to modify.

   The Action properties appear on the right.

Figure 10.8 Action Properties

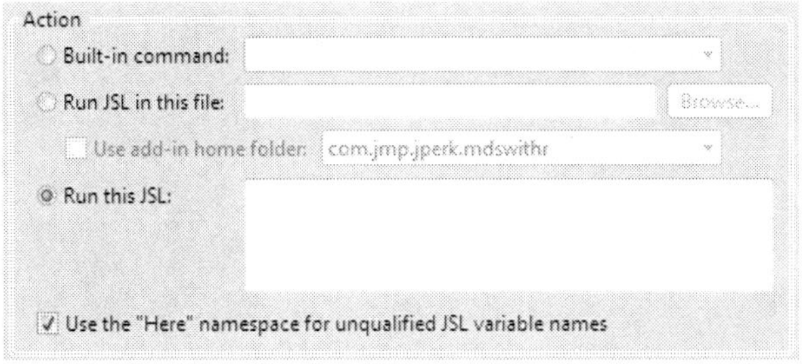

2. Select the action that you want to execute.
3. To run an internal JSL script, delete the placeholder text *print("Not implemented.");* in the **Run this JSL** area and enter the JSL.
4. (Optional) If your internal or external JSL script includes namespaces, click **Use the "Here" namespace for unqualified JSL variable names.**
5. Click **Save** to save your changes.

Step 4: Show an Icon on the Menu Item or Button (Optional)

New menu items have no icon next to the item. New buttons show up as blue squares on the toolbars. You can assign a descriptive icon to the menu item or button. The graphic can reside on your computer or in an add-in folder. .ico, .png, .jpg, and .bmp graphics are supported.

Assigning an icon to a button is particularly important. Otherwise, the user must view the button's tooltip to see what the button does.

To show an icon on the menu item or button, follow these steps:

1. With the Menu Editor still opened, select the menu item or button.

   The Icon properties appear on the right.

Figure 10.9 Icon Properties

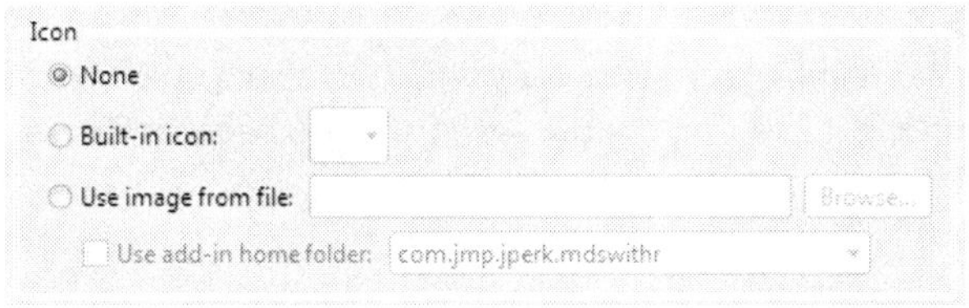

2. Select one of the following options:

   None Select this option to show a blue square instead of an icon.

   Built-in icon Select this option to show an icon that JMP provides, and then select the icon from the list.

   Use image from file Select this option to show an image that you created. Click **Browse**, select the graphic, and then click **OK**. You can also use an icon defined in an add-in. Select **Use add-in home folder**, and then select the add-in.

3. Click **Save** to save your changes.

Step 5: Assign a Shortcut Key (Optional)

A shortcut key executes an action so that you do not have to select the menu item or click the button on the toolbar. This option is also helpful when the button's toolbar is not shown, but you still want to execute the action.

Shortcuts begin with CTRL, CTRL+SHIFT, CTRL+ALT and end with a number, letter, or symbol that appears on the keyboard. The Function keys (such as F1) are also supported. Many CTRL + *letter* and Function shortcuts are already assigned in JMP, but you can reassign them to your buttons if you want.

Shortcut keys appear in menus next to each menu item.

To assign a shortcut key, follow these steps:

1. With the Menu Editor still opened, select the menu item or button.

   The Shortcuts properties appear on the right.

Figure 10.10 Shortcuts Properties

2. Do one of the following:
   - To assign a new shortcut, click in the **New shortcut** area and press the shortcut keys. If the keystrokes then appear in the **Currently assign to** box, press another combination of shortcut keys. (You do not have to delete the keystrokes before pressing the shortcut keys.)
   - To change a shortcut that you previously specified, click **Remove**, click in the **New shortcut** area and press the shortcut keys.
3. Click **Assign.**

   The shortcut appears in the **Current shortcuts** list.
4. Click **Save** to save your changes.

To remove a shortcut, select the shortcut and click **Remove**.

Figure 10.11 shows examples of a completed toolbar, menu, and submenu.

Figure 10.11 Examples of Buttons and Menu Items in the Menu Editor

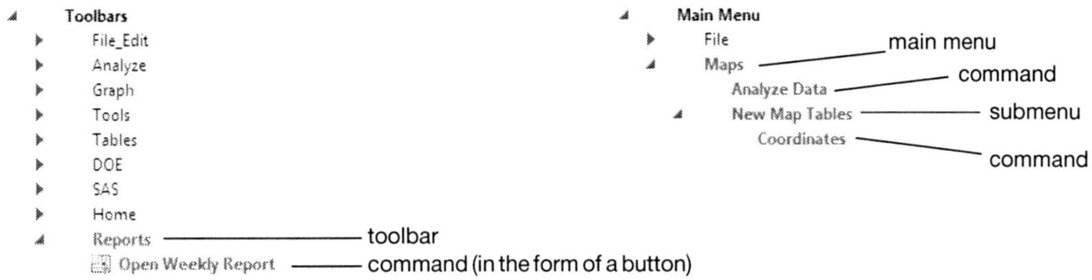

Rearrange Toolbars

On Windows, there are two ways to rearrange toolbars:

- To rearrange toolbars quickly for the current user, drag the toolbar above, below, to the left, or to the right of the adjacent toolbar.

- To specify the new location rather than drag the toolbar, right-click the toolbar and select a position from the **Location** menu. This method lets you move toolbars to the top or bottom of the window. You can also position the toolbars vertically on the left or right side of the window.

  In some windows, the toolbar is hidden by default. Specifying a new location shows the toolbar in all windows of that type. On Windows, you can change the toolbar preferences to always show toolbars. See "Windows Specific," p. 459 in the "JMP Preferences" chapter

Drag a Toolbar

1. Click the left corner of the toolbar until the Move  cursor appears.
2. Drag the toolbar to the new location. When a toolbar is too wide to show completely, click the arrow to show all of the buttons, as shown in Figure 10.12.

Figure 10.12 Expanding a Toolbar

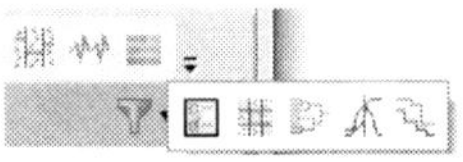

Specify the Toolbar Position

1. Right-click the toolbar that you want to move.
2. Select **Location**, and then select the new position.

Note: After you move a toolbar to the left, right, or bottom positions, you can add other toolbars to the same position. For example, you might relocate the **File_Edit** toolbar to the right side of the JMP window. To add other toolbars on the right side, right-click the bar and select the toolbar that you want to add.

Copy and Paste Menus, Menu Items, Toolbars, and Buttons

Copying existing items is a shortcut to creating new items in a menu or toolbar. This option lets you change only a few settings in the new item when possible.

Copy and Paste Menus and Menu Items

1. Select **View > Customize > Menus and Toolbars.**

   The Menu Editor appears.

2 (Optional) Change the customization set to control which users see your customizations. See "Change Customization Sets," p. 335 for details.

3. Right-click the menu or menu item that you want to copy and select **Copy.**
4. Right-click the menu or menu item before, after, or into which the item will appear and select **Paste.**

One of the following occurs:

– A list of possible locations appears.

– The menu or menu item appears below the selected item. Skip to step step 7.

5. Select the location of the menu by doing one of the following:

   – To paste the item *before* the selected menu, select **Paste before**.

   – To paste the menu *after* the selected menu, select **Paste after**.

   – To paste the menu *inside* the selected menu, select **Paste into**. (Only available when you select a menu as the new location.)

   – To cancel the action, select **Cancel**.

   The item appears as you indicated.

6. Modify the item.

7. Click **Save** to save your changes.

See "Create Main Menus," p. 340 and "Create Menu Items and Toolbar Buttons," p. 341 for details about modifying the items.

Copy and Paste Toolbars and Buttons

1. Select **View > Customize > Menus and Toolbars**.

   The Menu Editor appears.

2 (Optional) Change the customization set to control which users see your customizations. See "Change Customization Sets," p. 335 for details.

3. Right-click the toolbar or button that you want to copy and select **Copy**.

4. Do one of the following:

   – To paste a toolbar, right-click **Toolbar** and select **Paste**. The toolbar appears at the end of the list of toolbars.

   – To paste a button, right-click the button before or after which you want the new button, and select **Paste**. If you selected the first button on the toolbar, select the location of the button. Otherwise, the button is pasted below the selected button.

5. Modify the item.

6. Click **Save** to save your changes.

See "Create Toolbars," p. 337 and "Create Menu Items and Toolbar Buttons," p. 341 for details about modifying the items.

Rearrange Custom Menus, Menu Items, and Buttons

On Windows, you can rearrange the order of menus, menu items, and buttons that you create. For example, under the **File > New** menu, you could move your custom menu before the **Data Table** command.

In the item's General properties, the Source determines whether you can move the item. You can move items if the Source is *Custom Item*. Built-in items cannot be moved. Items that are defined in other customization sets can be moved only in that customization set. See "Change Customization Sets," p. 335 for details.

Note: The right-click menu has options for cutting and pasting items. To cut an item, you can also hold down the CTRL key and press X, or hold down the SHIFT key and press DELETE. To paste, hold down the CTRL key and press V, or hold down the SHIFT key and press INSERT.

Rearrange Custom Menus and Menu Items

1. Select **View > Customize > Menus and Toolbars.**

   The Main Menu list appears on the left (Figure 10.4).

2 Do the following:

   – Select the item that you want to move.

   – Verify that the Source is *Custom Item.*

   – If the Source is *All Users* or *JMP Add-In*, select the specified customization set and verify that the Source is *Custom Item.*

3. Right-click the custom menu or menu item that you want to move and select **Cut.**

4. Right-click the new location of the cut item and select **Paste.**

   One of the following occurs:

   – A list appears with the possible locations of the cut item. This occurs when you select a menu (such as **File** or **New**) or the first item in a menu (such as **Data Table** in the **File > New** menu).

   – The cut menu item appears below the selected menu item. Skip to step 6.

5. Select the location of the cut item by doing one of the following:

   – To paste the cut item *before* the selected item, select **Paste before.**

   – To paste the cut item *after* the selected item, select **Paste after.**

   – To paste the cut item *inside* the selected menu, select **Paste into**. (Only available when you select a menu as the new location.)

   – To cancel the move, select **Cancel.**

   The item appears as you indicated.

6. Click **Save** to save your changes.

Rearrange Custom Buttons

1. Select **View > Customize > Menus and Toolbars.**

   The Toolbars list appears on the left.

2 Do the following:
   - Select the custom button that you want to move.
   - Verify that the Source is *Custom Item*.
   - If the Source is *All Users* or *JMP Add-In*, select the specified customization set and verify that the Source is *Custom Item*.
3. Right-click the custom button that you want to move and select **Cut**.
4. Right-click the new location of the cut button.

   One of the following occurs:
   - A list appears with the possible locations of the cut button. This occurs when you select the first button on the toolbar (such as **New Data Table** in the **File_Edit** toolbar).
   - The button is pasted at the end of the toolbar. This occurs when you select the name of the toolbar. For example, when you select **File_Edit** and paste, the button appears after the last button, **Run Script**. Skip to step 6.
   - The cut button appears after the selected button. Skip to step 6.
5. Select the location of the cut button by doing one of the following:
   - To paste the cut button *before* the selected button, select **Paste before.**
   - To paste the cut button *after* the selected button, select **Paste after.**
6. Click **Save** to save your changes.

Delete Custom Items

Rather than temporarily showing or hiding toolbars, buttons, menus, and menu items, you can delete items that you created to remove them permanently from JMP. The Source determines whether you can delete the item and which customization set contains the item. Figure 10.13 shows examples of customization Source types.

Figure 10.13 Examples of Customization Sources

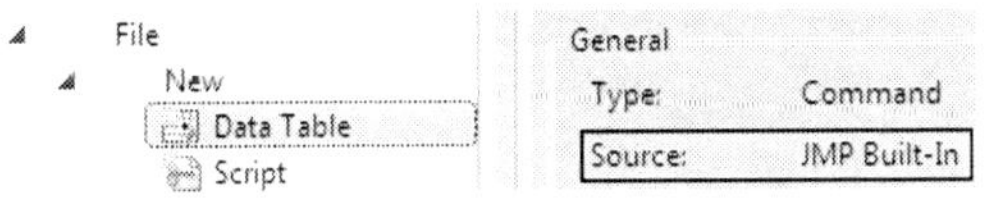

The item cannot be deleted.

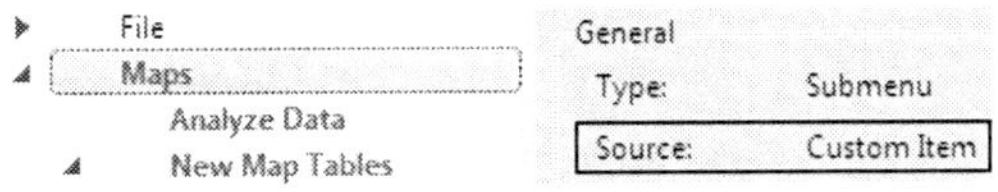

Delete the item in the current customization set.

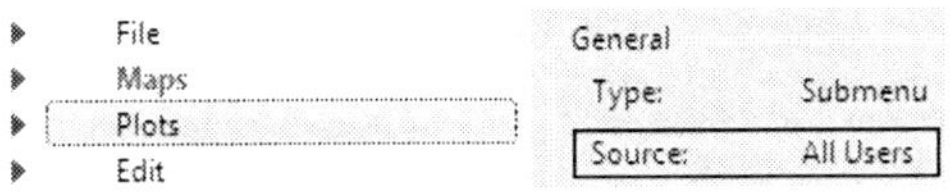

Delete the item from the *All Users* customization set.

Empty menus are not supported. When you delete the only item in a menu, the entire menu is deleted, not just the selected item.

Note: Make sure that you really want to delete custom items. They are removed immediately without confirmation.

Delete a Custom Item

1. Select **View > Customize > Menus and Toolbars.**

   The Menu Editor appears.

2 Select the item that you want to delete.

3. Verify whether the item can be deleted, and change the customization set, if necessary.

4. Right-click the selected item and select **Delete.**

   The item is immediately deleted.

5. Click **Save.**

   Your changes are saved.

Show and Hide Items

In the Menu Editor, you show or hide toolbars, buttons, menus, and menu items in specific customization sets. The item's Source determines where you show or hide the item. For example:

- When the Source is *All Users*, change the customization set to **All users** to hide the item from all users.
- When the Source is *JMP Add-In*, change the customization set to **JMP Add-In** followed by the name of the add-in. The item will be shown or hidden in all sets that include the selected JMP add-in.

If you do not change the customization set, the item is shown or hidden only in the currently selected set. See "Change Customization Sets," p. 335 for details about customization sets.

Note: To quickly hide or show toolbars for the current user, select or deselect them from the **View > Toolbars** list.

Show and Hide Items in a Customization Set

1. Select **View > Customize > Menus and Toolbars.**

   The Menu Editor appears.

2 Select the customization set that you want to modify.

3 Select the item that you want to show or hide.

4. Select or deselect **Hidden** in the General properties. See "Step 2: Specify the Caption and Internal Name," p. 339 for details about the General properties.

5. Click **Save** to save your changes.

Import Customizations

Changes to toolbars and menus are stored in customization sets, or plain text files. You open a customization set in JMP to import your modified toolbars and menus. The customizations are then shown in the Menu Editor.

Import Customizations

1. Select **File > Open**.
2. In the **File name** list (or the **Files of type** list in Windows XP), select **All JMP Files** (or **JMP Menu Files**).
3. Click **Open**.

   A confirmation is displayed, stating that the customizations have been imported. The Menu Editor is also opened.
4. Display the Menu Editor and click **OK** to save the changes, or click **Cancel** to discard them.

Remove Customizations

As you modify items, you discard unsaved changes by clicking the Menu Editor's **Cancel** button and clicking **Yes** to confirm. After saving customizations, you can also remove all customizations and revert to the original menus and toolbars.

Remove All of the Current User's Customizations

1. Select **View > Customize > Revert to Factory Defaults**.
2. Click **Yes** to remove the current user's customizations.

   When JMP finds customizations from a previous version of JMP, a confirmation appears. Do one of the following:

   – Select **No** to avoid adding those customizations to the current installation of JMP.

   – Select **Yes** to add those customizations to the current installation of JMP.

Note: Select **Do not ask me again about merging my old menus** for JMP to ignore customizations found in a previous JMP version *every time* you open JMP. However, when you revert toolbars and menus to the factory defaults, you always have the opportunity to merge old menus, whether you previously checked the merging old menus option.

Remove Customizations from Another Customization Set

1. Select **View > Customize > Menus and Toolbars**.

   The Menu Editor appears.
2. Click **Change**.
3. Select the customization set, and click **OK**.

4. Click **Revert All.**

   A confirmation window appears.

5. Click **Yes** to remove your customizations.

Remove Customizations from the Selected Item

1. Select **View > Customize > Menus and Toolbars.**

   The Menu Editor appears.

2. (Optional) Select the customization set that contains the item.
3. Select the item that you want to modify.
4. Do one of the following:
   – To restore the *previous* properties of an *unsaved* item, click **Reset.**
   – To restore the *original* properties of a *built-in* item, click **Revert All**, and then click **Yes** to confirm.
5. Click **Save** to save your changes.

Personalize Toolbars (Macintosh)

On the Macintosh, you can set up toolbars to display only the icons that you need. Icons are available based on the type of window. For example, in a script window, icons for options such as reformatting, encrypting, and running the script are available. Table icons are available only for data table toolbars.

You can specify whether the icon, icon and icon name, or just the name appear on the toolbar. Displaying small icons is also an option.

To Add, Remove, or Rearrange Toolbar Icons

1. Open the type of window whose toolbar you want to customize.
2. Select **View > Customize Toolbar.**

   A window appears that shows icons relevant to the current type of window.

3. Do any of the following:
   – To add an icon to the toolbar, drag the icon from the window onto the toolbar.
   – To move an icon on the toolbar, drag the icon to its new location onto the toolbar.
   – To remove an icon from the toolbar, drag the icon from the toolbar onto the window.

   Your changes are applied to the current window and all open windows of the same type.

Figure 10.14 Add, Rearrange, and Remove Toolbar Icons on the Macintosh

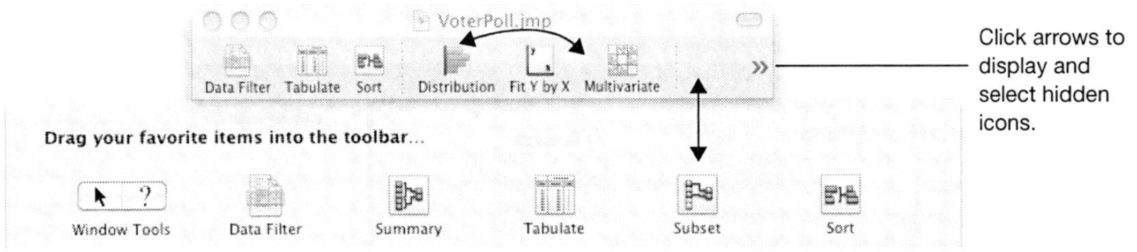

4. Click **Done**.

   When a toolbar is too wide to display completely, click the right arrows on the toolbar to show and then select other icons. Alternatively, you can click and drag the window until the entire toolbar appears.

Note: To display the original set of toolbar icons, drag the default set from the bottom of the toolbar customization window to the toolbar.

Change the Appearance of Toolbar Icons

1. Open the type of window whose toolbar you want to customize.
2. Select **View > Customize Toolbar**.

   A window containing toolbar icons appears.
3. At the bottom of the window, select one of the following options from the Show list:
   - To display only an icon, select **Icon**.
   - To display an icon and its name, select **Icon and Text.**
   - To display only the icon name, select **Text**.
4. To display small or standard icons, select or deselect **Use small size**.

   Your changes are applied to the current window and all open windows of the same type.
5. Click **Done**.

JMP Add-Ins

Experienced JSL script writers can create scripts that extend JMP in many ways. For example, add a custom analytical tool, new JSL functions, or a user interface to communicate with a database. The JMP add-in architecture makes deploying and using these complicated scripts simpler.

You can send co-workers a suite of scripts and tell them how to run them. Another option is sending a single add-in file that any JMP user can install and then use just like any other part of JMP.

As a JMP user, you might be given add-ins to use by co-workers. You can also find add-ins on the JMP Web site at http://www.jmp.com/addins.

Manage JMP Add-Ins

Install Add-Ins

A JMP add-in is a file that has the .jmpaddin extension. You can install it in one of two ways:

1. Select **File > Open**.
2. Navigate to the .jmpaddin file and select it.
3. Click **Open**.

Or, you can double-click the .jmpaddin file.

View Your Add-Ins

Select **View > Add-Ins** to see the add-ins that you have already installed.

Update Add-Ins

If you are given an update to an add-in you already have installed, you can just install the update as you did the original add-in. Doing so overwrites the old version with the new one.

Disable, Enable, and Remove Add-Ins

Temporarily disable an add-in without removing it

1. Select **View > Add-Ins**.
2. Select the add-in that you want to disable.
3. Deselect the **Enabled** check box.

Enable a disabled add-in

1. Select **View > Add-Ins.**
2. Select the add-in that you want to enable.
3. Select the **Enabled** check box.

Remove an add-in completely

1. Select **View > Add-Ins.**
2. Select the add-in that you want to remove.
3. Click **Unregister**.

Create JMP Add-Ins

Developing add-ins requires experience with JMP scripting, and the process is discussed at length in the *Scripting Guide*. Here is the basic workflow for creating an add-in:

1. Create a working folder.

   This folder contains all the files that are necessary for your add-in.

2. Create a text file that is named addin.def.

   This file contains the official, registered name of your add-in, and the name that you want your users to see. See the "Extending JMP" chapter in the *Scripting Guide* for more details about naming conventions and the file specification for addin.def.

Note: Each JMP add-in *must* have a unique ID. This is a name that is rarely seen by the end user of an add-in. We strongly recommend that you use the reverse DNS naming convention to create this unique name. For example, com.jmp.developer.myaddin.

3. Register your add-in by opening your addin.def file in JMP.

   Although your add-in has no functionality yet, developing and debugging your add-in is easier if it is registered just like it will be for your users. Also, you can create a custom menu for your add-in only if it is registered first.

   Registering an add-in saves the information about your add-in so that JMP knows where it is, what it is, where to place it on the menu, and so on.

4. Write your scripts and create anything else you might need (for example, data tables, DLLs, images, and so forth). Place everything in your add-in folder.

5. (Optional) If you want your users to be able to run your add-in from the JMP menu, create a custom menu for it. The menu is saved as a .jmpcust files that you bundle with your add-in. See "Personalize Toolbars and Menus (Windows Only)," p. 335 for information about customizing menus.

Note: If you are developing your add-in on a Macintosh, you must create your custom menu on Windows.

6. When your add-in is finished, compress your add-in folder into a .zip file.

7. Change the .zip extension to .jmpaddin.

Chapter 11

External Data and Analytical Sources

Connecting to and Working with Other Applications

From JMP, you can use other data sources and even other analytical applications directly:

- SAS
- R
- Excel
- databases

Contents

Working with SAS

You can connect to a SAS server and work directly with SAS data sets:

- Import whole SAS data sets or portions of data sets
- Make changes to imported SAS data in JMP and then export those changes as a SAS data set
- Run stored processes
- Submit SAS code from JMP

Access SAS options from the **File > SAS** menu:

Browse Data Browse and import data residing on a SAS Server.

Export Data to SAS Export JMP data tables to a SAS Server.

Browse SAS Folders Browse and run SAS stored processes or open Metadata-defined data tables.

SAS Add-ins Opens a window with links to additional JSL and SAS programs available.

New SAS Program Opens a script window for writing and submitting SAS code.

Open SAS Log Window Opens a SAS log window for the active SAS server.

Open SAS Output Window Opens a SAS output window for the active SAS server. This window shows recent SAS output.

Server Connections Administer connections to SAS servers.

You can also find shortcuts for SAS options on the SAS page of the JMP Starter, and there is a SAS toolbar. You can save certain settings pertaining to SAS Integration on the SAS Integration page of the Preferences window (**File > Preferences**). For more information about setting your SAS Integration preferences, see "SAS Integration," p. 468 in the "JMP Preferences" chapter.

Connecting to SAS

You can either connect to a SAS metadata server or directly to a SAS workspace server. Once connected to a SAS metadata server, you can browse through SAS servers, libraries, and data sets.

Note: The earliest supported release of the SAS metadata server is version 9.1.3 SP4. Connections to earlier releases of the SAS metadata server are experimental and are not supported.

To connect to a SAS server, click **File > SAS > Server Connections**. The SAS Server Connections window in Figure 11.1 appears. All connections are made in this window.

Figure 11.1 SAS Server Connections

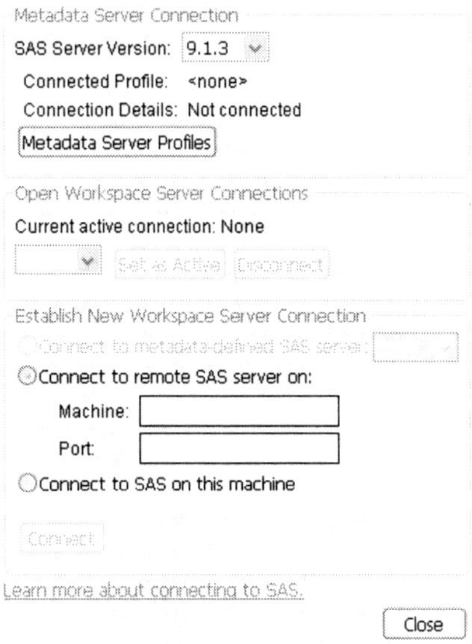

Connect to a SAS Metadata Server

Note: You can be connected to only one metadata server at a time. If you make a second connection, your first one is disconnected for you.

To connect to a metadata server

1. Select the version for the SAS Server. Your SAS metadata server administrator should have this information.
2. Click **Metadata Server Profiles.**

The **Manage Metadata Server Profiles** window in Figure 11.2 appears.

Figure 11.2 Manage Metadata Server Profiles

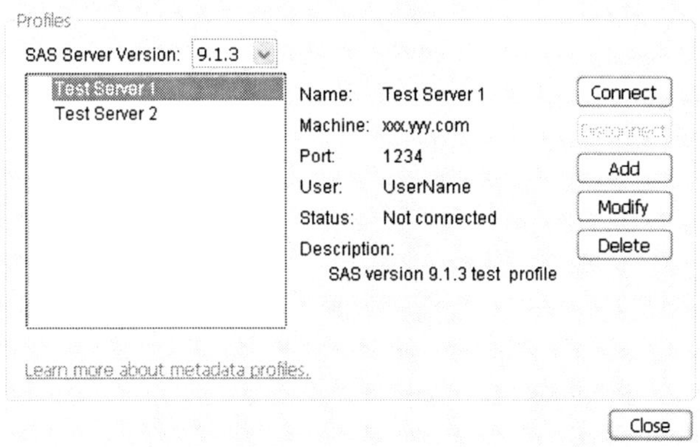

3. Select a profile in the list.

Note: If you need to add a profile, see "To create or modify a metadata server profile," p. 362.

4. Click **Connect**.

   If a connection is established, the Status information changes to **Connected**. If JMP is unable to establish a connection, an error message gives you the reason. Common reasons are invalid user names or passwords. If you need to update the information for the profile, see "To create or modify a metadata server profile," p. 362.

5. Click **Close**.

Once you are connected to a SAS Metadata Server, you can connect to any SAS Workspace Servers that the Metadata Server offers.

To connect to a SAS Workspace Server

1. In the SAS Server Connections window, select the workspace server to connect to. (See Figure 11.3.)

Figure 11.3 Open a Connection to a Workspace Server

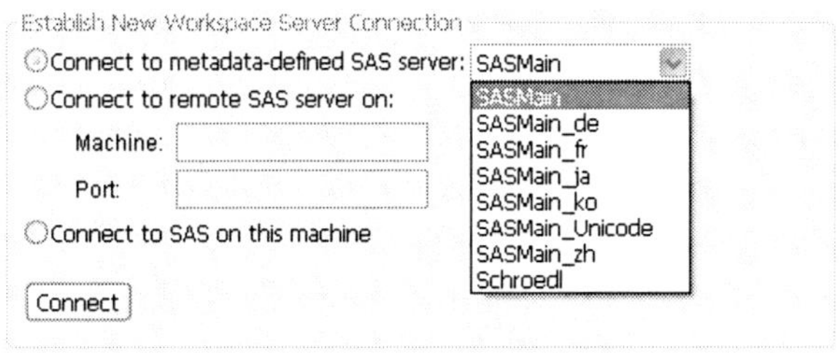

2. Click **Connect**.

   Under Open Workspace Server Connections, the workspace server is shown as the current active connection. See Figure 11.4.

Figure 11.4 Current Active Connection

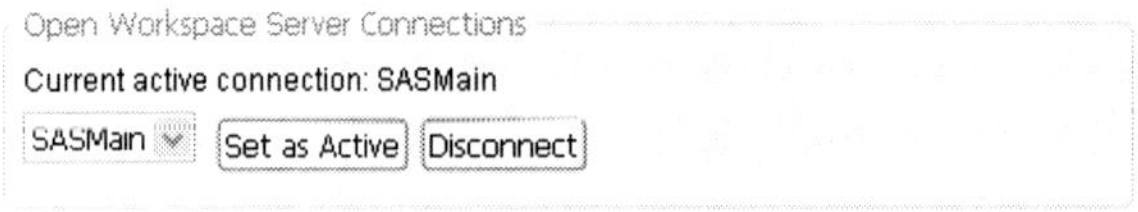

3. Click **Close**.

To change the active connection

Note: The active connection is what is used to submit SAS code or handle SAS script commands.

To change the active connection, you first need to be connected to more than one server. Follow the instructions in "To connect to a SAS Workspace Server," p. 361 to add two or more server connections.

1. In the Open Workspace Server Connections section, click the drop-down menu and select the desired server.
2. Click **Set as Active**.
3. Click **Close**.

Tip: You can change the active server at any time.

To disconnect from a workspace server

1. In the SAS Server Connections window, select the workspace server to disconnect under Open Workspace Server Connections.
2. Click **Disconnect**.

To disconnect from the metadata server

1. In the SAS Server Connections window, click **Metadata Server Profiles**.
2. Select the Metadata Server that you are connected to.
3. Click **Disconnect**.

To create or modify a metadata server profile

1. In the SAS Server Connections window, select the **SAS Server Version**.
2. Click **Metadata Server Profiles**.
3. Click **Add** to add a new profile, or click **Modify** to change a profile's settings.

The Create Profile or Modify Profile window appears. If you are adding a new profile, all fields are empty except the Authentication domain field, which contains DefaultAuth, and the Port field. If you are modifying a profile, the fields contain the current information.

Figure 11.5 Create or Modify a Metadata Server Profile

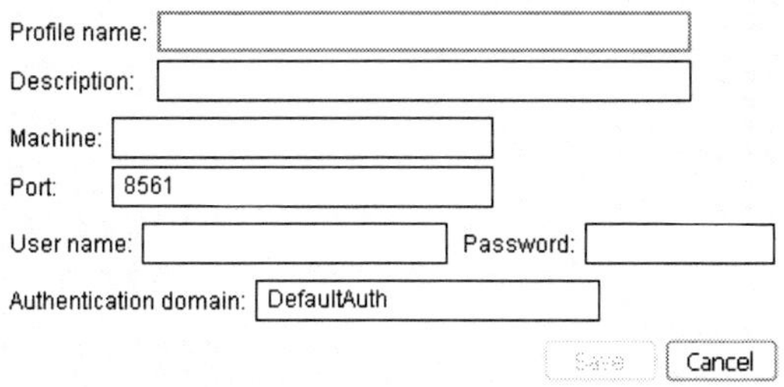

4. Fill in the information needed to connect to a SAS metadata server. Your SAS metadata server administrator should have this information.

   Profile name Select a name for this profile. This name is shown in the list of profiles.

   Description (Optional) You can enter a short description of this profile.

   Machine The name of the machine that hosts the metadata server. (Example: myserver.mycompany.com)

   Port The port through which you should connect to the machine. (Example: 8561)

   User name Your user name for the metadata server.

   Password Your password. This is always displayed as asterisks.

   Authentication domain The domain you, as a user, belong to.

5. Click **Save**.

Connect to a SAS Server on a Remote Machine

You can also connect directly to a SAS workspace server, instead of going through a metadata server.

To connect to a SAS server

1. Open the SAS Server Connections window (**File > SAS > Server Connections**).
2. Under Establish New Workspace Server Connection, select **Connect to remote SAS server on**. See Figure 11.6.

Figure 11.6 Open a Connection to a Remote SAS Server

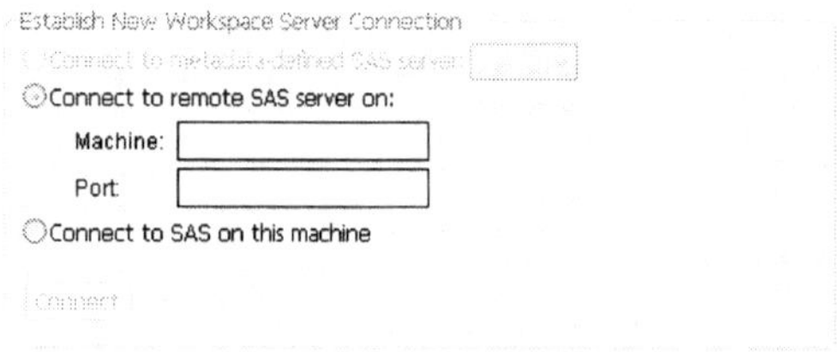

3. Enter the machine name and the port number. Your SAS server administrator has this information.
4. Click **Connect**.
5. Enter your user name and password in the window that appears.
6. Click **OK**.
7. Click **Close** in the SAS Server Connections window.

To disconnect from a SAS server

1. In the SAS Server Connections window, select the server to disconnect under Open Workspace Server Connections.
2. Click **Disconnect**.

Connect to SAS on Your Local Machine (Windows Only)

You can also connect directly to SAS on your local machine.

To connect to SAS on your computer

1. Open the SAS Server Connections window (**File > SAS > Server Connections**).
2. Under Establish New Connection, select **Connect to SAS on this machine**.
3. Click **Connect**.
4. Click **Close** in the SAS Server Connections window.

To disconnect from SAS on your computer

1. In the SAS Server Connections window, select **Local** under Open Connections.
2. Click **Disconnect**.

Opening SAS Data Sets through a SAS Server

Once you connect to a SAS workspace server, you can browse through the SAS libraries on that server and import data into JMP.

Note: On Windows, you can open SAS data sets directly through the **File > Open** command. Another option is connecting to a SAS server by selecting **File > SAS > Browse Data** (as described in this chapter). On Macintosh, you must connect through a SAS server first. For details about opening SAS data sets without connecting to a SAS server, see "Open SAS Data Sets," p. 39 in the "Getting Started" chapter.

To browse the data sets on the SAS server, select **File > SAS > Browse Data.** The Browse SAS Data window appears. See Figure 11.7.

Figure 11.7 Browse SAS Data

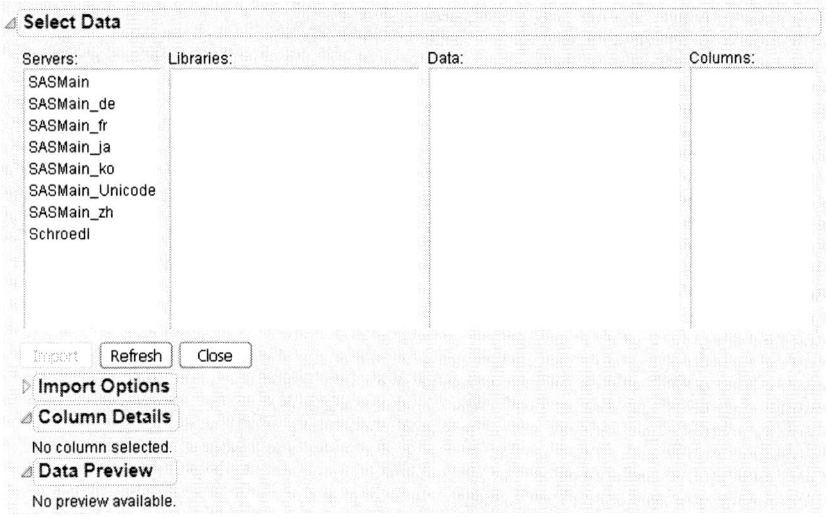

The window is initially populated with a list of servers the SAS metadata server provides (if connected). Any physical and local connections are also shown (as listed in Figure 11.6 "Open a Connection to a Remote SAS Server," p. 364).

- Select a server to see a list of libraries that server contains.
- Select a library to see a list of data sets within that library.
- Select a data set to see a list of columns within that data set.

At any time, you can select a different data set, library, or server.

Tip: If a server is unavailable, or if the connections failed, the server's name is shown in light, italic text. Click it to try to re-establish the connection.

Browsing SAS Data Information

You can select a SAS data set and see information about its contents before opening it using Column Details and Data Preview.

Data Preview

When you select a data set, the Data Preview outline shows you the first ten rows and columns in the data set. See Figure 11.8.

Figure 11.8 Data Preview

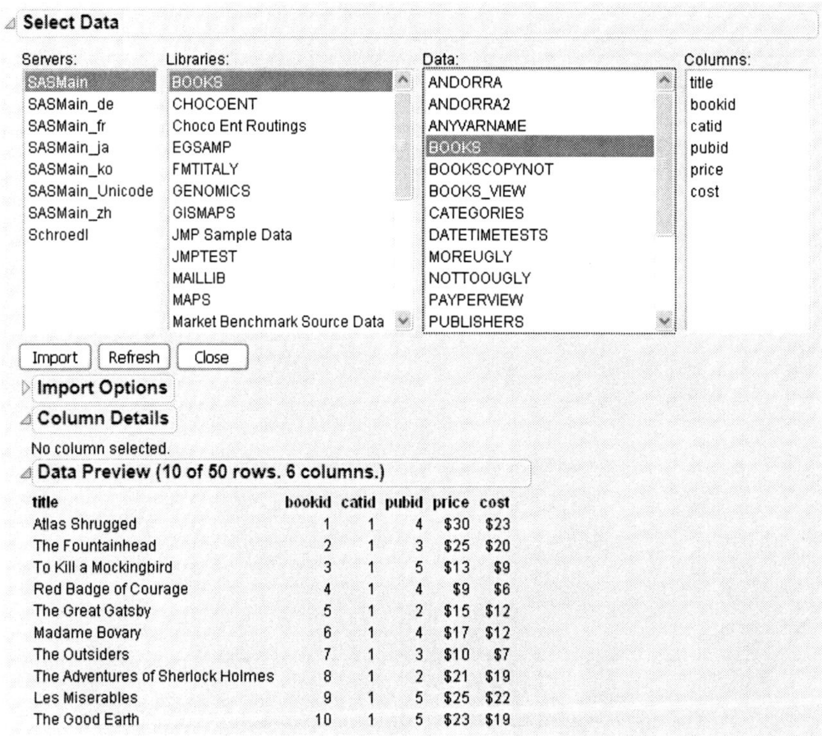

Column Details

To see information about a particular column in the data set, select it. The Column Details outline shows you some basic information about the data column. See Figure 11.9.

Figure 11.9 Column Details

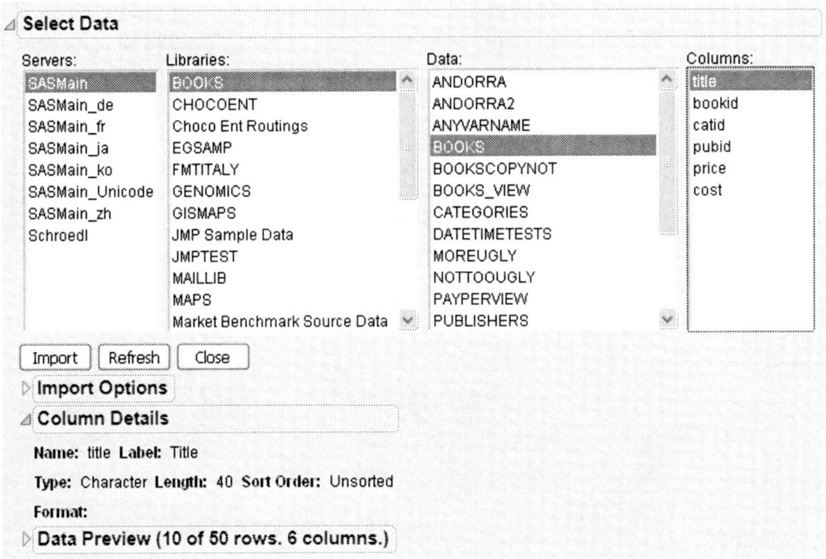

Name Column name from the SAS data set.

Label Descriptive column label. The label can be longer than the name, and is often helpful to determine what the column name means.

Type Specifies whether the column has a character or numeric data type.

Length The length in bytes of data in the column.

Sort Order How the column is sorted in SAS.

Format The format for the SAS column, such as DOLLAR. This format field also contains information about the width of formatted values and the number of decimal places.

Open a SAS Data Set in JMP

You can import SAS data sets directly into JMP.

1. Select a data set.
2. Click **Import.**

The SAS data set is imported into a JMP data table. When SAS data is imported, JMP attempts to make the best match to the SAS format.

If you want to import only a portion of a data set, you can do any of the following:

- Select a subset of the columns. See "To select a subset of columns," p. 368.
- Construct a WHERE clause to filter the data. See "To import using a WHERE clause," p. 369.
- Take a random sample of the data. See "Importing a random sample of the data," p. 369.

To select a subset of columns

1. Open the Import Options outline. See Figure 11.10.

Figure 11.10 Import Options

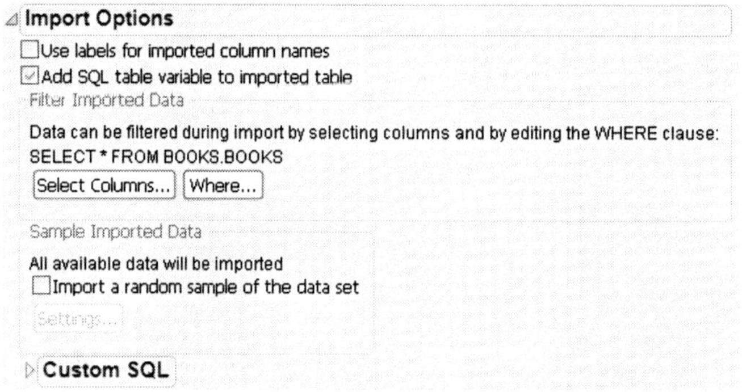

2. Click **Select Columns.**

   The Select Columns window appears. See Figure 11.11.

Figure 11.11 Select Columns

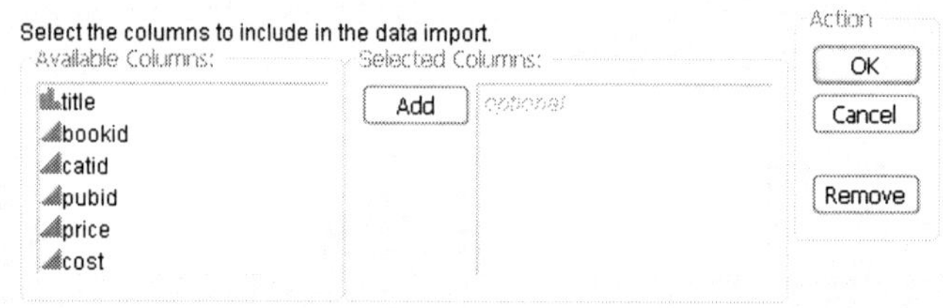

3. Select the columns that you want to import.

   To select more than one column at a time, Control-click (Windows) or ⌘-click (Macintosh) each column.

4. Click **Add**.
5. When you have added all the columns that you want, click **OK.**
6. In the Browse SAS Data window, click **Import**.

Only the columns that you selected from the SAS data set are imported into a JMP data table.

To import using a WHERE clause

1. Click **Where**.
2. Use the WHERE clause editor to construct your WHERE clause.
3. Click **OK** to return to the Browse SAS Data window.
4. Click **Import**.

Only the data that matches your WHERE clause are imported into a JMP data table.

For information about constructing WHERE clauses and using the WHERE clause editor, see "Using the WHERE Clause Editor," p. 387.

Note: If you import data using both a WHERE clause and sampling, the WHERE clause is applied first, and then a sample of the filtered data is taken.

You can also write your own SQL statements.

Custom SQL

You can also open a SAS data set using a custom SQL statement.

1. Open the Custom SQL outline under the Import Options outline. See Figure 11.10.

Figure 11.12 Custom SQL

2. Enter your SQL statement in the window.
3. Click **Execute Custom SQL**.

Note: Your SQL is run on the selected server but is not restricted to any selected library or data set.

Importing a random sample of the data

You can also import a random sample of the rows of the SAS data set.

Note: The sampling feature requires that the SAS server has the SAS/STAT product licensed and installed. If SAS/STAT is not present, sampling is disabled.

In the Sample Imported Data area of the Import Options outline, select the **Import a random sample of the data set** check box. By default, 5% of the rows are imported. To change the random sample import settings, click the **Settings** button.

Figure 11.13 Sampling Settings

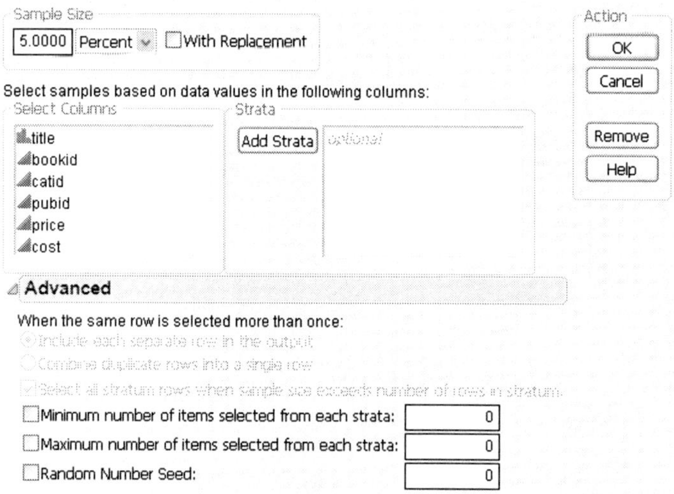

In this window, you specify any of the following:

Sample Size You can set the sample size be percentage or by number of rows. To ensure that each row is sampled only once, de-select the **With Replacement** option. To ensure that any row can be sampled and appear more than once in the imported data, select the option.

Selecting by Column You can select strata by moving columns into the Strata list.

Handling Multiple Row Sampling If **With replacement** is selected, you can specify to either add each duplicated row as a separate row or combine all duplicated rows into one row. If the second option is selected, a column is added to the table that contains a count of how many times each row was sampled.

Setting minimum and maximum numbers of items selected Select the option and enter a number.

Setting the random number seed Select the option and enter a seed. Specifying the seed lets you reproduce the exact same sample multiple times.

Note: If you import data using both a WHERE clause and sampling, the WHERE clause is applied first, and then a sample of the filtered data is taken.

Import Options

There are additional options you can use to specify how SAS data is imported into JMP.

Use labels for imported column names When selected, this option switches the column name, which has a limited length and might be difficult to decipher, with the column label. This option is turned off by default. To use the SAS data column names as column names in JMP, uncheck this box.

Add SQL table variable to imported table When selected, this option adds SQL queries to the data table panel as a variable. This option is turned on by default. If you turn off this option, only two variables are added when you import the data table: the SAS server and the data set.

Tip: If your data is password-protected, you might want to turn this option off, because your password might be shown in the SQL.

Table variables

After you import the JMP data table, table variables appear in the upper left panel of the data table. These variables show the SAS server, data set, and the SQL query and sampling settings if applicable. There is also a source script added that lets you re-do the import at any time.

Opening Password-Protected Data Sets

On Windows, JMP can open SAS version 7 or higher data sets that are password protected.

To open password-protected data sets:

1. Select **File > Open**.
2. Select **SAS Data Sets** from the **Files of type** drop-down menu.
3. Click the box beside **Dataset is Password Protected**.
4. Select the file.
5. Click **Open**.

   A window appears that asks you to type the password and choose the appropriate type of password (**Read**, **Alter**, or **Write**)

Running Stored Processes

Stored processes are SAS DATA step code saved on the SAS server that you are connected to. You can run them from JMP and see the results of the script in JMP.

Note: Depending on the preferences that you have set for SAS, error messages are sent either to the JMP log or to a separate SAS log window.

You must be connected to a metadata server to view and run stored processes. If you select **File > SAS > Browse SAS Folders** without such a connection, you are prompted to either make a connection or cancel your action.

To select and run a stored process

1. Select **File > SAS > Browse SAS Folders.**

   The Browse SAS Folders window appears. See Figure 11.14.

Figure 11.14 Browse SAS Folders

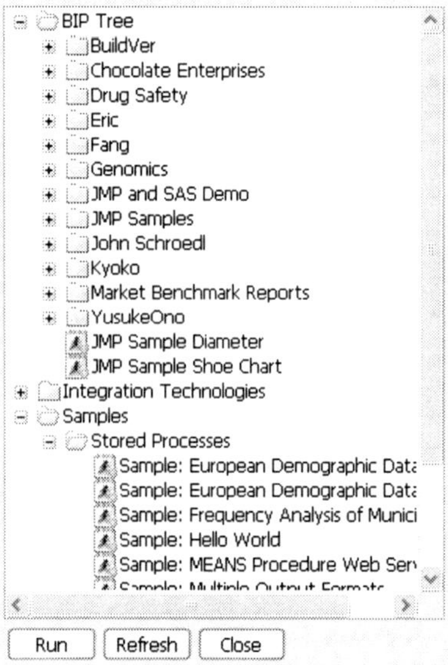

2. Browse through the stored processes to find the one that you want to run.
3. Select it.
4. Click **Run.**

On Windows, you can also right-click a stored process and select **Run.**

On Windows, you can right-click a stored process and select **Copy Metadata Path**, which copies the path to the clipboard. You can then paste it into a script window to include it as a parameter for the JSL operator `Meta Get Stored Process()`. For more information, see the *Scripting Guide.*

Note: Static graphs might not appear in the results returned from a SAS stored process when streaming output is selected. The "Sample:" programs shown in Figure 11.14 are provided for the SAS Information Delivery Portal, and produce streaming output by default. The graphs in these examples might not display correctly in JMP.

Stored processes send reports to HTML by default, but you can select RTF or PDF instead on the SAS Integration page of the JMP preferences. Select **File > Preferences** (Windows) or **JMP > Preferences** (Macintosh) to view the JMP preferences.

Submitting SAS Code

You can submit SAS code directly from JMP to the currently active SAS server. If the submitted SAS code generates SAS Listing output, that output is automatically retrieved from the SAS server and displayed in JMP. Also, the generated SAS Log is retrieved, and, if there are any errors in the submitted code, the SAS Log is automatically displayed in the SAS Log window.

Figure 11.15 SAS Code Submission Example

SAS Code	JMP Code to Submit SAS Code

`data a;`
`x=1;`
 	
`proc print;`
`run;`
	`SAS Connect("sasmain");`
	`SAS Submit("data a; x=1; proc print; run;");`

To run SAS code directly from JMP

1. Either open an existing SAS program using **File > Open**, or create a new SAS program. (Create a new SAS program by selecting **File > SAS > New SAS Program** and typing in the SAS code.)
2. From the **Edit** menu, select **Submit to SAS.**

   You can also right-click in the Program Editor window and select **Submit to SAS**. The menu item also includes the name of the active SAS server that the SAS code will be submitted to.

   You can also press the F8 key (⌘-shift-R on Macintosh).

To run SAS code using a JSL script

Write and run a JSL script that uses either the `SAS Submit` or `SAS Submit File` JSL functions. For more information about writing JSL scripts that submit SAS code, see the *Scripting Guide*.

To view the SAS Listing Output

If the submitted SAS code generates SAS Listing (textual) output, that output is automatically be displayed in a SAS Output window when the job is completed. If you need to view the SAS Listing output again later in the JMP session, select **File > SAS > Open SAS Output Window**. The SAS Output Window retains the listing output from the previous 25 submits to the active SAS server.

To view the SAS Log

If the submitted SAS code contained errors, the SAS Log window for the active SAS server is automatically opened, displaying the SAS Log for the job. However, you can view the SAS Log for the most recent 25 submits to the active server at any time by selecting **File > SAS > Open SAS Log Window**.

If you prefer that SAS Log information is appended to the JMP log after a submit completes:

1. Select **File > Preferences** (Windows) or **JMP > Preferences** (Macintosh).
2. Open the SAS Integration category.
3. In the Show SAS Log section, select **JMP Log** rather than **Separate Window**.

Also in the Show SAS Log section, you can set whether the SAS Log should be displayed **Always**, **Never**, or **On Error** (the default).

Generating ODS Results

The SAS Output Delivery System (ODS) is a powerful mechanism for generating reports in HTML, RTF, PDF, and other formats. ODS output is generally much more attractive and customizable than plain-text SAS Listing output. You can set your submitted SAS code generate ODS results rather than SAS Listing output using Preferences.

To generate ODS results from your submitted SAS code

1. Select **File > Preferences** (Windows) or **JMP > Preferences** (Macintosh).
2. Open the SAS Integration category and find the large SAS Submit Options group, as shown in Figure 11.16.

Figure 11.16 SAS Submit Options in Preferences

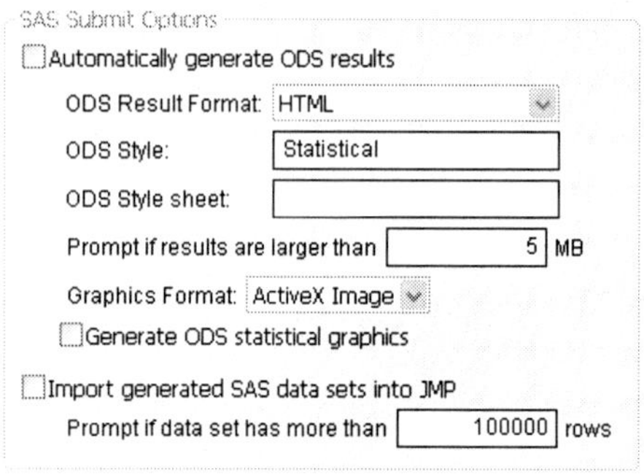

3. Check the **Automatically generate ODS results** option.
4. From the **ODS Result Format** menu, select the format in which to generate the ODS results: HTML, PDF, or RTF.
5. (Optional) You can use other options to specify a style or style sheet to format the results or set the format for generated graphics. For more details, see "SAS Integration," p. 468 in the "JMP Preferences" chapter.

Performing the previous steps causes JMP to generate additional SAS code, including an ODS statement, that is wrapped around the SAS code that you submit. The SAS code that you submit then automatically generates ODS results in the specified format. Those results are downloaded to your computer and displayed either within JMP, when possible, or in an appropriate external application.

Retrieving Generated SAS Data Sets

SAS code that you submit might generate SAS data sets. You can have them automatically imported into JMP for further analysis.

1. Select **File > Preferences** (Windows) or **JMP > Preferences** (Macintosh).
2. Open the SAS Integration category.
3. Select the **Import generated SAS data sets into JMP** option.

Exporting JMP Data Tables to SAS

You can export JMP data tables to a SAS workspace server.

1. Connect to the SAS workspace server.
2. Select **File > SAS > Export Data to SAS.**

   If you are not connected to a SAS workspace server, you are prompted to do so.

Figure 11.17 Export Data to SAS

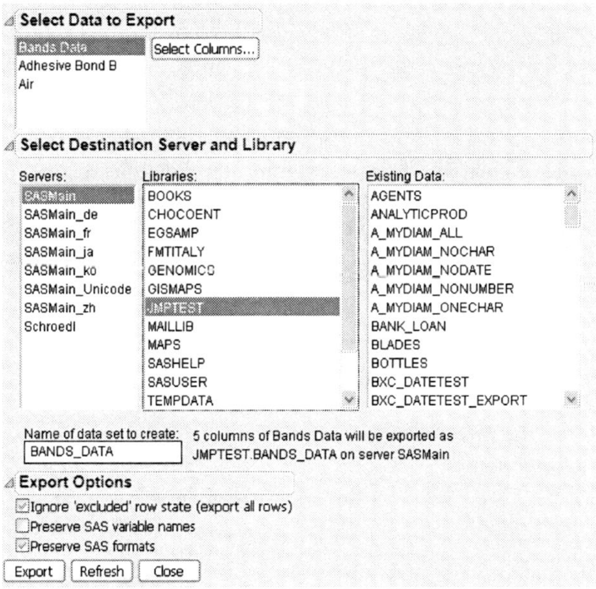

3. Select the data table that you want to export to SAS from the list of open data tables under Select Data to Export.
4. (Optional) To export only some of the columns in the data table, click **Select Columns**. See "Select Columns to Export," p. 376 for details.
5. Select the Destination Server.
6. Select the Library.

   A list of the data sets in the library appears.
7. Enter the name as you want it to appear in the SAS library.
8. (Optional) Set the export options that you want to use. See "Export Options," p. 376 for details.
9. Click **Export**.

Select Columns to Export

1. To export only some of the columns in the data table, click **Select Columns**.
2. In the window that appears, select the columns to export and click **Add**.
3. When all the columns have been added to the Selected Columns list, click **OK**.

Export Options

The available export options are as follows:

Ignore 'excluded' row state (export all rows) Select this option to export all rows in the data table. Deselect this option to export only those rows that are not excluded. This option is on by default.

Preserve SAS variable names This option is useful for data tables that were imported originally from SAS. When importing a SAS data set, the original SAS variable name is saved in a column property for each column. Select this option to use the SAS variable name for each column when exporting to SAS. Deselect this option to export the JMP variable names. This option is off by default.

Preserve SAS formats This option is useful for data tables that were imported originally from SAS. When importing a SAS data set, the original SAS format and informat is saved in a column property for each column. Select this option to use the SAS format and informat for each column when exporting to SAS. Deselect this option to export the JMP formats instead. This option is on by default.

Working with R

JMP provides a scripting interface to R. See the *Scripting Guide* for details.

Working with Excel (Windows Only)

The add-in for Excel provides new capabilities to JMP and Excel users. JMP automatically installs the JMP Add-In for your version of Excel.

- Transfer selected cells in Excel to JMP data tables with one click.
- Use the profiler in JMP with calculation models in Excel workbooks.

Use the **Transfer To JMP** button in the Excel add-in to create a new data table in JMP and populate it with whatever data you selected in your Excel worksheet. Empty cells are brought in as missing data, and dates, numbers, and strings are recognized correctly. In addition, JMP column names can be created automatically, using the preference window to specify how many rows to use as the column name. See "Transferring Excel Data to JMP," p. 379.

The profiler tool is designed to bring the power of the JMP profiler to models that reside in Excel spreadsheets. You do not have to recreate your Excel models in JMP, verify that they are correct, and maintain the model in both JMP and Excel. Use the add-in to specify the input and output cells of the model and save that information within the Excel workbook. Once an Excel file has the model inputs and outputs defined, you can profile the model using the **Excel Profiler** in the **Graph** menu. Information about using the JMP profiler with Excel is in the chapter on Profiling in the *Modeling and Multivariate Methods* book.

Requirements

- JMP 9 running on the Windows XP or later operating system
- Microsoft Excel 2003, Microsoft Excel 2007, or Microsoft Excel 2010

Installing the Excel Add-In

JMP automatically installs the Excel Add-In for your version of Excel. If you have a previous version of this add-in, uninstall it:

1. Open the Windows Add or Remove Programs utility (in the Control Panel).
2. Select **JMP_ExcelAddinSetup** and click **Uninstall.**

Uninstalling the Excel Add-In

1. Open the Windows Add or Remove Programs utility (in the Control Panel).
2. Select **JMP Profiler Core** and click **Uninstall.**
3. Select **JMP Profiler GUI** and click **Uninstall.**

About the JMP Add-In for Excel

The JMP add-in for Excel is installed in the following areas, depending on the version of Excel that you have:

- In Excel 2010, all options are on the JMP ribbon.
- In Excel 2007, all options are on the JMP ribbon.
- In Excel 2003, the menu is in the toolbar on a JMP button that produces a menu of options.

The functionality for all commands is the same for all versions of Excel.

Information about using the JMP profiler with Excel is in the chapter on Profiling in the *Modeling and Multivariate Methods* book.

The JMP Add-In for Excel 2007 and 2010

The JMP add-in commands for Excel 2007 and Excel 2010 are in two groups:

Transfer to JMP

Preferences Set preferences for transferring data from Excel to JMP.

Data Table Transfer the selected data in your Excel file to a JMP data table.

Graph Builder Transfer the selected data in your Excel file to a JMP data table and launch the Graph Builder platform.

Distribution Transfer the selected data in your Excel file to a JMP data table and launch the Distribution platform.

Profile in JMP

Edit/New Model Set up preferences for using the JMP profiler with Excel data.

Run Model Run the JMP profiler.

The JMP Add-In for Excel 2003

The JMP add-in commands for Excel 2003 are on the JMP menu on the toolbar:

Preferences Set preferences for transferring data from Excel to JMP.

Transfer to JMP Transfer the selected data in your Excel file to a JMP data table.

Graph Builder() Transfer the selected data in your Excel file to a JMP data table and launch the Graph Builder platform.

Distribution() Transfer the selected data in your Excel file to a JMP data table and launch the Distribution platform.

Profiler Configuration Set up preferences for using the JMP profiler with Excel data.

Run Profiler Run the JMP profiler.

Transferring Excel Data to JMP

1. In your Excel worksheet, show the JMP add-in on the ribbon. (For Excel 2003, this is not necessary, since the add-in is on the toolbar.)
2. Open the Preferences window by selecting **Preferences**.

Figure 11.18 Preferences for Transfer to JMP

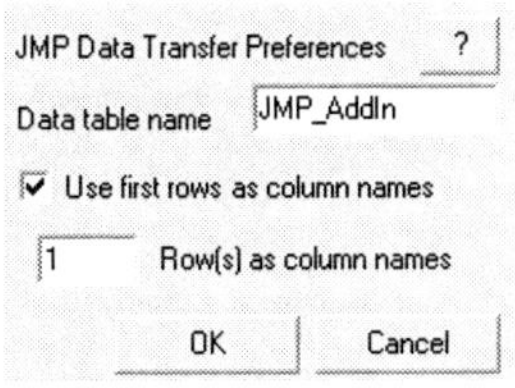

3. Enter the name for the new JMP data table.
4. To use one or more rows of Excel data as JMP columns names, select the **Use First Rows as Column Name** option and enter the number of rows to use.

   The cells that contain column names do not have to be the first rows in the spreadsheet. They need only be the first rows in the group of cells that you select to transfer to JMP.
5. Click **OK**.
6. Select the cells to transfer into JMP, including any cells that you want to use as column names.

   If you are using cells as column names, they need to be the first rows in your selection.
7. Click **Data Table** in Excel 2007 and Excel 2010, or select **Transfer to JMP** in Excel 2003.

If JMP is not running, it is started, and the selected data is placed in a new JMP data table.

8. (Optional) Click **Graph Builder** or **Distribution** to both transfer the data to JMP and to launch either the Graph Builder or Distribution platform.

Note: Your JMP windows might be hidden behind your Excel window, especially if you maximize Excel.

For more information about using Graph Builder and Distribution, see the *Basic Analysis and Graphing* book.

Profiling in JMP

See the chapter on Profiling in the *Modeling and Multivariate Methods* book.

Connecting to Databases

You can import data from a database if you have an ODBC (Open Database Connectivity) driver for the database.

Your operating system provides an interface for JMP to communicate with databases using ODBC data sources. Data sources can be created and configured with operating system software: For example, on Windows XP, use **Control Panel > Administrative Tools > Data Sources**; on the Macintosh, use **Applications > Utilities > ODBC Administrator**.

Note: When you import data from an ODBC database, a table variable is added that might contain user ID and password information. There is a JSL-only preference that can be set to prevent including this possibly sensitive information. See the *Scripting Guide* for more details.

Opening Data from a Database

To import data from a database:

1. Select **File > Database > Open Table** to display the window shown in Figure 11.19. The Connections box lists data sources to which JMP is connected. The Schemas box lists schemas for those databases that support them. The Tables box lists database tables for the currently selected data source connection.

Figure 11.19 Database Open Table Window

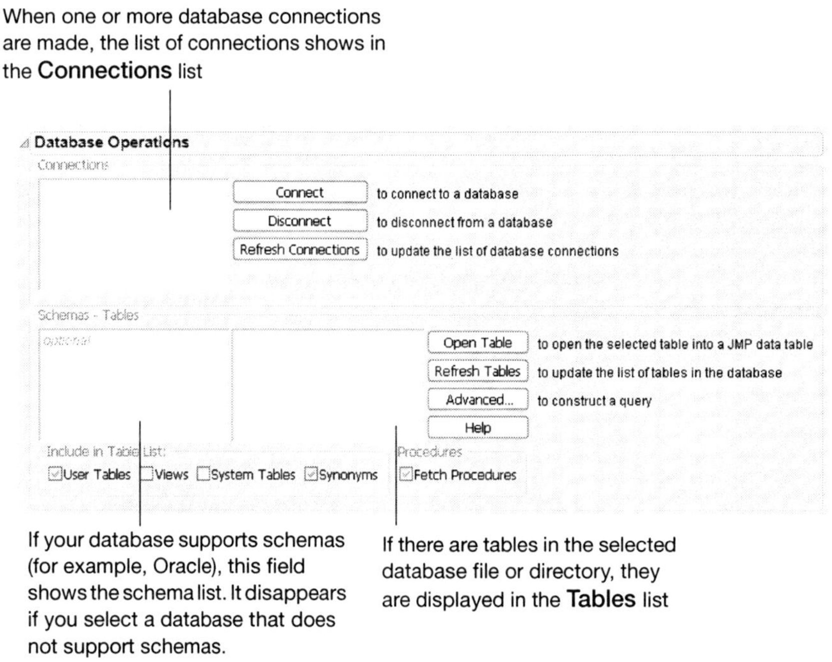

2. If the desired data source is *not* listed in the Connections box, click **Connect** to choose a data source. The method of choosing a data source depends on your operating system. Figure 11.20 shows the data source chooser for Windows. Select a data source and click **OK**.

Figure 11.20 Select a Database Source (Windows)

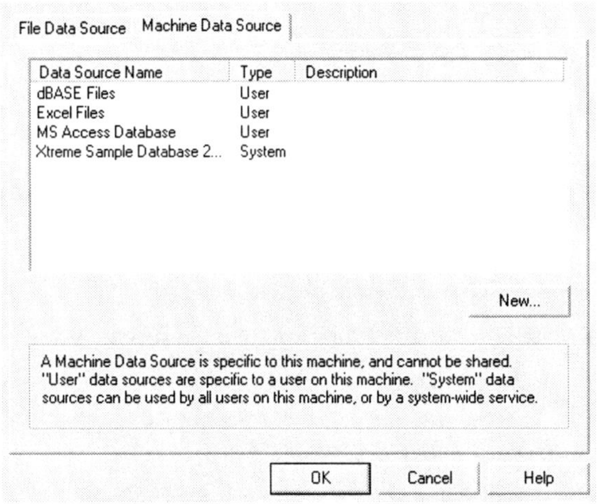

3. Select the desired data source in the Connections box. The tables list in the Tables box updates accordingly. The update might take a several seconds, depending on the number of tables and the speed of the connection to the database. If your database supports schemas, tables are loaded for the first schema in the list, and on other schemas as you click on them.
4. To control which tables are listed by choosing the options in the **Include in Table List** group of check boxes. Different drivers interpret these labels differently. Your options are:

   User Tables When clicked, displays all available user tables in the **Tables** list. User tables are specific to which user is logged on to the computer.

   Views When clicked, displays "views" in the **Tables** list along with all other file types that can be opened. "Views" are virtual tables that are query result sets updated each time you open them. They are used to extract and combine information from one or more tables.

   System Tables When clicked, displays all available system tables in the **Tables** list. System tables are tables that can be used by all users or by a system-wide service.

   Synonyms When clicked, displays all available ORACLE synonyms in the **Tables** list.
5. Select the desired table from the Tables list.

Note: If you are connected to a dBase database, select the database *folder* to which you would like to connect. Individual files are grayed out and cannot be selected.

6. Click **Open Table** to import all the data in the selected table, or click **Advanced** to specify a subset of the table to be imported.

Note: Some databases require that you enter the user ID and password to access the data.

Retrieving Data Using SQL Statements

You can use Structured Query Language (SQL) statements to control what you import from a database. When you open a database file in JMP, you are actually sending an SQL statement to the database. By default, this statement gets all files and records in the database table. In some cases, this is too much data. When you are interested only in a subset of the table's data, you can customize the SQL request to only request the data that you want. After you execute an SQL query, the code for the query is stored in the data table in the SQL table variable.

Note: The SQL Query that you run in this window operates only on the tables and procedures that are displayed in the left panes of the window. Running unrelated SQL here has no results.

To retrieve data using SQL statements:

1. Connect to a database by following the steps in "Opening Data from a Database," p. 380.
2. From the **Database Open Table** window, click the **Advanced** button to open specific subsets of a table.
3. Either type in a valid SQL statement, or modify the default statement. Figure 11.21 shows a default SQL `Select` statement appropriate for the selected file. See "Structured Query Language (SQL): A Reference," p. 383, for a description of SQL statements you can use.

   Alternately, you can add expressions by clicking the **Where** button and using the WHERE Clause editor to create expressions. See "Using the WHERE Clause Editor," p. 387, for details.

Figure 11.21 Reading All Variables From the Solubility Table Stored in an Excel File

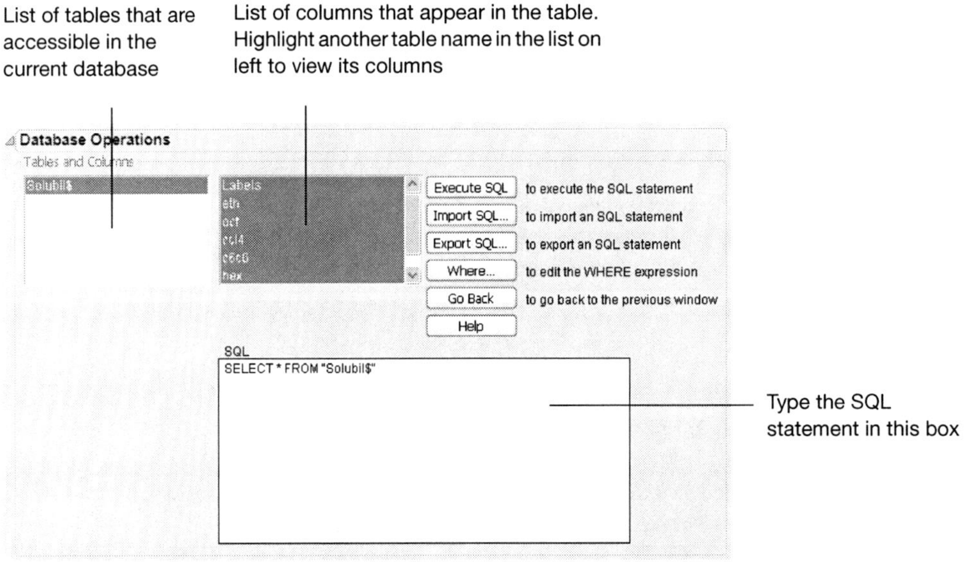

4. Click **Execute SQL**. A JMP data table appears with the columns that you selected. The SQL statement becomes an SQL table variable in the JMP data table. (See "Adding Table Variables," p. 81, for details.)

Note that you can enter any valid SQL statement and click **Execute SQL** to execute the command. Valid SQL varies with the data source and ODBC driver.

Structured Query Language (SQL): A Reference

The following sections are a brief introduction to SQL. They give you insight to the power of queries, and they are not meant to be a comprehensive reference.

Using the SELECT Statement

The fundamental SQL statement in JMP is the SELECT statement. It tells the database which rows to fetch from the data source. When you completed the process in "Opening Data from a Database" on page 380 with the Solubility.jmp sample data table, you were actually sending the following SQL statement to your data source.

```
SELECT * FROM Solubil
```

The * operator is an abbreviation for "all columns." So, this statement sends a request to the database to return all columns from the specified data table.

Rather than returning all rows, you can replace the * with specific column names from the data table. In the case of the Solubility data table example, you could select the ETH, OCT, and CCL4 columns only by submitting this statement:

```
SELECT ETH, OCT, CCL4 FROM Solubil
```

Note: JMP does not require you to end SQL statements with a semicolon.

JMP provides a graphical way of constructing simple SELECT statements without typing actual SQL. To select certain columns from a data source, highlight them in the list of columns (Figure 11.21).

To highlight several rows,

- Shift-click to select a range of column names
- Ctrl-click (Windows) or Command-click (Macintosh) to select individual column names.

Note that the SQL statement changes appropriately with your selections.

Sometimes, you are interested in fetching only unique records from the data source. That is, you want to eliminate duplicate records. To enable this, use the DISTINCT keyword.

```
SELECT DISTINCT ETH, OCT, CCL4 FROM Solubil
```

Sorting Results

You can have the results sorted by one or more fields of the database. Specify the variables to sort by using the ORDER BY command.

```
SELECT * FROM Solubil ORDER BY LABELS
```

selects all fields, with the resulting data table sorted by the LABELS variable. If you want to specify further variables to sort by, add them in a comma-separated list.

```
SELECT * FROM Solubil ORDER BY LABELS, ETH, OCT
```

Using the WHERE Statement

With the WHERE statement, you can fetch certain rows of a data table based on conditions. For example, you might want to select all rows where the column ETH has values greater than 1.

```
SELECT * FROM Solubil$ WHERE ETH > 1
```

The WHERE statement is placed after the FROM statement and can use any of the following logical operators.

Table 11.1 WHERE Operators

Operator	Meaning
=	Equal to
!= or < >	Not equal to
>	Greater than
<	Less Than
>=	Greater than or equal to
<=	Less than or equal to
NOT	Logical NOT
AND	Logical AND
OR	Logical OR

When evaluating conditions, NOT statements are processed for the entire statement first, followed by AND statements, and then OR statements. Therefore

```
SELECT * FROM Solubil WHERE ETH > -2 OR OCT < 1 AND CCL4 > 0
```

is equivalent to

```
SELECT * FROM Solubil WHERE ETH > -2 OR (OCT < 1 AND CCL4 > 0)
```

Using the IN and BETWEEN Statements

To specify a range of values to fetch, use the IN and BETWEEN statements in conjunction with WHERE. IN statements specify a list of values and BETWEEN lets you specify a range of values. For example,

```
SELECT * FROM Solubil  WHERE LABEL IN ('Methanol', 'Ethanol', 'Propanol')
```

fetches all rows that have values of the LABEL column Methanol, Ethanol, or Propanol.

```
SELECT * FROM Solubil WHERE ETH BETWEEN 0 AND 2
```

fetches all rows that have ETH values between 0 and 2.

Using the LIKE Statement

With the `LIKE` statement, you can select values similar to a given string. Use % to represent a string of characters that can take on any value. For example, you might want to select chemicals out of the Solubil data that are alcohols, that is, have the –ol ending. The following SQL statement accomplishes this task.

```
SELECT * FROM Solubil WHERE LABELS LIKE '%OL'
```

The % operator can be placed anywhere in the `LIKE` statement. The following example extracts all rows that have labels starting with M and ending in OL:

```
SELECT * FROM Solubil WHERE LABELS LIKE 'M%OL'
```

Using Aggregate Functions

Aggregate functions are used to fetch summaries of data rather than the data itself. Use any of the following aggregate functions in a SELECT statement.

Table 11.2 SELECT Statement Functions

Function	Meaning
`SUM( )`	Sum of the column
`AVG( )`	Average of the column
`MAX( )`	Maximum of the column
`MIN( )`	Minimum of the column
`COUNT( )`	Number of rows in the column

Some examples include:

- The following statement requests the sum of the ETH and OCT columns:
  `SELECT SUM(ETH), SUM(OCT) FROM Solubil`
- This statement returns the number of rows that have ETH values greater than one:
  `SELECT COUNT(*) FROM Solubil WHERE ETH > 1`

- The following statement lets you know the average OCT value for the data that are alcohols:

  `SELECT AVG(OCT) FROM Solubil WHERE LABELS LIKE '%OL'`

Note: When using aggregate functions, the column names in the resulting JMP data table are Expr1000, Expr1001, and so on. You probably want to rename them after the fetch is completed.

The GROUP BY and HAVING Commands

The `GROUP BY` and `HAVING` commands are especially useful with the aggregate functions. They enable you to execute the aggregate function multiple times based on the value of a field in the data set.

For example, you might want to count the number of records in the data table that have ETH=0, ETH=1, and so on, for each value of ETH.

- `SELECT COUNT(ETH) FROM Solubil GROUP BY (ETH)` returns a single column of data, with each entry corresponding to one level of ETH.
- `SELECT COUNT(ETH) FROM Solubil GROUP BY (ETH) WHERE OCT > 0` does the same thing as the above statement, but only for rows where OCT > 0.

When using GROUP BY with an aggregate function of a column, include the column itself in the SELECT statement. For example,

```
SELECT ETH, COUNT(ETH) FROM Solubil GROUP BY (ETH)
```

returns a column containing the levels of ETH in addition to the counts.

Using Subqueries

Aggregate functions are also useful for computing values to use in a `WHERE` statement. For example, you might want to fetch all values that have greater-than-average values of ETH. In other words, you want to find the average value of ETH, and then select only those records that have values greater than this average. Remember that `SELECT AVG(ETH) FROM Solubil` fetches the average that you are interested in. So, the appropriate SQL command uses this statement in the `WHERE` conditional:

```
SELECT * FROM Solubil WHERE ETH > (SELECT AVG(ETH) FROM Solubil)
```

Saving and Loading SQL Queries

After constructing a query, you might want to repeat the query at a later time. You do not have to hand-type the query each time you want to use it. Instead, you can export the query to an external file. To do this, click the **Export SQL** button in the window shown in Figure 11.21. This brings up a window that lets you save your SQL query as a text file.

To load a saved query, click the **Import SQL** button in the window shown in Figure 11.21. This brings up a window that lets you navigate to your saved query. When you open the query, it is loaded into the window.

Using the WHERE Clause Editor

JMP provides help building `WHERE` clauses for SQL queries during ODBC import. It provides a `WHERE` clause editor that helps you build basic expressions using common SQL features, allowing vendor-specific functions. For example, you do not need to know whether SQL uses '=' or '==' for comparison, or `avg()` or `average()` for averaging.

In addition, string literals should be enclosed by single quotes (`'string'`)rather than double quotes ("`string`").

To open the WHERE Clause Editor:

1. Connect to a database by following the steps in "Opening Data from a Database," p. 380.
2. From the **Database Open Table** window, shown in Figure 11.19, click the **Advanced** button.
3. Click the **Where** button.

USE the WHERE Clause Editor to add any of the following from the work panel: expressions, functions, and terms. They are applied to the highlighted red box.

1. Click the Table Name Browser to select a table. The columns in that table appear in the list.
2. Click the SQL Vendor Name Browser to select the type of SQL that you want to use: GenericSQL, Access, DB2, MySQL, Oracle, SQL Server, or all of the above. Perform an action by clicking a function or operator in the list and selecting an operator from the list that appears.
3. Select an empty formula element in the formula editing area by clicking it. It is selected when there is a red outline around it. All terms within the smallest nesting box relative to the place that you clicked become selected. The subsequent actions apply to those combined elements.
4. Add operators to an expression by clicking buttons on the keypad.
5. (Optional) To customize your WHERE clause, select one of the options from the red triangle menu above the keypad:

   Show Boxing Show or hide boxes around the WHERE clause terms.

   Larger Font Increase the font size of the formula.

   Smaller Font Decrease the font size of the formula.

   Simplify Simply the WHERE clause statement as much as possible.

The WHERE clause editor works similarly to the formula editor, which is described in "Formula Editor," p. 301.

Figure 11.22 The WHERE Clause Editor

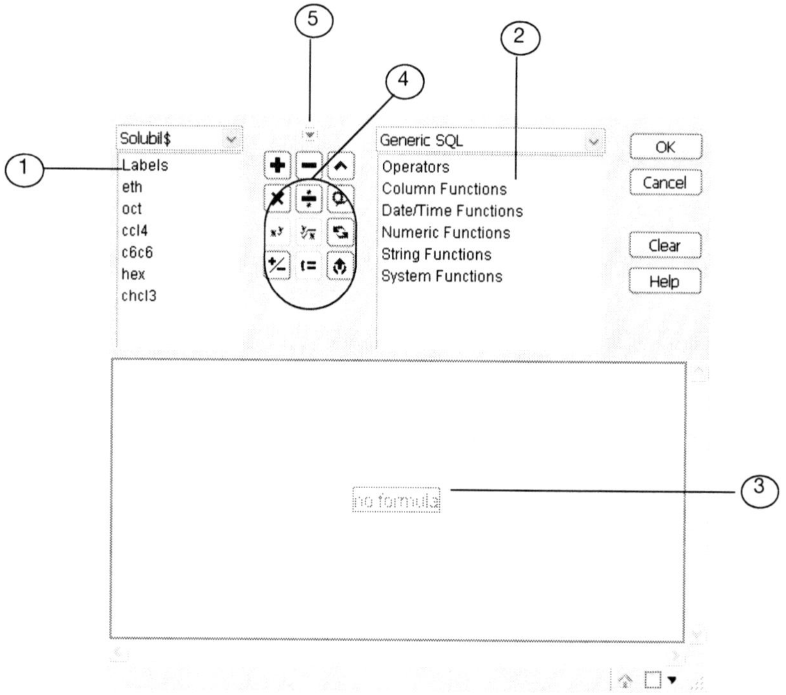

Reading in Real-Time Data (Windows Only)

The term *live data feed* describes the way an external data source sends information via a physical or a logical communication link to another device. You can connect JMP to a live data feed through the serial port of your Windows computer to read a stream of incoming data in real time. Remember the following:

- The data feed must come through a standard nine-pin serial port. Data cannot be read through a USB port unless there is a driver that can simulate a serial port
- You need to know the exact baud rate, parity, stop bits, and data bits for the attached device.

Once you obtain the numbers for your device, enter them into the `Open Datafeed()` command in the script below. (The `4800`, `even`, `2`, and `7` in the script below are examples, so replace them with your information). Then connect the data feed to your computer and open and run the script:

```
streamScript = expr( line = feed <<Get Line;show(line);
             len = length(line); show(len);
   if (length(line)>=1, show("Hi"); show(line);
                  field = substr(line,5,8); show(field);
       x = Num(field); show(x);
       if (!IsMissing(x), current data table()<<add row({:Column1=x});
     show(x);
```

```
   )));
feed = open DataFeed(Baud Rate(4800),parity(even),Stop bits(2), Data bits(7));
feed<<Set Script(streamScript);
feed<<Connect;
```

To ensure harmony between the communications settings for JMP and the instrument reading data from an external source, select **File > Preferences > Communications.** Refer to the documentation for your instrument to find the appropriate settings.

For more details about scripting a datafeed, see the *Scripting Guide*.

Appendix A

JMP Starter

A Review of Categories and Buttons

When the JMP application opens, you see the JMP Starter on Macintosh. On Windows, you can open the JMP Starter at any time through the **View** menu or set you preferences to open the Starter by default.

The JMP Starter, as illustrated below, is a good way to get started if you have not used JMP before. It gives alternative access to most commands found on the main menu or on toolbars.

This chapter gives an overview of the JMP Starter and briefly describes the items and commands on each page of the Starter.

Figure A.1 The JMP Starter

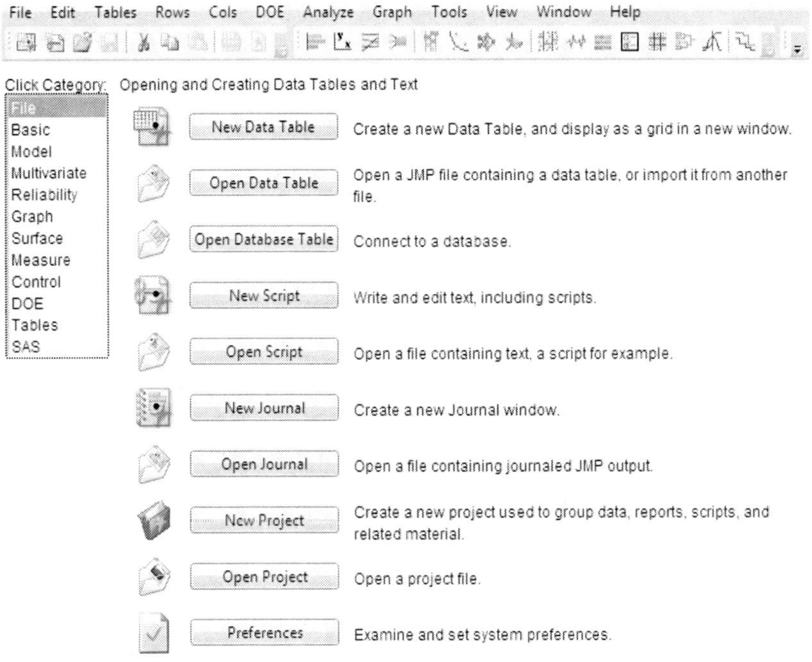

Contents

Overview of the JMP Starter Window

When JMP opens, the JMP Home Window appears behind the Tip of the Day window by default. You can set the JMP preferences to open the JMP Starter window instead. The JMP Starter gives alternative access to most commands found on the main menu or on toolbars. You can close the JMP Starter if you want; it is not required for running JMP.

To open and close the JMP Starter, select **View > JMP Starter** on Windows or **Window > JMP Starter** on Macintosh.

To display the JMP Starter automatically when you open JMP, select **File** (**JMP** on the Macintosh) **> Preferences** and then select **JMP Starter** from the Initial JMP Window drop-down list.

The JMP Starter categories are organized in a way similar to the main menu, as illustrated in Figure A.2. The four categories that follow the **File** category correspond with options available under the **Analyze** menu. They lead you step by step from your data to an appropriate analysis. The next four categories correspond with the **Graph** menu. The next two categories correspond with the **DOE** and **Tables** menus. The last category is also listed in the **File** menu.

Figure A.2 JMP Starter Categories

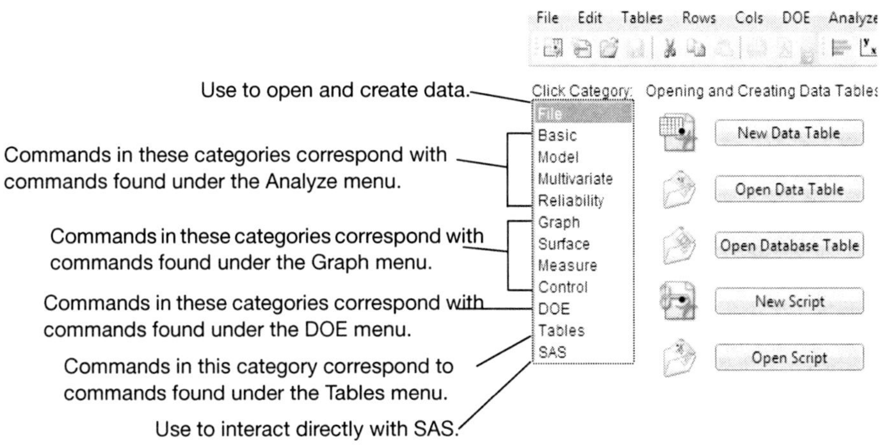

The File Category

The JMP Starter first appears with the **File** category showing, as in Figure A.3. Most commands in the **File** category correspond to **File** menu commands on the main menu bar.

The commands in the **File** category open or create new JMP data tables or other types of JMP windows, which is often what you need to do first.

Figure A.3 JMP Starter File Page

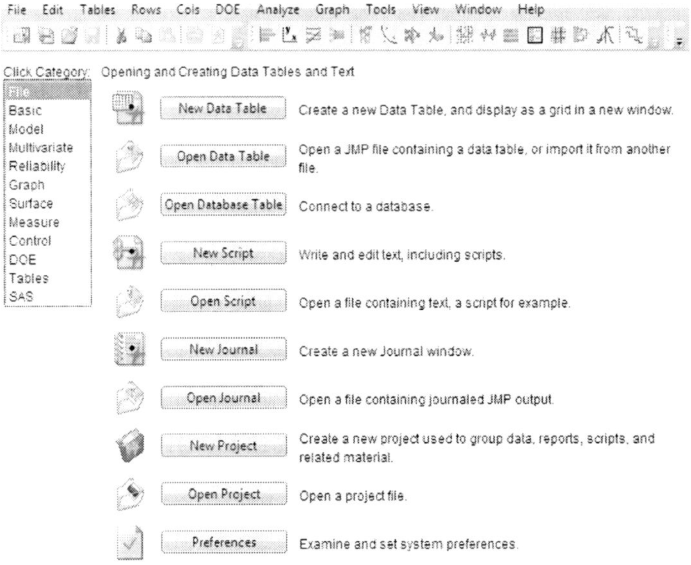

New Data Table Opens an empty untitled data table. See "Create New Data Tables," p. 23, for details.

Open Data Table Lets you select the data table that you want to open and select the file format of the file if necessary. See "Open Existing JMP Files," p. 24 for details.

Open Database Table Opens a connection to any database on your system that has an ODBC driver. See "Opening Data from a Database," p. 380 in the "External Data and Analytical Sources" chapter for details.

New Script Opens a new empty script window, which is an editable text window in which you enter and submit JMP Scripting Language (JSL) commands. See the *Scripting Guide* for details about JSL.

Open Script Displays a filtered open file window to locate the script that you want to open. See "Open Existing JMP Files," p. 24, for details.

New Journal Opens an empty journal window. See "Create Journals," p. 116, for details.

Open Journal Displays a filtered open file window to locate the journal that you want to open. See "Open Existing JMP Files," p. 24, for details.

New Project Opens an empty project. See "Creating a JMP Project," p. 125, for details.

Open Project Displays a filtered open file window to locate the project that you want to open. See "Open Existing JMP Files," p. 24, for details.

Preferences Enables you to specify general and specific settings and save the settings so they are associated with your JMP installation. See "To change preferences," p. 443, for details.

The Basic Category

The **Basic** category (Figure A.4) addresses univariate and bivariate analyses. You see how to examine variables one at a time by looking at distributions and comparing them to known distributions. When there are two variables, a single response (*y*) and a single factor (*x*), JMP performs the appropriate bivariate analysis according to whether the variables are continuous or categorical. These analyses can be run by clicking their buttons on the **Basic** category, as shown in Figure A.4.

Figure A.4 JMP Starter Basic Page

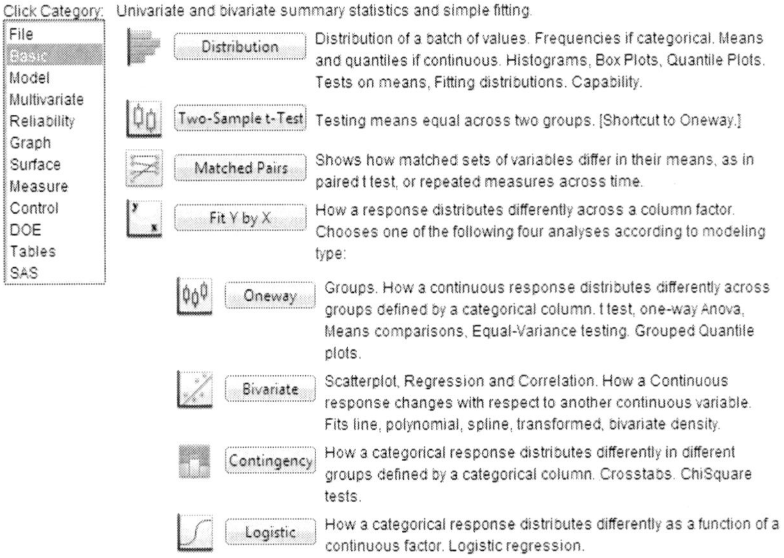

Distribution Clicking the **Distribution** button, or choosing **Analyze > Distribution,** launches the Distribution platform. Distribution reports show the distribution of column values using histograms and other graphical and text reports. See the *Basic Analysis and Graphing* book for details about using this platform.

Two-Sample *t*-Test Clicking the **Two-Sample t-Test** button takes you directly to a window for setting up a one-way analysis. Use this button rather than clicking **Analyze > Fit Y by X**. See the *Basic Analysis and Graphing* book for details about using this platform.

Matched Pairs Clicking the **Matched Pairs** button, or choosing **Analyze > Matched Pairs**, looks at the mean difference between two (or more) columns. It plots the points in a rotated scatterplot that presents them in terms of the differences and means of two variables.

See the *Basic Analysis and Graphing* book for details about using this platform.

Fit Y by X Clicking the **Fit Y by X** button, or choosing **Analyze > Fit Y by X**, launches a report platform for each pair of columns in the active data table that are assigned *x* and *y* roles.

See the *Basic Analysis and Graphing* book for details about using this platform.

The four sub-buttons under **Fit Y by X** lead directly to one of the four models within the Fit Y by X platform.

Oneway If *x* has a nominal or ordinal modeling type and *y* is continuous, clicking the **Oneway** button lets you plot the distribution of *y* values for each discrete value of *x* and computes a one-way analysis of variance table.

Bivariate If both *x* and *y* have continuous modeling types, clicking the **Bivariate** button and making your selections in the window displays a scatterplot.

Contingency If both *x* and *y* have a nominal or ordinal modeling type, clicking the **Contingency** button and making your selections displays a contingency table and a mosaic bar chart.

Logistic If *x* is continuous and *y* is nominal or ordinal, clicking the **Logistic** button and making your selections performs a logistic regression and displays a family of logistic probability curves.

The Model Category

The **Model** category (Figure A.5) gives choices for fitting all types of models—from simple regression and analysis of variance to complex nonlinear fits.

Figure A.5 JMP Starter Model Page

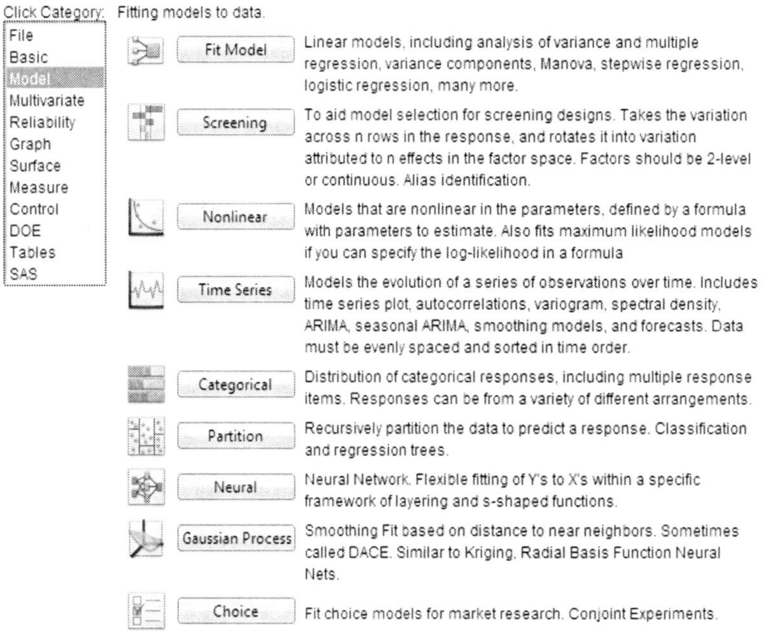

Fit Model Clicking the **Fit Model** button, or choosing **Analyze > Fit Model**, fits one or more *y* variables to a model of *x* variables.

See the *Modeling and Multivariate Methods* book for details about using this platform.

Screening Clicking the **Screening** button, or choosing **Analyze > Modeling > Screening**, helps select a model to fit to a two-level screening design by showing which effects are large.

See the *Modeling and Multivariate Methods* book for details about using this platform.

Nonlinear Clicking the **Nonlinear** button, or choosing **Analyze > Modeling > Nonlinear**, lets you specify nonlinear models, which are models that are nonlinear in their parameters. You orchestrate the fitting process as a coordination of three important parts of JMP: the data table, the formula editor, and the nonlinear fit platform.

Details about the JMP data table and formula editor are in the chapters "Entering, Editing, and Managing Data," p. 45, "Reshaping Data," p. 229, and "Formula Editor," p. 301 in this book.

See the *Modeling and Multivariate Methods* book for details about using this platform.

Time Series Clicking the **Time Series** button, or choosing **Analyze > Modeling > Time Series**, looks at the relationship between a value and its lagged values. The data must be sorted and have equal intervals between time periods. **Time Series** plots the time series and autocorrelations, shows a variogram, and shows the spectral density.

See the *Modeling and Multivariate Methods* book for details about using this platform.

Categorical Clicking the **Categorical** button, choosing **Analyze > Modeling > Categorical**, does tabulation and summarization of categorical response data, including multiple response data, and calculates test statistics.

See the *Modeling and Multivariate Methods* book for details about using this platform.

Partition Clicking the **Partition** button, or choosing **Analyze > Modeling > Partition**, recursively partitions rows into groups according to *x*-values that associate with *y*-values. This partitioning creates a tree of partitions. The factor columns (*x*) can be either continuous or categorical (nominal or ordinal).

See the *Modeling and Multivariate Methods* book for details about using this platform.

Neural Clicking the **Neural** button, or choosing **Analyze > Modeling > Neural**, fits a one- or two-layer neural network. The neural network is a set of nonlinear equations that predict output variables (*y*) from input variables (*x*) in a flexible way using layers of linear regressions and S- and Gaussian-shaped functions. JMP fits the neural net using standard nonlinear least squares regression methods.

See the *Modeling and Multivariate Methods* book for details about using this platform.

Gaussian Process Clicking the **Gaussian Process** button, or choosing **Analyze > Modeling > Gaussian Process**, models the relationship between a continuous response and one or more continuous predictors. The Gaussian Process platform fits a spatial correlation model to the data. The correlation of the response between two observations decreases as the values of the independent variables become more distant.

See the *Modeling and Multivariate Methods* book for details about using this platform.

Choice Clicking the **Choice** button, or choosing **Analyze > Modeling > Choice**, makes probabilistic predictions regarding human choice. It is particularly valuable in marketing research and product design.

See the *Modeling and Multivariate Methods* book for details about using this platform.

The Multivariate Category

The **Multivariate** category provides ways to look at continuous variables when they are considered as responses only; there are no factor or independent variables. Multivariate exploration with correlations and cluster analysis lets you look at many variables at the same time.

Figure A.6 JMP Starter Multivariate Page

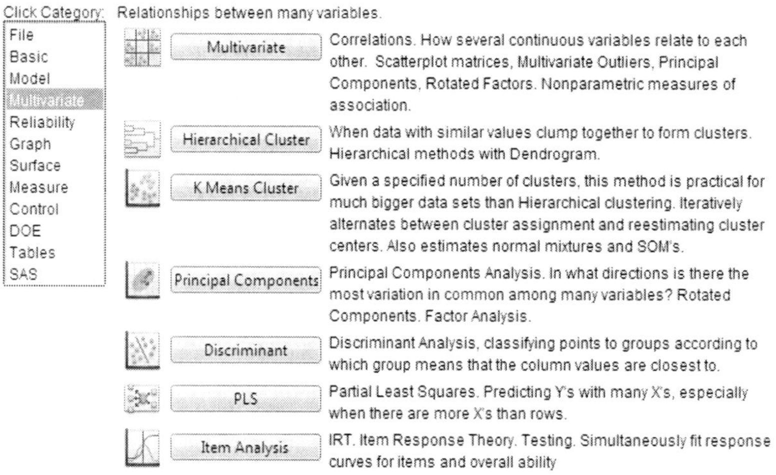

Multivariate Clicking the **Multivariate** button, or choosing **Analyze > Multivariate Methods > Multivariate**, gives you ways to explore how several variables relate to each other. The platform begins with a standard correlation matrix and gives additional options for correlations and other techniques for looking at several variables.

See the *Modeling and Multivariate Methods* book for details about using this platform.

Hierarchical Cluster and K-Means Cluster Clustering is a technique of grouping rows together that share similar values across a number of variables. It is an exploratory technique to help you understand the clumping structure of your data. Clicking the **Hierarchical Cluster** or **K-Means**

Cluster buttons, or choosing **Analyze > Multivariate Methods > Cluster**, lets you select whether to run a K-means cluster or a hierarchical cluster.

- Hierarchical clustering is appropriate for small tables, up to several hundred rows. It combines rows in a hierarchical sequence portrayed as a tree. In JMP, the tree, also called a *dendrogram*, is a dynamically responding graph. You can specify the number of clusters after the tree is built.
- K-means clustering is suitable for larger tables, up to hundreds of thousands of rows. It begins with a cluster seed point. Then the clustering process iterates between assigning points to the number of clusters that you specify and recalculating cluster centers. After the clustering process is complete, you can save the cluster assignments to the data table or use them to set colors and markers for the data table rows.

See the *Modeling and Multivariate Methods* book for details about using this platform.

Principal Components Clicking the **Principal Components** button, or choosing **Analyze > Multivariate Methods > Principal Components**, helps derive a small number of independent linear combinations (principal components) of a set of variables. This captures as much of the variability in the original variables as possible.

See the *Modeling and Multivariate Methods* book for details about using this platform.

Discriminant Clicking the **Discriminant** button, or choosing **Analyze > Multivariate Methods > Discriminant**, starts a discriminant analysis. The technique is based on how close a set of measurement variables is to the multivariate means of the levels being predicted. Also, you could do a stepwise discriminant analysis.

See the *Modeling and Multivariate Methods* book for details about using this platform.

PLS Clicking the **PLS** button, or choosing **Analyze > Multivariate Methods > PLS**, fits models using the partial least squares (PLS) method. This balances the objectives of explaining response variation and explaining predictor variation. The PLS techniques work by extracting successive linear combinations of the predictors, called *factors* (also called *components* or *latent vectors*) that address one or both of these two goals. The PLS platform in JMP also lets you select the number of extracted factors by cross validation. This involves fitting the model to part of the data and minimizing the prediction error for the unfitted part.

See the *Modeling and Multivariate Methods* book for details about using this platform.

Item Analysis Clicking the **Item Analysis** button, or choosing **Analyze > Multivariate Methods > Item Analysis**, provides analysis of test items using the Item Response Theory.

See the *Modeling and Multivariate Methods* book for details about using this platform.

The Reliability Category

Reliability data contain duration times until the occurrence of a specific event and are sometimes referred to as event-time or survival data. In survival data, the event can be failure, such as the failure of an engine or death of a patient.

Figure A.7 JMP Starter Reliability Page

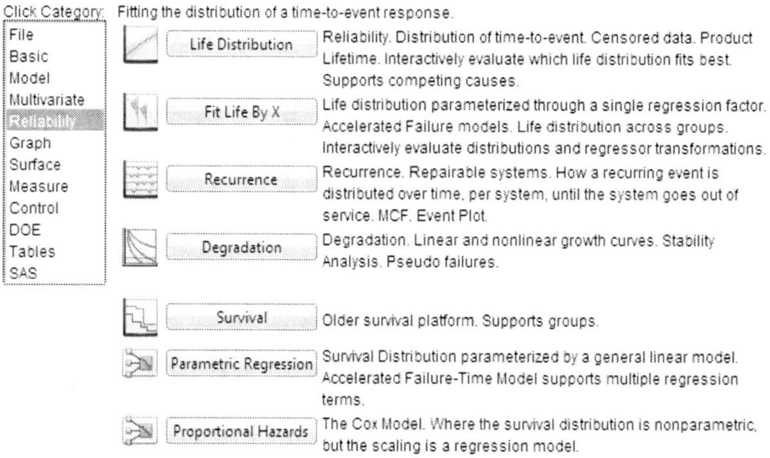

The buttons on the **Reliability** category, also found under **Analyze > Reliability and Survival**, can help you analyze survival data several ways:

Life Distribution Is used to find the most suited distributional fit for your data, and to make predictions. Weibull, Lognormal, Fréchet, Extreme Value, and other common distributions are included.

See the *Quality and Reliability Methods* book for details about using this platform.

Fit Life By X Is used to analyze lifetime events when only one factor is present. Transformations include Arrhenius, Erying, voltage, and linear. This platform also enables you to create a custom transformation of your data. You can also compare different distributions at the same factor level or the same distribution across different factor levels.

See the *Quality and Reliability Methods* book for details about using this platform.

Recurrence Starts an analysis that looks at the age of a system when it requires a repair. A system can have multiple repairs, each with its associated age, and is followed until it is no longer in service. A typical system might be some component of an engine or appliance.

See the *Quality and Reliability Methods* book for details about using this platform.

Degradation Analyzes degradation over time to help predict failure rates.

See the *Quality and Reliability Methods* book for details about using this platform.

Survival (Univariate survival analysis) is an older survival platform for a univariate survival analysis using product-limit (Kaplan-Meier) life table survival computations with estimation of Weibull, lognormal, and exponential parameters.

See the *Quality and Reliability Methods* book for details about using this platform.

Parametric Regression Is used for an analysis that tests the fit of an exponential, Weibull, or lognormal distribution. The accelerated failure-time model supports multiple regression terms.

See the *Quality and Reliability Methods* book for details about using this platform.

Proportional Hazards Is used for a survival analysis that fits a Cox model.

See the *Quality and Reliability Methods* book for details about using this platform.

Note: You can also use the nonlinear platform to handle nonlinear models with loss functions for other parametric survival modeling.

The Graph Category

The **Graph** category, shown in Figure A.8, corresponds with commands in the main menu. These commands produce plots and charts of summarized data, a three-dimensional plots, contour plots, and ternary plots.

Figure A.8 JMP Starter Graph Page

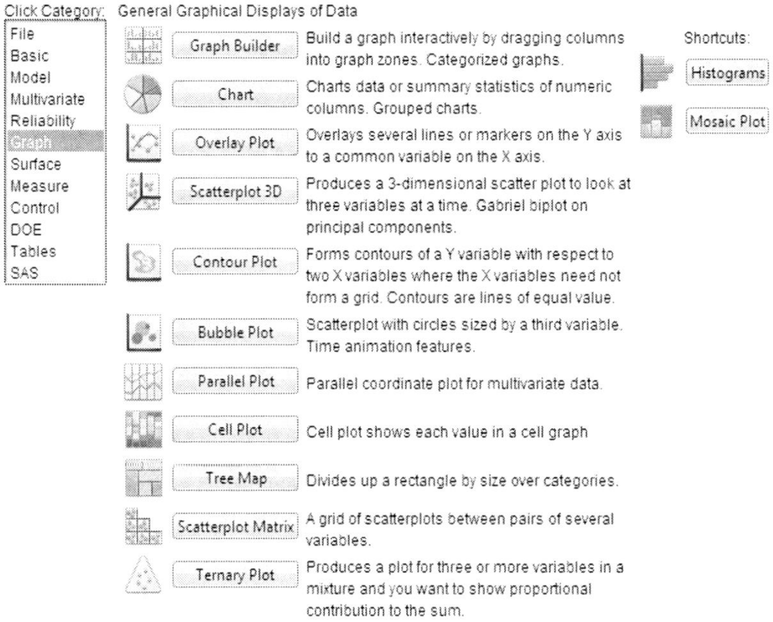

Graph Builder Clicking the **Graph builder** button (or choosing **Graph > Graph Builder**) enables the interactive display of both continuous and categorical data. Its graphs are built through drag and drop. Each drag shows a preview of the resulting report, enabling quick visualization of many chart types. A variety of graph elements are available (scatterplots, box plots, histograms, trend lines, bar charts, smoothers for studying univariate and multivariate relationships). Data can be partitioned by both continuous and categorical variables to easily see interactions.

See the *Basic Analysis and Graphing* book for details about using this platform.

Chart Clicking the **Chart** button (or choosing **Graph > Chart**) and making your selections gives a chart for every numeric *y* specified where the *y*'s are statistics to chart. The *x*-values are always treated as discrete values. **Chart** automatically updates when values change in the current data table and you then activate the chart window.

See the *Basic Analysis and Graphing* book for details about using this platform.

Overlay Plot Clicking the **Overlay Plot** button (or choosing **Graph > Overlay Plot**) and selecting options shows an overlaid plot of a single numeric or categorical *x* column and all numeric *y* variables. The plots for each *y* can be shown separately, with or without a common *x*-axis. Plots can be modified with range and needle options, color, log axes, and grid lines. Overlay plots automatically update when you change values in the current data table and then activate the overlay plot report window.

See the *Basic Analysis and Graphing* book for details about using this platform.

Scatterplot 3D Clicking the **Scatterplot 3D** button (or choosing **Graph > Scatterplot 3D**), displays a three-dimensional view of data and an approximation of higher dimensions through principal components. The plot is a rotatable display of the values of numeric columns in the current data table. The Scatterplot 3D platform displays three variables at a time from the columns that you select.

To help capture and visualize variation in higher dimensions, the Scatterplot 3D platform displays a biplot representation of the points and variables when you request principal components.

See the *Basic Analysis and Graphing* book for details about using this platform.

Contour Plot Clicking the **Contour Plot** button (or choosing **Graph > Contour Plot**) and selecting options constructs a contour plot for a response variable, *y*, for the values of two *x* variables. A contour plot requires two variables for the *x*- and *y*-axes and a third variable (denoted *y* in the Contour Plot window) for contours.

See the *Basic Analysis and Graphing* book for details about using this platform.

Parallel Plot Clicking the **Parallel Plot** button (or choosing **Graph > Parallel Plot**) draws a parallel coordinate plot, which shows connected line segments representing each row of a data table.

See the *Basic Analysis and Graphing* book for details about using this platform.

Cell Plot Clicking the **Cell Plot** button (or choosing **Graph > Cell Plot**) produces a rectangular array of cells drawn with a one-to-one correspondence to data table values.

See the *Basic Analysis and Graphing* book for details about using this platform.

Tree Map Clicking the **Tree Map** button (or choosing **Graph > Tree Map**) shows tree maps. Tree maps are similar to bar charts that have been folded over in two dimensions, leaving no unused space. It is useful when there are a lot of categories. Tree maps are a graphical technique of observing patterns among groups that have many levels. They are especially useful in cases where histograms are ineffective.

See the *Basic Analysis and Graphing* book for details about using this platform.

Bubble Plot Clicking the **Bubble Plot** button (or choosing **Graph > Bubble Plot**), displays a scatter plot that represents its points as circles (bubbles). The bubbles can be sized according to a another column, colored by another column, aggregated across groups defined by one or more other columns, and dynamically indexed by a time column. With the opportunity to see up to five dimensions at once (*x* position, *y* position, size, color, and time), bubble plots can produce dramatic visualizations and make interesting discoveries easy. You can also create a Flash (.swf) version of the Bubble Plot for viewing outside of JMP.

See the *Basic Analysis and Graphing* book for details about using this platform.

Scatterplot Matrix Clicking the **Scatterplot Matrix** button (or choosing **Graph > Scatterplot Matrix**) explores how multiple variables relate to each other and how points fit that relationship. This platform helps you see correlations between two or more response (*y*) variables, look for points that are outliers, and examine principal components to look for factors.

See the *Basic Analysis and Graphing* book for details about using this platform.

Ternary Plot Clicking the **Ternary Plot** button (or choosing **Graph > Ternary Plot**) creates a ternary plot, which shows the distribution and variability of three-part compositional data. Examples of compositional data are the proportion of sand, silt, and clay in soil, or the proportion of three chemical agents in a trial drug. You can use data expressed in proportions, or in absolute measures. The Ternary Plot platform converts absolute values to proportions.

The ternary plot in JMP also has a contour feature, which means you can use a response measure at each point to form a contour plot within the ternary structure. The contour feature must be specified using a column formula.

See the *Basic Analysis and Graphing* book for details about using this platform.

Histograms Clicking the **Histograms** button (or choosing **Analyze > Distribution**) lets you construct an analysis that shows a distribution of values with histograms and other graphical and textual reports. This is a shortcut to the Distribution platform.

See the *Basic Analysis and Graphing* book for details about using this platform.

Mosaic Plot Clicking the **Mosaic Plot** button when you have nominal or ordinal *x* and *y* values produces a contingency table and a mosaic bar chart. This is a shortcut to the Contingency personality of the Fit Y by X platform.

See the *Basic Analysis and Graphing* book for details about using this platform.

The Surface Category

The **Surface** category, shown in Figure A.9, corresponds with commands in the main menu that produce multi-dimensional graphs, such as profilers and surface plots.

Figure A.9 JMP Starter Surface Page

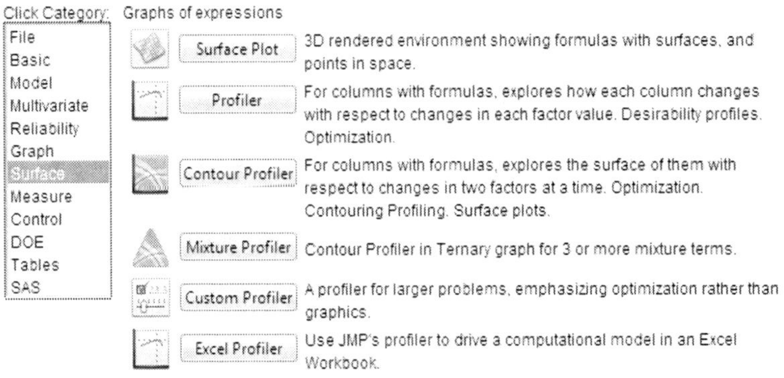

Surface Plot Clicking the **Surface Plot** button, or choosing **Graph > Surface Plot**, produces plots with smooth, three-dimensional surfaces.

See the *Modeling and Multivariate Methods* book for details about using this platform.

Profiler Clicking the **Profiler** button, or choosing **Graph > Profiler**, and selecting options produces a *prediction profiler* for the effect variables in a model, with one or more *y* variables. The prediction profiler displays *prediction traces* for each *x* variable. A prediction trace is the predicted response as one variable is changed while the others are held constant at the current values. If you save the prediction equations for a model, you can access the prediction profiler for that data table directly with **Graph > Profiler** without rerunning the model. You can also use this command to create a Flash (.swf) version of the Profiler for viewing outside of JMP.

See the *Modeling and Multivariate Methods* book for details about using this platform.

Contour Profiler The report that results from clicking the **Contour Profiler** button, or choosing **Graph > Contour Profiler**, does interactive contour profiling. This is useful for optimizing response surfaces graphically. This interactive plot lets you adjust both multiple responses and multiple factors simultaneously and shows acceptable factor setting regions for response values that you specify. If you save the prediction equations for a model, **Graph > Contour Profiler** accesses the contour profiler for that data table directly without rerunning the model.

See the *Modeling and Multivariate Methods* book for details about using this platform.

Mixture Profiler Lets you profile models involving mixture experiments. It respects constrained models and enables you to explore alternatives inside feasible design regions.

See the *Modeling and Multivariate Methods* book for details about using this platform.

Custom Profiler Clicking the **Custom Profiler** button, or choosing **Graph > Custom Profiler**, displays an interface for interacting with and optimizing a prediction formula.

See the *Modeling and Multivariate Methods* book for details about using this platform.

Excel Profiler Provides a method for transferring Excel data to JMP for analysis in JMP's profiler.

See the *Modeling and Multivariate Methods* book for details about using this platform.

The Measure Category

The **Measure** category, shown in Figure A.10, accesses the commands on the **Graph** menu that are used in statistical quality control.

Figure A.10 The JMP Starter Measure Page

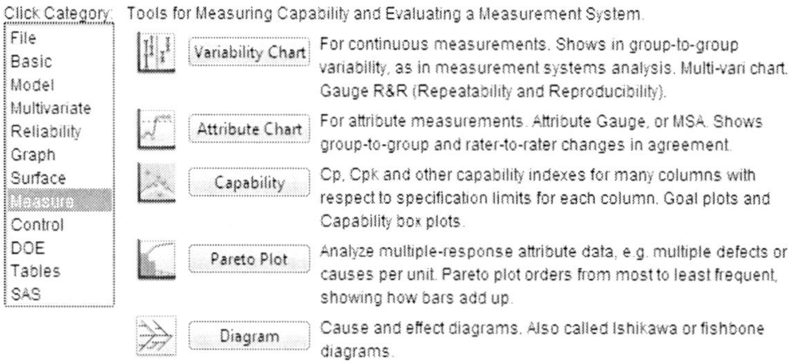

Variability Chart Clicking the **Variability Chart** button, or choosing **Graph > Variability/Gauge Chart**, performs a variability analysis, also called a *Gauge R&R analysis*. In a variability analysis, a number of parts assumed to be identical are taken from a production line. Each one is measured several times by a number of operators using different measuring instruments. You want to know the magnitudes of the variation due to operators, parts, and instruments.

See the *Quality and Reliability Methods* book for details about using this platform.

Attribute Chart Click the **Attribute Chart** button, or select **Graph > Variability/Gauge Chart** and choose **Attribute** from the drop-down menu, when your response values are binary or categorical. For example, a circuit might be measured as pass or fail. Because different reports are generated on the raters, raters each need to be in different columns.

See the *Quality and Reliability Methods* book for details about using this platform.

Capability Clicking the **Capability** button, or choosing **Graph > Capability**, gives a platform option for capability analysis for quality control applications. A capability study measures the deviation of a process from given specification limits. A window prompts you for the lower specification limit, upper specification limit, and target. You can also enter a known value for sigma, the process standard deviation.

See the *Quality and Reliability Methods* book for details about using this platform.

Pareto Plot Clicking the **Pareto Plot** button, or choosing **Graph > Pareto Plot**, creates a bar chart (Pareto chart) that displays the severity (frequency) of problems in a quality-related process or operation. A Pareto chart is a statistical quality improvement tool that shows frequency, relative frequency, and cumulative frequency of problems. The defining characteristic of Pareto plots is that the bars are in descending order of values, which visually emphasizes the most important measures or frequencies.

See the *Quality and Reliability Methods* book for details about using this platform.

Diagram Clicking the **Diagram** button, or choosing **Graph > Diagram**, constructs *Ishikawa* charts, also called *fishbone* charts, or *cause-and-effect diagrams*. These charts are useful to organize the sources (causes) of a problem (effect), perhaps for brainstorming, or as a preliminary analysis to identify variables in preparation for further experimentation.

See the *Quality and Reliability Methods* book for details about using this platform.

The Control Category

The **Control** category, shown in Figure A.11, accesses the commands that are used in statistical quality control.

Figure A.11 The JMP Starter Control Page

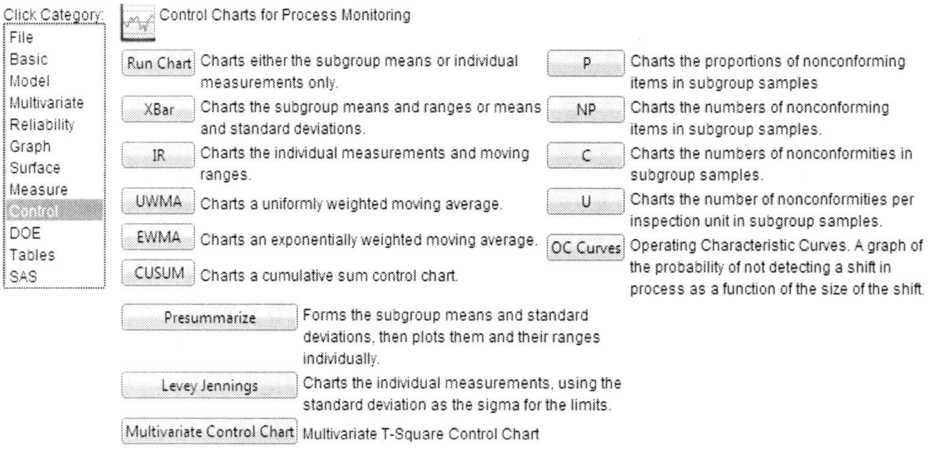

Run Chart Clicking the **Run Chart** button, or clicking **Graph > Control Chart > Run Chart**, gives an overlay of a single numeric or categorical *x* column. All specified numeric *y* variables are also included.

See the *Quality and Reliability Methods* book for details about using control charts.

XBar Clicking the **XBar** button, or clicking **Graph > Control Chart > XBar**, offers a combination of $\bar{X}$-, R-, and S-charts. The $\bar{X}$-chart plots subgroup means (averages), the R-chart plots subgroup ranges, and the S-chart plots subgroup standard deviations.

See the *Quality and Reliability Methods* book for details about using control charts.

IR Clicking the **IR** button, or clicking **Graph > Control Chart > IR**, creates two graphs. The individual measurement chart shows individual measurements. The moving range chart shows the moving ranges of two or more successive measurements.

See the *Quality and Reliability Methods* book for details about using control charts.

UWMA Clicking the **UWMA** button, or clicking **Graph > Control Chart > UMWA**, gives a Uniformly Weighted Moving Average (UWMA) chart, also called a Moving Average chart. Each point on a UWMA chart is the average of the *w* most recent subgroup means, including the present subgroup mean.

See the *Quality and Reliability Methods* book for details about using control charts.

EWMA Clicking the **EWMA** button, or clicking **Graph > Control Chart > EWMA**, gives an Exponentially Weighted Moving Average (EWMA) chart, also referred to as a Geometric Moving Average (GMA) chart. Each point on an EWMA chart is the weighted average of all the previous subgroup means, including the mean of the present subgroup sample. The weights decrease exponentially going backward in time.

See the *Quality and Reliability Methods* book for details about using control charts.

CUSUM Clicking the **CUSUM** button, or clicking **Graph > Control Chart > CUSUM**, creates a Cumulative Sum control chart. This chart shows cumulative sums of the deviations of measurements or subgroup means from a target value.

See the *Quality and Reliability Methods* book for details about using control charts.

Presummarize Clicking the **Presummarize** button, or clicking **Graph > Control Chart > Presummarize**, summarizes the process column into sample means or standard deviations. The summary is based on either the sample size or sample label chosen. Then it charts the summarized data in individuals or moving range charts.

See the *Quality and Reliability Methods* book for details about using control charts.

Levey Jennings Clicking the **Levey Jennings** button, or clicking **Graph > Control Chart > Levey Jennings**, gives a plot that shows a process mean with control limits based on a long-term sigma. The control limits are placed at 3*s* distance from the center line.

See the *Quality and Reliability Methods* book for details about using control charts.

Multivariate Control Chart Clicking the **Multivariate Control Chart** button, or clicking **Graph > Control Chart > Multivariate Control Chart**, creates a t-squared control chart. This chart measures how far a multivariate point is from the target with respect to the covariance structure.

See the *Quality and Reliability Methods* book for details about using control charts.

P Clicking the **P** button, or clicking **Graph > Control Chart > P**, gives *P*-charts, which display the proportion of nonconforming (defective) items in subgroup samples, which can vary in size. Each subgroup for a *P*-chart consists of N_i items. Each item is also judged as either conforming or nonconforming. As a result, the maximum number of nonconforming items in a subgroup is N_i.

See the *Quality and Reliability Methods* book for details about using control charts.

NP Clicking the **NP** button, or clicking **Graph > Control Chart > NP**, gives *NP*-charts, which display the number of nonconforming (defective) items in subgroup samples. Each subgroup for a *NP*-chart consists of N_i items. Each item is also judged as either conforming or nonconforming. As a result, maximum number of nonconforming items in subgroup *i* is N_i.

See the *Quality and Reliability Methods* book for details about using control charts.

C Clicking the **C** button, or clicking **Graph > Control Chart > C**, gives *C*-charts, which display the number of nonconformities (defects) in a subgroup sample that usually consists of one *inspection* unit. The maximum number of defects is unlimited.

See the *Quality and Reliability Methods* book for details about using control charts.

U Clicking the **U** button, or clicking **Graph > Control Chart > U**, gives *U*-charts, which display the number of nonconformities (defects) per unit in subgroup samples that can have a varying number of inspection units. The maximum number of defects is unlimited.

See the *Quality and Reliability Methods* book for details about using control charts.

OC Curves Clicking the **OC Curves** button, or clicking **Graph > Control Chart > OC Curves**, accesses a set of JSL scripts that compute and plot operating characteristic curves.

See the *Quality and Reliability Methods* book for details about using control charts.

The DOE Category

The **DOE** category, shown in Figure A.12, corresponds to the commands in the **DOE** main menu. These commands construct classical and custom experimental designs and save them in a JMP table. Selecting a design type presents an environment for describing the factors, responses, and other specifications needed to make a design of that type.

Figure A.12 JMP Starter DOE Page

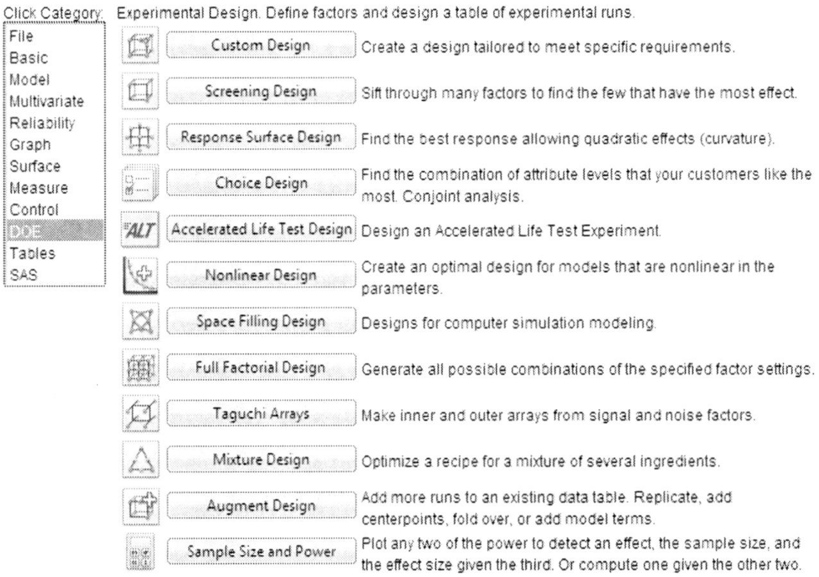

Custom Design Clicking the **Custom Design** button, or choosing **DOE > Custom Design**, generates a *D*-optimal design for the design that you specify, and an optional *I*-optimal design for response surface designs. Custom designs can have continuous factors and categorical factors with arbitrary numbers of level, interactions, and mixture ingredients. A design can also have covariates, which are factors that already have unchangeable values; a design is built around the covariate. You can construct custom design models, including interaction terms and polynomial terms for continuous factors. Factors can have inequality constraints. You can also specify the number of experimental runs, which can be any number greater than or equal to the number of terms in the model.

See the *Design of Experiments* book for details about this platform.

Screening Design Clicking the **Screening Design** button, or choosing **DOE > Screening Design**, enables you to specify continuous factors, two- and three-level factors, and categorical factors. It creates a table of classical screening designs from which you can pick.

See the *Design of Experiments* book for details about this platform.

Response Surface Design Clicking the **Response Surface Design** button, or choosing **DOE > Response Surface Design**, lets you define a number of continuous factors. This environment then offers a table of classical response surface designs from which to select.

See the *Design of Experiments* book for details about this platform.

Choice Design Clicking on the Choice Design button, or choosing **DOE > Choice Design**, lets you design Choice experiments.

See the *Design of Experiments* book for details about this platform.

Accelerated Life Test Design Clicking on the **Accelerated Life Test Design** button, or choosing **DOE > Accelerated Life Test Design**, lets you design tests based on one or two accelerating factors with either a main effects model or an interaction model for the latter.

See the *Design of Experiments* book for details about this platform.

Nonlinear Design Clicking the **Nonlinear Design** button, or choosing **DOE > Nonlinear Design**, lets you create an optimal design for models that are nonlinear in the parameters.

See the *Design of Experiments* book for details about this platform.

Space Filling Design Clicking the **Space Design Filling** button, or choosing **DOE > Space Filling Design**, lets you run experiments for modeling systems that are deterministic or near deterministic. You can choose to emphasize spreading points apart, mimic the uniform distribution, or choose a compromise between the two.

See the *Design of Experiments* book for details about this platform.

Full Factorial Design Clicking the **Full Factorial Design** button, or choosing **DOE > Full Factorial Design**, lets you specify a set of continuous and categorical factors with arbitrary numbers of levels. JMP creates the design containing all possible combinations of those factors.

See the *Design of Experiments* book for details about this platform.

Taguchi Arrays Clicking the **Taguchi Arrays** button, or choosing **DOE > Taguchi Arrays**, guides you through the definition of signal and noise factors. The signal factors form the inner array and the noise factors form the outer array. The inner and outer array designs are the traditional Taguchi orthogonal arrays such as L4, L8, L16, and so on.

See the *Design of Experiments* book for details about this platform.

Mixture Design Clicking the **Mixture Design** button, or choosing **DOE > Mixture Design**, lets you define a set of factors that are ingredients in a mixture. JMP creates a new window for choosing among several classical mixture design approaches such as simplex, extreme vertices, and lattice. For the extreme vertices approach, you can supply a set of linear inequality constraints limiting the geometry of the mixture factor space.

See the *Design of Experiments* book for details about this platform.

Augment Design Clicking the **Augment Design** button, or choosing **DOE > Augment Design**, lets you modify existing designs. You can add center points, replicate the design, create a foldover design, and add runs to the design using a model with more terms than the original design.

See the *Design of Experiments* book for details about this platform.

Sample Size and Power Clicking the **Sample Size and Power** button, or choosing **DOE > Sample Size and Power**, computes power, sample size, or the effect size that you want to detect for a given alpha and error standard deviation. You supply two of these values and the sample size and power feature computes the third. If you supply only one of these values, the result is a plot of the other two. This feature is available for the single sample, two sample, and *k* sample situations.

See the *Design of Experiments* book for details about this platform.

The Tables Category

The **Tables** category (Figure A.13) corresponds to the commands in the **Tables** main menu.

Figure A.13 JMP Starter Tables Page

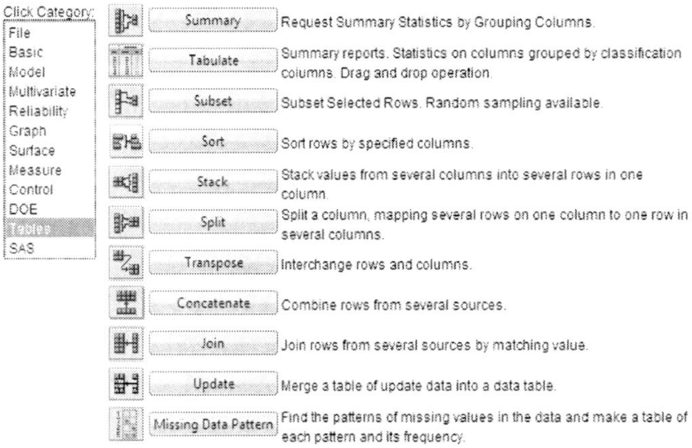

The buttons on the **Tables** page can help you analyze data in several ways:

Summary Creates a *summary table*, which summarizes columns from the active data table, called its *source table*. The summary table has a single row for each level of a grouping variable that you specify. Also, you can add columns of summary statistics to this table and request summary statistics.

See “Summarizing Columns,” p. 277 in the “Summarizing Data” chapter for details.

Tabulate Creates a tabulated table from the active data table. It calculates, sorts, and displays totals and subtotals for you in a report format, rather than in a data table.

See “Tabulating Data,” p. 285 in the “Summarizing Data” chapter for details.

Subset Creates a new data table formed by the selected rows and columns or by a random sample in the active data table.

See “Creating a Subset Data Table,” p. 231 in the “Reshaping Data” chapter for details.

Sort Sorts a data table by one or more columns.

See “Sorting Data Tables,” p. 234 in the “Reshaping Data” chapter for details.

Stack Creates a new data table from the active table by stacking specified columns into a single column.

See “Stacking Columns,” p. 237 in the “Reshaping Data” chapter for details.

Split Creates a new data table from the active table by dividing one or more columns to form multiple columns.

See “Splitting Columns,” p. 243 in the “Reshaping Data” chapter for details.

Transpose Creates a new data table that has the columns of the active table as its rows and the rows of the active table as its columns.

See “Transposing Rows and Columns,” p. 248 in the “Reshaping Data” chapter for details.

Concatenate Creates a new data table from two or more open data tables by combining them end to end.

See “Concatenating Data Tables,” p. 254 in the “Reshaping Data” chapter for details.

Join Creates a new data table by merging (joining) two tables side by side.

See “Joining Data Tables,” p. 256 in the “Reshaping Data” chapter for details.

Update Updates one table with data from a second table without producing a new data table.

See “Updating a Data Table,” p. 271 in the “Reshaping Data” chapter for details.

Missing Data Pattern Shows a pattern that your missing data creates. The pattern might help you make discoveries about your data.

See “Viewing Patterns of Missing Data,” p. 70 in the “Entering, Editing, and Managing Data” chapter for details.

The SAS Category

The **SAS** category (Figure A.14) corresponds to the commands in the **File > SAS** menu. See “External Data and Analytical Sources,” p. 357 for details.

Figure A.14 JMP Starter SAS Page

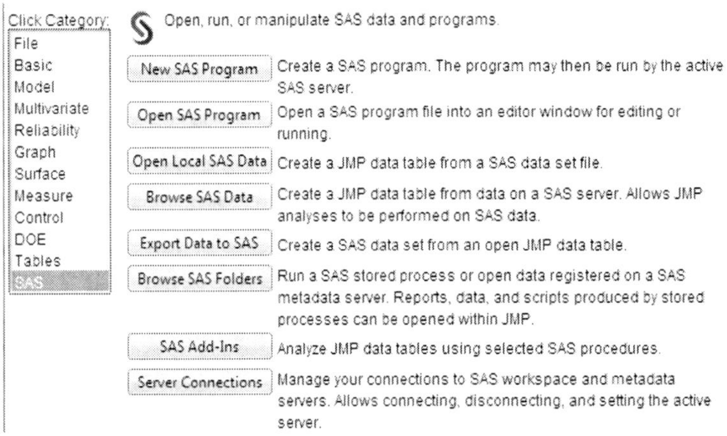

New SAS Program Creates a new SAS program to be run by the default SAS server.

Open SAS Program Opens a SAS program file into the JMP script editor.

Open Local SAS Data Imports a SAS data set into a JMP data table.

Browse SAS Data Creates a JMP data table using data that resides on a SAS server.

Export Data to SAS Exports a JMP data table to a SAS server.

Browse SAS Folders Lets you open reports, data, and scripts produced by SAS programs in JMP. The items reside on a SAS metadata server.

SAS Add-Ins Runs pre-defined SAS procedures using JMP data tables.

Server Connections Connects to and disconnects from SAS servers and sets the default server.

Appendix B

Main Menu

A Description of Commands

This appendix gives an overview of commands in the JMP main menu bar, which is located at the top of the screen.

Contents

The JMP Menu (Macintosh Only)

About JMP

Displays product version, registration, user and system information, the JMP Web page link, and more. (*Pro* is displayed in the menu item when you install JMP Pro.)

U.S. Government Notice

Displays the U.S Government restricted rights notice.

Preferences

Lets you select options and settings for start-up items, reports, tables, analysis platforms, import and export settings, graphic formats, fonts, SAS integration, and more. Also lets you reset default preferences. See "JMP Preferences" chapter, for details about the preferences.

Services

Shows available Mac OS X services. No services are available in JMP.

Hide JMP

Removes the JMP application from view.

Hide Others

Removes all other applications from view, leaving only JMP visible.

Show All

Brings all open applications into view.

Quit JMP

Closes all JMP windows and quits JMP. Prompts you to save new or changed journals, scripts, and data tables. Also asks whether you want to save the session script.

The File Menu

The **File** menu has commands that perform file management or affect the JMP environment.

New

Reveals a submenu with options to create a new data table, script, journal, or project (Windows only). The submenu contains:

Data Table Opens an empty untitled data table. See "Create New Data Tables," p. 23, for details.

Script Opens a new empty script window in which you enter and submit JMP Scripting Language (JSL) commands.

Journal Opens an empty journal window. See "Create Journals," p. 116, for details.

Project (Windows only) Opens the Project window and creates an empty, untitled project in the Projects list. See "Working with JMP Projects (Windows Only)," p. 125 in the "Saving Tables, Reports, and Sessions" chapter.

Open

Displays an open file window to locate the file that you want to open and specify the file format of the incoming file. The Open window displays information appropriate for the incoming file type. See "Open Existing JMP Files," p. 24, for details of the **Open** command.

Open Recent (Macintosh only)

Reveals a submenu that lists the most recently opened files. When you select a table name in this list, JMP opens the table. When you select a script name in this list, JMP opens the script. Some scripts automatically run. To open these scripts without running them, hold down the OPTION key while making the selections.

Close

Closes the active window. Works the same as clicking the close box of the active window. On Macintosh, holding the OPTION key and selecting **File** displays the **Close All** command, which closes all windows (except for the JMP Starter). Additionally, you can hold down the OPTION key and click the close box to close all windows.

Import as Data

Imports text-formatted data from a script window. On the Macintosh, this option is listed on the **Edit** menu. See "Import Text from the Script Window," p. 34, for details.

Save

Writes the active window to a file, providing a choice of appropriate extensions. See the "Saving Tables, Reports, and Sessions" chapter for details.

Save As

Writes the active data table, journal, or layout to a file after prompting you for a name and file type. See the "Saving Tables, Reports, and Sessions" chapter for details.

Save Selection As (Macintosh Only)

Saves selected content in a data table, report, journal, or layout window as a .PNG graphic. The option does not appear on the menu until you select the content, hold down the CTRL and SHIFT keys, and then select **File**. See "The Edit Menu," p. 418 for details about saving selected content on Windows.

Revert (Windows) or Revert to Saved (Macintosh)

Restores the current data table, script, journal, or layout to its last saved condition.

Export (Macintosh Only)

Exports data into text, Microsoft Excel, and SAS Transport file formats. Exports reports into text, images, HTML, and RTF formats. See the "Saving Tables, Reports, and Sessions" chapter for details.

Database

Opens a connection to any database on your system that has an ODBC driver, and opens or saves a data table. See "Opening Data from a Database," p. 380 in the "External Data and Analytical Sources" chapter, and "Saving Data Tables to a Database," p. 111, for details.

Selecting **File >Database** reveals a submenu with options to open or save to any database on your system that has an ODBC driver.

SAS

Connects to a SAS server; browses and opens SAS data files and folders; provides an editor window; opens SAS Log and Output windows; exports data to SAS; connects to SAS add-ins.

Internet Open

Opens remote files in a browser or text editor window, and opens an HTML table as a data table. Also lets you open JMP tables and run scripts from a Uniform Resource Locator (URL). See "Open Remote Files and Web Pages," p. 35, for details.

Preferences (Windows only)

Lets you specify options and settings for start-up items, reports, tables, analysis platforms, import and export settings, graphic formats, SAS integration, and more. Also lets you reset defaults. On Macintosh, this option is listed on the **JMP** menu. For more information about customizing the preferences, see the "JMP Preferences" chapter.

Print

Prints the active window. The appearance of the window depends on your operating system and printer driver. See "Printing Reports," p. 184, for details.

Print Preview (Windows only)

Displays the active window in print view. **Print Preview** is dimmed on the **File** menu unless there is an open window.

Print Setup (Windows) or Page Setup (Macintosh)

Displays the standard window for setting printed page properties. The form of the window depends on your operating system and current printer driver.

Send (Windows only)

Submits the open window to the default e-mail program. Prompts you to save new windows first. See "E-mail Tables and Reports (Windows Only)," p. 41, for details.

Recent Files (Windows only)

Displays a list of the twenty most recently opened files for easy re-opening.

Save Session Script

Creates a JMP Scripting Language (JSL) script to re-open all currently open files and re-run all currently open analyses and graphs. See "Saving JMP Sessions," p. 123 for details.

Save Projects (Windows only)

Saves JMP projects with a .jmpprj extension. This option is available only when a project is open.

Exit JMP (Windows only)

Closes all JMP windows and quits JMP. Prompts you to save new or changed journals, scripts, data tables, and projects. Also asks whether you want to save the session script.

The Edit Menu

The **Edit** menu contains standard commands that operate on selected rows, columns, areas of reports, scripts, text, and formula elements in the formula editor. The commands operate on entire rows if no columns are selected. Likewise, they operate on whole columns if no rows are selected. When both rows and columns are selected, **Edit** commands affect the subset of values defined by the intersection of those rows and columns.

The **Edit** menu in conjunction with JMP tools can be used to copy all or part of active analysis report windows. See "The Tools Menu," p. 433, for details.

Undo

Cancels the effect of the most recent reversible **Edit, Rows,** or **Columns** commands.

Redo

Reverses the most recent **Undo** command.

Cut

Copies selected data table cells to the clipboard and replaces the data with missing values. In reports and other windows, copies selected content to the clipboard.

Copy

Copies selected data table cells to the clipboard. In reports and other windows, copies selected content to the clipboard.

Copy as Text

Copies all text (but no graphs), or text selected using the selection tool, from the active report window to the clipboard. On the Macintosh, the **Copy as Text** command appears only when applicable.

Copy with Column Names

Copies selected data table cells and column names to the clipboard. Select **Paste with Labels** (**Paste with Column Names** on the Macintosh) to paste the information into a new column with the column name intact. See "Copying, Cutting, and Pasting," p. 75, for details.

Paste

Copies data from the clipboard into the active data table.

Paste with Column Names

Copies data from the clipboard into the active data table. Converts text on the first line of the clipboard to column names.

Clear (Windows) or Delete (Macintosh)

Removes selected data table cell contents and adds missing values. In journal, layout, and script windows, removes selected content.

Select All

Selects all rows in a data table. In reports and other windows, selects all content when one item is selected.

Save Selection As

Saves selected portions of an analysis window in a graphic format. On Macintosh, hold down the CTRL and SHIFT keys and select **File > Save Selection As** to save the selected portion of an analysis window in a graphic format. See "Saving Parts of a Report in a Graphic Format," p. 116, for details.

Reformat Script

Formats JSL code for easier reading.

Encrypt Script

Places encoding in a script so that you must have the password to open or run the script. When an encrypted script is in the active window on Macintosh, this option changes to **Decrypt Script**. The option lets you open or run the encrypted script.

Import as Data

Imports text-formatted data, using the first row as column names. Before invoking the **Import as Data** command, you must paste data into a script window. That window must also be active. On Windows, this option is listed on the **File** menu. See "Import Text from the Script Window," p. 34, for details.

Decrypt Script

Lets you open or run an encrypted script. Only available if an encrypted script is in the active window.

Run Script

Processes JSL commands entered in the script window. For example, if the script calls for a data table and none are open, you are prompted to select a data table. The script then runs on that data table. Or scripts that do not require a data table are run instantly.

For other examples of how scripts are run, see the *Scripting Guide* book.

Stop Script

Stops running a JSL script.

Search

Finds and replaces text in data tables and scripts. See "Using the Search and Find Commands," p. 93, for details.

Go to Line

Scrolls to a specific line number in a script.

Balance

Selects all text within a pair of brackets that include the cursor. Only available if a script window is active.

Shift Right (Macintosh only)

Shifts selected lines in a script to the right.

Shift Left (Macintosh only)

Shifts selected lines in a script to the left.

Journal

Copies output from a data table or report window to a journal window. Lets you save reports, add notes, and organize output. See "Create Journals," p. 116 in the "Saving Tables, Reports, and Sessions" chapter, for details.

Layout

Copies output from a data table or report window to a layout window, where you can rearrange report elements for printing or presentations. When a layout window is the active (front-most) window, the **Layout** menu appears between the **Edit** and **Tables** menus. See "Saving Using the Layout Command," p. 114, and "The Layout Menu," p. 441, for details.

Submit to SAS

Connects to a specified SAS server and submits SAS code directly from JMP.

Special Characters (Macintosh only)

Opens the Character Palette and lets you insert special characters into report, journal, or layout windows.

The Tables Menu

The **Tables** menu commands modify or create a new JMP table from one or more existing tables. This section gives you an overview of each **Tables** menu command.

Summary

Creates a table that contains user-specified summary statistics calculated from the active data table. See "Summarizing Columns," p. 277, for details.

Subset

Creates a new data table that is a subset of the active data table and can be linked to the original table. To quickly create a subset table, highlight the rows you would like to subset, hold down the SHIFT key and select **Tables > Subset**. JMP creates a linked subset table without asking you to complete the Subset window. See "Creating a Subset Data Table," p. 231, for details about creating subsets.

Sort

Sorts a JMP data table by one or more columns. This creates a new data table, or you can replace the original table with the sorted table. See "Sorting Data Tables," p. 234, for details.

Stack

Stacks specified columns into one or more new columns in a new data table. The values in other columns are preserved in the new data table. **Stack** does the reverse of **Split**. See "Stacking Columns," p. 237, for details.

Split

Splits one or more columns, according to values in a Split By column, to form multiple columns in a new data table. **Split** requires one or more columns whose combined values identify each row in the new table. See "Splitting Columns," p. 243, for details.

Transpose

Creates a new data table where the data in the selected columns of the original table are arranged in rows. This equates to switching the rows and columns in the original selection. See "Transposing Rows and Columns," p. 248, for details.

Concatenate

Stacks two or more tables on top of each other by matching column names. See "Concatenating Data Tables," p. 254, for details.

Join

Adds columns from a second table to the main table, using row number or matching columns, or by forming a Cartesian join. See "Joining Data Tables," p. 256, for details.

Update

Replaces data in the active table with data from a second table without producing a new table. See "Updating a Data Table," p. 271, for details.

Tabulate

Constructs tables of descriptive statistics using an interactive interface. See "Tabulating Data," p. 285, for details.

Missing Data Pattern

Describes the missing data in your table in terms of the columns where it is missing. In the original data table, missing character data is represented with a blank and missing numerical data is represented with a black dot. See "Viewing Patterns of Missing Data," p. 70, for details.

The Rows Menu

You can access commands that affect selected rows from the **Rows** menu. Most of these commands are also on the red triangle menu:

- Click the red triangle icon on the rows panel to the left of the data table.
- Click the red triangle icon in the upper left corner of the data table above the row numbers.

Exclude/Unexclude

Excludes selected rows from analyses, or includes selected rows in analyses. See "Excluding Rows and Columns," p. 135, for details.

Hide/Unhide

Hides or shows selected points in all plots. See "Hiding Rows and Columns," p. 136, for details.

Label/Unlabel

Labels or unlabels selected points in all plots. See "Labeling Rows and Columns," p. 136, for details.

Colors

Changes selected points in all plots to a specified color. See "Giving Rows a Color," p. 137, for details.

Markers

Replaces selected points in all plots with symbols that you select. See "Adding Markers to Rows," p. 137, for details.

Next Selected

Locates and flashes the first selected row in a data table after the current row. See "Locating Next and Previously Selected Rows," p. 94, for details.

Previous Selected

Locates and flashes the first selected row in a data table before the current row. See "Locating Next and Previously Selected Rows," p. 94, for details.

Row Selection

Provides several options for selecting rows based on one or more conditions. See "Selecting Rows and Columns," p. 88, for details.

Clear Row States

Clears all active row states in the data table. All rows become included, visible, unlabeled, and show in plots as black dots. It does not affect row states saved in row state columns.

Color or Mark by Column

Assigns colors or markers to rows based on values in a selected column. See "Assigning Colors or Markers to Rows According to Column Values," p. 138, for details.

Row Editor

Opens a window that lets you edit cells in a row. Helps you select rows and create new rows. See "Using the Row Editor," p. 78, for details.

Delete Rows

Removes all selected rows from the data table. Use the **Undo** command on the **Edit** menu to undo an accidental deletion. See "Adding and Deleting Rows," p. 63, for details.

Add Rows

Displays a window that prompts for the number of rows to add and to specify their location in the table. See "Adding and Deleting Rows," p. 63, for details.

Move Rows

Moves selected rows to the location that you specify in the data table.

Data Filter

Selects, shows, or includes rows based on conditions on the values of specified columns. See "The Data Filter," p. 95 in the "Entering, Editing, and Managing Data" chapter.

The Cols Menu

You can access commands that affect selected columns from the several locations:

- **Cols** main menu
- Columns panel to the left of the data table
- Red triangle icon in the upper left corner of the data table.

New Column

Adds a new column, and lets you name the column and change attributes and properties. See "Adding and Deleting Columns," p. 64, for details.

Add Multiple Columns

Adds several columns of the same data type at one time to a specified location in the data table. See "Adding and Deleting Columns," p. 64, for details.

Go to

Highlights and scrolls to a specified column in the data table.

Column Info

Displays and allows changes to column name, attributes, and properties for selected column or columns. See "Assigning Properties to Columns," p. 145, for details.

Preselect Role

Assigns an analysis role, which can be saved with the data table, to selected column or columns. See "Giving Columns a Preselected Analysis Role," p. 143, for details.

Formula

Shows the formula editor window for selected column or columns. See "Giving Columns a Formula to Compute Values," p. 146, and "Formula Editor," p. 301, for details.

Validation

Places restrictions on the values that a column can hold, and enables you to specify their display order on graphics. See "Validating Column Data," p. 148, for details.

Label/Unlabel

Tells JMP whether to use values from the selected column or columns to identify points in plots when the cursor is placed near them. To display the labels for points at all times, the corresponding rows must have a label row state. See "Labeling Rows and Columns," p. 136, for details.

Scroll Lock/Unlock

Locks or unlocks selected columns into or from the left-most position in the table. Locking a column keeps its contents visible when you scroll horizontally. See "Locking Columns in Place," p. 143, for details.

Hide/Unhide

Hides or shows selected columns without removing them from the table. Retains them in analyses. See "Hiding Rows and Columns," p. 136, for details.

Exclude/Unexclude

Excludes selected rows from analyses, or includes selected rows in analyses. See "Excluding Rows and Columns," p. 135, for details.

Standardize Attributes

Lets you apply attributes (which include data type, modeling type, numeric format, and so on). Also lets you apply properties (which include formulas, lists, notes, list and range checks, and so on) to multiple columns. See "Standardizing Attributes and Properties Across Columns," p. 169, for details.

Compress Selected Columns

Compresses data in a selected column based on the data type. Also adds a List Check column property when the column contains fewer than 255 unique values. See "Compress Selected Columns," p. 172 in the "Properties and Characteristics of Data" chapter for details.

Reorder Columns

Lets you move columns according to your selection from its submenu:

Move Selected Columns Moves the selected columns **To first** (left-most in the data table), **To last** (right-most in the data table), or **After:** (after a column that you identify) in the Move Selected Columns window.

Original Order Returns the columns to the order they were in at the time the data table was last saved.

Reorder by Name Arranges the columns (except for row state columns) from left to right in alphabetical order by column name.

Reorder By Data Type Arranges the columns from left to right in alphabetic order by data type (Row State, Character, Numeric).

Reorder By Modeling Type Arranges the columns from left to right in alphabetic order by modeling type (continuous, ordinal, nominal). Row State columns have no modeling type, and are shown last.

Reverse Order Reverses the order of the data table columns.

If you mistakenly move one or more columns, use the **Undo** command in the **Edit** menu to restore the previous order. See "Reordering Columns," p. 74 for details.

Delete Columns

Removes selected columns from the data table. See "Adding and Deleting Columns," p. 64, for details.

Recode

Lets you correct data entry errors or inconsistencies or change old values to new values. The changes can be made in the original column, in a new column, or in a new column that tracks the changes with a formula. See "Recoding Data," p. 69, for details.

Group Columns

Groups columns under one heading to manage large numbers of columns and facilitate analysis role assignment. Grouped columns are shown in outline view under a gray disclosure icon in the columns panel. See "Grouping Columns," p. 286 in the "Summarizing Data" chapter for details.

Ungroup Columns

Ungroups the selected group of columns.

The DOE Menu

The Design of Experiments (**DOE**) menu provides an environment for describing the factors, responses, and other specifications needed to create a designed experiment. Your then and save the information in a JMP data table. The specific designs are described briefly here and covered in detail in the *Design of Experiments* book.

Custom Design

Provides the most flexibility of all design choices. With this option you can have continuous factors, categorical factors with any number of levels, interactions, mixture ingredients, and covariates (factors that have unchangeable fixed values). You can construct custom-designed models that include interaction terms and polynomial terms for continuous factors. You can also specify inequality constraints on the factors and the number of experimental runs. The number can be any number greater than or equal to the number of terms in the model.

Screening Design

Generates a list of classical screening designs for the specified list of 2-level continuous factors and 2- and 3-level categorical factors.

Response Surface Design

Generates a list of classical response surface designs for the specified number of continuous factors.

Full Factorial Design

Creates a design using all combinations of factor levels for a set of continuous and categorical factors that have up to nine levels.

Mixture Design

Creates a design that is appropriate when the factors are components in a mixture. Allows specification of both linear and regional constraints. Several classical mixture design approaches are offered.

Choice Design

Creates an optimal discrete choice conjoint experimental design. These designs are useful in studies where users are asked to specify preferences relative to combinations of product or service features.

Space Filling Design

Creates a design that is appropriate when modeling deterministic or nearly deterministic systems, such as computer simulations. Design points are distributed throughout the experimental region and are not replicated.

Accelerated Life Test Design

Designs tests based on one or two accelerating factors with either a main effects model or an interaction model for the latter.

Nonlinear Design

Creates an optimal design for estimating the parameters of a nonlinear model.

Taguchi Arrays

Constructs a Taguchi design. Signal and noise factors, a choice of inner and outer array designs, and signal-to-noise ratios are supported.

Augment Design

Lets you modify existing designs. You can add center points, replicate the design a specified number of times, and create a foldover design. This feature also lets you add runs to the design using a model with more terms than the original design.

Sample Size and Power

Computes power, sample size, and effect size for means, variance, proportions, and counts. Also computes sigma quality level.

The Analyze Menu

Each **Analyze** command launches a platform. A *platform* is an interactive window that you use to analyze data, work with points on plots, and save results. The reports in a JMP analysis are organized hierarchically. Methods unfold that suit the context of your data. Many results appear automatically, and more are offered through drop-down menus.

Distribution

Provides a histogram for continuous data and a bar graph for nominal or ordinal data, along with relevant summary statistics. Presents options for many one-sample analyses, based on modeling type. See the *Basic Analysis and Graphing* book for details.

Fit Y by X

Shows plots that describe the relationship between any two variables. Provides two-sample analyses based on the modeling types of the two variables, such as bivariate, oneway, logistic, and contingency analysis. See the *Basic Analysis and Graphing* book for details.

Matched Pairs

Handles bivariate data in the special situation where the two responses form a pair of measurements coming from the same experimental unit or subject. For example, a matched pair might be a before-and-after blood pressure measurement from the same subject. The responses are correlated, and the statistical method called the paired t-test takes that into account.

The Matched Pairs platform displays the data as a scatterplot of the difference between each pair (y) by the mean of each pair (x). Includes the standard paired t-test, which is equivalent to testing that the mean difference between the paired values is zero. See the *Basic Analysis and Graphing* book for details.

Fit Model

Fits models involving one or more Y variables and multiple X variables. Techniques include standard least squares, stepwise, MANOVA, loglinear variance, logistic, proportional hazards, parametric survival, and generalized linear models. See the *Modeling and Multivariate Methods* book for details.

Modeling

Offers various modeling techniques: screening (for designs with many effects), nonlinear, neural net, Gaussian process, partition analysis, time series, categorical response analysis, and choice. See the *Modeling and Multivariate Methods* book for details.

Multivariate Methods

Offers techniques for exploring relationships among multiple variables: multivariate fitting, clustering, principal components, discriminant analysis, partial least squares, and item analysis. See the *Modeling and Multivariate Methods* book for details.

Reliability and Survival

Offers techniques for fitting survival and reliability data: life distribution, fit life by x, recurrence analysis, degradation, product-limit survival fit, parametric survival distributions, and proportional hazards modeling.

You can also use the Nonlinear platform to handle nonlinear models with loss functions for other parametric survival modeling. See *Modeling and Multivariate Methods* for details.

The Graph Menu

Graph menu commands show windows that contain specialized graphs or plots with supporting tables and statistics.

Graph Builder

Provides an interactive interface to create flexible trellis displays that are useful for visualizing multivariate data. Plots can be conditioned on discrete and continuous variables and include dot plots, box plots, lines, bar graphs, and histograms. See the *Basic Analysis and Graphing* book for details.

Chart

Plots bar, line, pie, needle, and point charts of user-specified summary statistics for multiple selected Ys across the values of up to two X variables. See the *Basic Analysis and Graphing* book for details.

Overlay Plot

Plots one or more numeric Y variables across the values of a single X variable. Constructs several plots in one window using grouping variables. By default, the values of the *x* variable are in ascending order, and the points are plotted in that order. You have the option of plotting the *x* values as they are encountered in the data table. See the *Basic Analysis and Graphing* book for details.

Note: For scatterplots of two variables with regression fitting options, use the **Fit Y by X** command instead of **Overlay Plot**.

Scatterplot 3D

Creates three-dimensional rotatable plots for multiple Y variables. Displays and saves principal components, standardized principal components, and rotated components. Also displays normal contour ellipsoids, nonparametric density contours, and biplot rays.

See *Basic Analysis and Graphing* for details about the **Scatterplot 3D** command and examples of plotting data. See *Modeling and Multivariate Methods* for examples of computing principal components.

Contour Plot

Constructs contour plots for one or more numeric response variables, Y, across a grid defined by two numeric X variables. See the *Basic Analysis and Graphing* book for details.

Bubble Plot

Displays circles according to positional specification. Circles can be sized and colored according to selected variables. When there is a time variable, selected variables can be viewed dynamically over time. See the *Basic Analysis and Graphing* book for details.

Parallel Plot

Draws a plot that connects the values of selected Y variables for each row in the data table. See the *Basic Analysis and Graphing* book for details.

Cell Plot

Creates a rectangular array of cells drawn with a one-to-one correspondence to data table values, and colored by the values in the cells. See the *Basic Analysis and Graphing* book for details.

Tree Map

Creates a rectangular tiling for a nominal or ordinal variable where you can tile categories to be proportional in size to a selected variable. A positional specification is optional. Useful when there are a lot of categories or when histograms are ineffective. See the *Basic Analysis and Graphing* book for details.

Scatterplot Matrix

Creates an ordered collection of bivariate graphs that help you assess the relationships between multiple variables simultaneously. Displays density ellipses, shaded ellipses, and nonparametric density contours. You can also add shaded density ellipses for each level of a grouping variable. See the *Basic Analysis and Graphing* book for details.

Ternary Plot

Illustrates the distribution of three-part (numeric) compositional data, using a two-dimensional plot. See the *Basic Analysis and Graphing* book for details.

Diagram

Constructs *Ishikawa charts*, also called *fishbone charts*, or *cause-and-effect diagrams*. These charts are useful when organizing the sources (causes) of a problem (effect), perhaps for brainstorming, or as a preliminary analysis to identify variables in preparation for further experimentation. See the *Quality and Reliability Methods* book for details.

Control Chart

Creates run charts as well as a wide range of dynamic, updatable control charts. These are plots of time-ordered data, along with control limits, that bound the variability of the plotted statistic. See the *Quality and Reliability Methods* book for details.

Variability/Gauge Chart

Creates variability charts, and runs Gauge R&R and variance components analysis. The Attribute Chart Type option gives measures of agreement across nominal responses. See the *Quality and Reliability Methods* book for details.

Pareto Plot

Creates a bar chart that displays causes of defects or problems in order of decreasing frequency. See the *Quality and Reliability Methods* book for details

Capability

Conducts capability studies on multiple responses. Displays a goal plot for normalized means and standard deviations, and provides other statistical and graphical output. See the *Quality and Reliability Methods* book for details.

Profiler

Displays prediction traces for each X variable, based on saved prediction formulas for each numeric response and linear constraints on the X variables. Allows simultaneous optimization of responses using desirability functions, estimation of response distributions using the simulator, and accesses and links to other profilers. Can be saved as an Adobe Flash file (`.swf`) to display in a web browser. See the *Modeling and Multivariate Methods* book for details.

Contour Profiler

Provides an interface for simultaneous optimization of responses, similar in function to the profiler. Allows linear constraints and a user-defined criterion for optimization, and displays predicted performance relative to benchmarks. Accesses the simulator, and links to other profilers. See the *Modeling and Multivariate Methods* book for details.

Surface Plot

Creates a three-dimensional, rotatable display of points or of a surface defined by a saved formula. See the *Modeling and Multivariate Methods* book for details.

Mixture Profiler

Displays a ternary plot and contours of the prediction formula for models involving mixture factors. Also provides links to other profilers. See the *Modeling and Multivariate Methods* book for details.

Custom Profiler

Provides an interface for simultaneous optimization of responses, similar in function to the profiler. Allows linear constraints and a user-defined criterion for optimization, and displays predicted performance relative to benchmarks. Accesses the simulator, and links to other profilers. See the *Modeling and Multivariate Methods* book for details.

Excel Profiler

Provides a method for transferring Excel data to JMP. You can then analyze the data using JMP's profilers. See the *Modeling and Multivariate Methods* book for details.

The Tools Menu

The **Tools** menu are palettes of special tools whose form the cursor can take. There are general tools and specific graphics tools that operate on points in plots. The default tool is the arrow. Click a tool in the menu to activate it.

Arrow

The default tool.

- Lets you select points in plots and analysis options. Drag the cursor to select multiple points.
- Rotates the plot in Surface Plot and Scatterplot 3D.

\? *Help*

Opens context-sensitive Help when you click the help tool in any part of a display or report.

Selection

Lets you select rows or regions of rows when you click and drag. Selects portions of graphical output or analysis results.

Click and drag to highlight multiple rows and columns, and Ctrl-click (Command-click on the Macintosh) to select discontinuous rows or columns. Clicking near the edge of the report highlights the entire report. To deselect, click a second time in a highlighted area.

Scroller

Scrolls a report, journal, or layout window in any direction with more precision than when using scroll bars. The scroller is available in report windows, journal, and layout windows.

Press and hold the SHIFT key while selecting the scroller to continue to use the scroller after multiple clicks.

Grabber

The hand (grabber) tool is for direct manipulation of plots, charts, axes, and formula components.

- Changes axis scaling when you hover with grabber and click over an axis. The hand changes orientation to indicate direction of change.
- Changes both axes simultaneously when you click and drag within a plot.
- In histograms, changes the number of bars or shifts the boundaries of the bars on the axis when you click and drag.
- In any report table, rearranges columns.
- In Surface Plot and Scatterplot 3D, click and drag an axis to move it. When you click the plot, the grabber turns to an arrow and lets you rotate the plot.

Brush

Lets you select one or more points when you brush over them. When you click, a rectangle appears. Move the rectangle over points to highlight them. Shift-click to extend the selection. Hold down the ALT key (Windows) or OPTION key (Macintosh) and click to change the size of the selection rectangle and also extend the selection. See “Selecting a Rectangular Area of Points,” p. 191, for details.

Lasso

Lets you circle points with a curve. Drag the lasso around any set of points. When you release the lasso, it automatically closes and highlights the points within the enclosed area. Hold down the SHIFT key and circle more points to extend a selection. See “Selecting an Irregular-Shaped Area of Points,” p. 192, for details.

Magnifier

Lets you automatically zoom in on any area of a plot. When you click the magnifier, the point or area where you click becomes the center of a new view of the data. Hold down the ALT key (Windows) or COMMAND key (Macintosh) and click to restore the original plot. On a ternary plot, drag the magnifier tool to zoom the triangular axes.

Crosshairs

Displays a moveable set of axes to locate points and estimate distances on plots. The values where the crosshairs intersect the vertical and horizontal axis appear automatically as you drag the crosshairs within a plot. On a ternary plot, this tool displays triangular crosshair lines.

Annotate

Lets you add editable text notes to JMP reports, journals, or layouts. The annotate tool reverts to the arrow after one click. Hold down the SHIFT key while selecting the annotate tool to have the cursor continue to use the tool after multiple clicks. See "Annotations," p. 217, for details.

Line, Polygon, and Simple Shape

Lets you draw lines, polygons, and shapes on JMP reports, journals, and layouts. Right-click a line or shape for a menu of options. See "Adding Shapes," p. 220, for details.

The shape tools revert to the arrow after one click. Hold down the SHIFT key while selecting the shape tool to have the cursor continue to have the shape form after multiple clicks.

The View Menu

The **View** menu lets you view or hide certain windows or panes.

On Windows

On Windows, you find these commands in the **View** menu:

JMP Starter

Opens and closes the JMP Starter window, which gives access to most commands found on the main menu or on toolbars. You can also close the JMP Starter with **File > Close** when it is the active window or by clicking its close box. On the Macintosh, the **JMP Starter** command is located in the **Window** menu. For more details, see "The JMP Starter Window," p. 21.

Home Window

Opens or closes a window that provides quick access to recently opened files. Also lists all open windows in outline view (the same list that appears when you select **View > Windows List**).

Window List

Opens or closes a window that lists open windows. These windows are shown in an outline view. Right-click the title of a window in the list to close, rename, redraw, or hide it. For more details, see "Work with Open Windows," p. 20.

Projects

Opens or closes a window that lets you create, view, and edit JMP projects.

Log

Opens a window that shows information such as JSL statements as they execute and script errors. You can use the log to locate errors in a script, correct the code, copy and paste it back into the script window, and then submit the corrected script.

Status Bars

Opens and closes the status bar at the bottom of the JMP window.

Thumbnail Panel

Opens and closes a panel that displays small previews of all open reports for the current data tables.

Toolbars

Lets you add, remove, or revert to the original toolbars. See "Personalize Toolbars and Menus (Windows Only)," p. 335, for details.

Add-Ins

Displays add-ins that you wrote in JSL and registered in JMP.

Customize

Lets you customize the menus and toolbar and revert to the original settings.

On Macintosh

On the Macintosh, you find these commands in the **View** menu:

New Data View

Displays a duplicate view of the active data table. The new view is linked to the original view and all corresponding reports. Changes made to a new view reflect on the original view immediately. On Windows, the **New Data View** command is located in the **Window** menu.

Redraw Window

Refreshes the active window. It is useful for cleaning up both data table views and graphs that have accumulated stray imperfections resulting from high-speed, dynamic handling of windows.

Make Text Bigger, Make Text Smaller

The **Make Text Bigger** and **Make Text Smaller** commands increase and decreases the size of text in all open windows.

Show Toolbar or Hide Toolbar

Show Toolbar adds the toolbar to the active window if it is not displayed. Only available when toolbars are hidden.

Hide Toolbar removes the toolbar from the active window if it is displayed. Only available when toolbars are shown.

Customize Toolbar

Lets you drag new toolbars, platform icons, and miscellaneous icons to the toolbar and change their display size. Also lets you display these items as icons, text, or both and revert to the original toolbars.

Add-Ins

Displays add-ins that you wrote in JSL and registered in JMP.

The Window Menu

The **Window** menu helps you organize the windows produced during a JMP session.

On Windows

On Windows, you find these commands in the **Window** menu:

New Data View

Displays a duplicate view of the active data table. The new view is linked to the original view and all corresponding reports. Changes made to a new view reflect on the original view immediately.

Close All Data Tables

Closes all windows of the specified type. Changes to **Close All Reports**, **Close All Journals**, **Close All Layouts**, or **Close All Scripts** based on the active window.

Close All

Closes all open windows.

Minimize All

Minimizes all open windows.

Restore All

Displays all minimized windows.

Bring All Forward

Brings all JMP windows to the front.

Arrange

The **Arrange** command helps you organize the open windows within JMP.

Arrange Selected Windows After you select windows, this option appears to let you show all selected windows side-by-side.

Arrange With Data Table \After you select a report, this option appears to let you show the report next to its data table.

Select For Arrange Selects the current window. A checkmark appears in the **Select for Arrange** box in the lower right corner of the window.

Select All For Arrange Selects all open JMP report, data table, script, and journal windows. A checkmark appears in the **Select for Arrange** box in the lower right corner of the windows.

Clear All Selections Deselects all selected windows.

Redraw

The **Redraw** command refreshes a window after it has accumulated stray imperfections from high-speed, dynamic handling.

Font Sizes

Change the font size in all open windows. See “Increasing Font Sizes,” p. 183 for details.

Set Title

Changes the name of the active JMP window. This is useful if you generate multiple output reports and need to distinguish among them during the JMP session. For options on changing the title of data tables, see “Changing Table Names,” p. 79. For options on changing the title of JMP output reports, see “Renaming a Report,” p. 182.

Move to Back

Moves the active window behind all other windows generated by the current JMP session and shows the next window in the sequence.

Hide

The **Hide** command suppresses the display of the active window but does not close it. Hidden windows are not in the list of open JMP windows found on the bottom of the **Window** menu.

Unhide

The **Unhide** command lets you select which hidden windows you want to make visible again. Choosing **Unhide All** from the **Unhide** menu displays all JMP windows.

On Macintosh

On the Macintosh, you find these commands in the **Window** menu:

Minimize

Shrinks the active window into an icon on the dock. To unroll or expand the window, select **Minimize** again. Hold down the OPTION key and select **Minimize All** to minimize all open JMP windows.

Zoom

Enlarges the window to use the available monitor screen. To reduce the window size, select **Zoom** again. Hold down the OPTION key and select **Zoom All** to resize all open JMP windows.

Bring All to Front

Moves all JMP windows to the front of the screen. Hold down the OPTION key and select **Arrange in Front** to display the windows one in front of another.

Set Report Title

The **Set Title** command lets you change the name of any active JMP window. This is useful if you generate multiple output reports and need to distinguish among them during the JMP session.

JMP Starter

Brings the JMP Starter to the front. You can also close the JMP Starter with **File > Close** when it is the active window or by clicking its close box.

Log

Opens a window that shows information such as JSL statements as they execute and script errors. You can use the log to locate errors in the script, correct the code, copy it and paste it back into the script window, submit the corrected script.

The last section of the **Window** menu lists all unhidden available windows. Bring any window to the front by clicking on it in this list.

The Help Menu

The Help menu provides links to information about JMP, from the online Help system to JMP tutorials.

Contents, Search, and Index (Windows), Search and JMP Help (Macintosh)

Lets you search for and open JMP menu items. Also lets you search the JMP Help system for key words.

Tip of the Day

Opens the Tip of the Day window, which provides tips about using JMP. See "Before You Start," p. 17 in the "Getting Started" chapter for details.

Tutorials

Opens a set of JMP tutorials.

Books

Opens .pdf files of the JMP documentation.

Sample Data

Opens a journal file with links to sample data and sample script files. Many sample data files include scripts that you can run to see examples of JMP tools and analyses. You can also run the sample scripts themselves to see examples of JMP tools and analyses.

Statistics Index

Defines statistical terms, opens online Help, launches analysis platforms, and lets you run scripts that illustrate the terms.

Click the **Topic Help** button to go to the place in the online Help that describes the current topic. Click **Launch** to run the script that corresponds to the item that you selected in the list. Click **Example** to view an example of the highlighted topic.

JSL Functions

Lists JSL functions, opens online Help, and lets you run scripts that illustrate the functions. Select a function name, and a description appears in the window on the right. Click the **Topic Help** button to locate the topic in the online Help. Click the **Run Example** button to view an example of the highlighted topic.

Object Scripting

Lists JSL objects with messages recognized by each object, opens online Help, and lets you run scripts that illustrate the objects. Select an object name, and messages the object recognizes appear in the window on the right. Click the **Topic Help** button to locate the topic in the online Help. Click the **Run Example** button to view an example of the highlighted topic.

DisplayBox Scripting

Lists display box elements with messages recognized by each element, opens online Help, and lets you run scripts that illustrate the objects. Click the **Topic Help** button to locate the topic in the online Help. Click the **Run Example** button to view an example of the highlighted topic.

JMP.com

Opens the JMP Web page www.jmp.com.

JMP User Community

Opens the JMP user community web page www.jmp.com/community.

About JMP

Displays product version, registration, user and system information, the JMP Web page link, and more. On Macintosh, **About JMP** is on the **JMP** menu.

The Layout Menu

To copy a JMP window to a separate window and manipulate pieces of it, you select **Edit > Layout**. See "Saving Using the Layout Command," p. 114, for details.

Group

Combines selected objects into a single object or group.

Ungroup

Ungroups selected objects. The first time you use **Ungroup**, the title of the analysis becomes an object, and is the outermost outline level of the analysis. The remaining nesting levels become an object. You can continue to select any portion of the layout window and use the **Ungroup** command repetitively to ungroup the analysis into its smallest elements.

Align

Aligns selected objects according to the type of alignment that you select from the **Align** submenu. You can select as many objects as you want for alignment. However, objects overlap if they are positioned on the layout window in such a way as to conflict with the alignment request.

Clean Up

Separates overlapping objects. Layout objects can be moved to any position in the Layout window, which means objects can overlap. When objects overlap, the front-most object hides some or all of the objects beneath it. The **Clean Up** command separates overlapping objects so that all are showing.

Move Forward, Move To Front, Move Backward, Move To Back

The **Move Forward** and **Move Backward** commands move the selected object or objects forward or backward one layer at a time. **Move to Front** moves the selected object or objects in front of all other objects. **Move to Back** moves the selected object or objects behind all other objects. Note that when objects overlap, the front most objects can hide objects beneath them.

Appendix C

JMP Preferences

Setting Preferences Using the Preferences Window

JMP preferences enable you to specify general and specific settings and save the settings.

To change preferences

1. Select **File > Preferences** (Windows) or **JMP > Preferences** (Macintosh). The window in Figure C.1 appears with the **General** preferences category showing.
2. Click a category and make selections. Click **Apply** to see the results, and then click **OK**.

Figure C.1 The General Preferences Window (Windows)

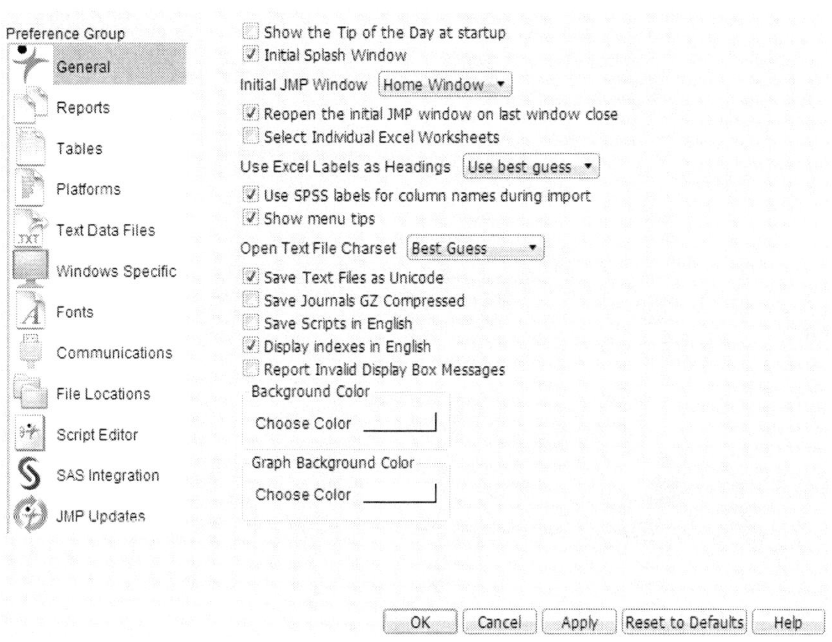

Contents

Overview

The **Preferences** command on the **File** menu (on the **JMP** menu on Macintosh) displays the Preferences window. See Figure C.2. Each category is described in this chapter.

Note: The **Reset to Defaults** button is on every page of the preferences. Clicking the button on any of the first three pages resets preferences on all of those pages. On the remaining pages, clicking the button resets preferences only on the current page.

General

The **General** page is initially displayed when you open Preferences. Most **General** preferences customize your JMP session at start-up. Others set preferences for file-handling and the general appearance of the JMP workspace.

Figure C.2 General Preferences

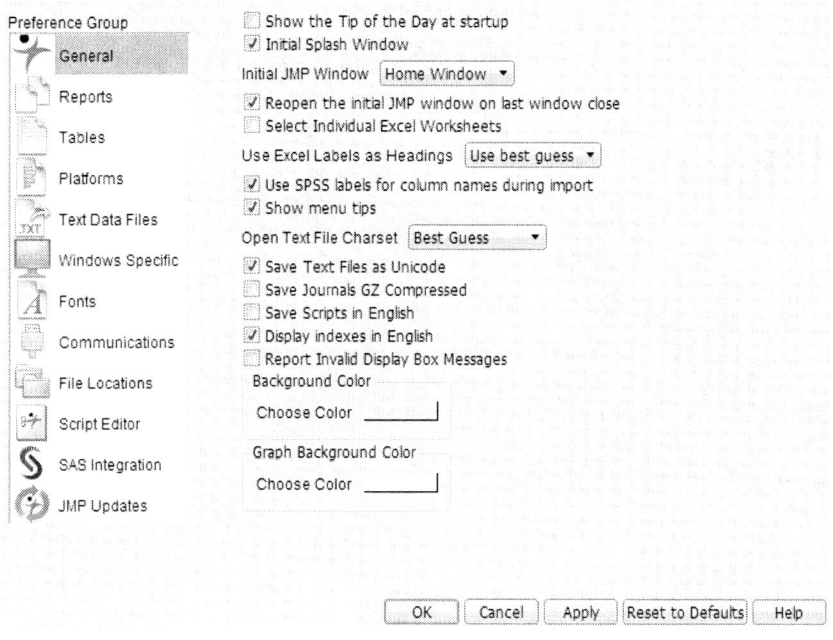

Table C.1 Preferences on the General Page

Preference	Description
Show Tip of the Day at startup	Select this option to show the Tip of the Day window at start-up. This option is selected by default. Clear this option to prevent the Tip of the Day window from appearing at start-up.
Initial Splash Window	Select this option to show the initial splash window at start-up. This option is selected by default. Clear this option to prevent the initial splash window from appearing at start-up.
Initial JMP Starter Window	(Macintosh only) Select this option to show the JMP Starter window every time you start JMP. This option is on by default.
Initial JMP Window	(Windows only) Select one of the listed windows as the default window. When JMP starts this window appears.
Reopen the initial JMP window on last window close	(Windows only) Select this option to open whichever window you have set as the Initial JMP Window when you close the last JMP window. If this option is turned off, closing the last JMP window also quits JMP. This option is on by default.
Select Individual Excel Worksheets	Select this option to always be prompted to select specific worksheets when you open a Microsoft Excel workbook. This option is not selected by default. Note that when you open an Excel file on Macintosh, an option to select worksheets is always available. Under Vista and subsequent Windows operating systems, the interactive method to select individual worksheets is on the **Open** button menu.

Table C.1 Preferences on the General Page *(Continued)*

<table>
<tr><th>Preference</th><th>Description</th></tr>
<tr><td>Use Excel Labels as Headings</td><td>Select one of the options from the menu to determine how the first line of your Excel worksheet is imported into a JMP data table. Use best guess is the default setting.

Use best guess Lets JMP to determine whether the first line in your Excel worksheet should be column names or a row of data. JMP assumes that the first row is a column name unless one of the following is true:
• The first row is entirely text and there is only one other row, or no other rows.
• The first row is entirely text and all of the variables are text
• The first row is not all text and none of the variables are text

Always Always use the first line in your Excel worksheet as column names.

Never Never use the first line in your Excel worksheet as column names. The first line is imported into JMP as the first row of data.

Note that when you open an Excel file in Windows, these options are available in the Open window. Therefore, you can use any of these options for any single file no matter what your preference setting is. On Macintosh, the option is available on the Open window, but you cannot choose the method. The method selected in the Preferences is always used.</td></tr>
<tr><td>Use SPSS labels for column names during import</td><td>Select this option to use SPSS labels as column names. This option is selected by default.</td></tr>
<tr><td>Show menu tips</td><td>Select this option to see tips on options listed in red triangle menus when you hover your mouse briefly over the option. This option is selected by default.</td></tr>
<tr><td>Open Text File Charset</td><td>Select one of the options from the menu to determine what character encoding JMP uses to open files. Best Guess is the default.</td></tr>
<tr><td>Save Text Files as Unicode</td><td>JMP uses the Unicode character set, which supports special characters such as é and ½. It saves files without special Unicode characters as plain text automatically. This option is selected by default.

Clear this check box to save all your files as plain text.

Note that versions of JMP earlier than 5.1.2 cannot read Unicode files.</td></tr>
</table>

Table C.1 Preferences on the General Page *(Continued)*

Preference	Description
Save Journals GZ Compressed	Select this option to save JMP journals in a compressed format to save disk space. Clear this option to save journals normally. This option is not selected by default.
Save Scripts in English	Non-English versions of JMP can save scripts in either English or your local language. User-specified values (such as column names and text strings) in these scripts appear as they do in the data table. However, command words (such as `Distribution` and `Set Value`) appear in English rather than the local language in order for the script to run on JMP in English. JMP properly displays non-Roman characters (such as Japanese variable names) in JMP in English when the fonts support the necessary characters. Select this option to save scripts in English no matter what language JMP is using. Clear this option to save scripts in your local language. Note that these scripts run correctly *only* if JMP is running in the same language. This option is cleared by default.
Display indexes in English	Shows sections of the JMP Indexes in English or the current locale's language. The preference is selected by default. When you deselect it, the following items are displayed in the current locale's language: • The object list, message list, and prototype in the Object Scripting Index. • The message list and prototype in the Display Box Index. • The category list in the JSL Functions Index. **Note**: The description of each index item is always displayed in the current locale's language. Examples are always in English.
Report Invalid Display Box Messages	Sends information about invalid display box messages to the log. This option is off by default. This option can be useful during script development, but can cause unwanted log messages for existing scripts.
Background Color	Click the color box to select a background color for windows.
Graph Background Color	Click the color box to select a background color for all graphs.

Reports

Report preferences customize the appearance of reports, including markers and colors for graphs.

Figure C.3 Reports Preferences

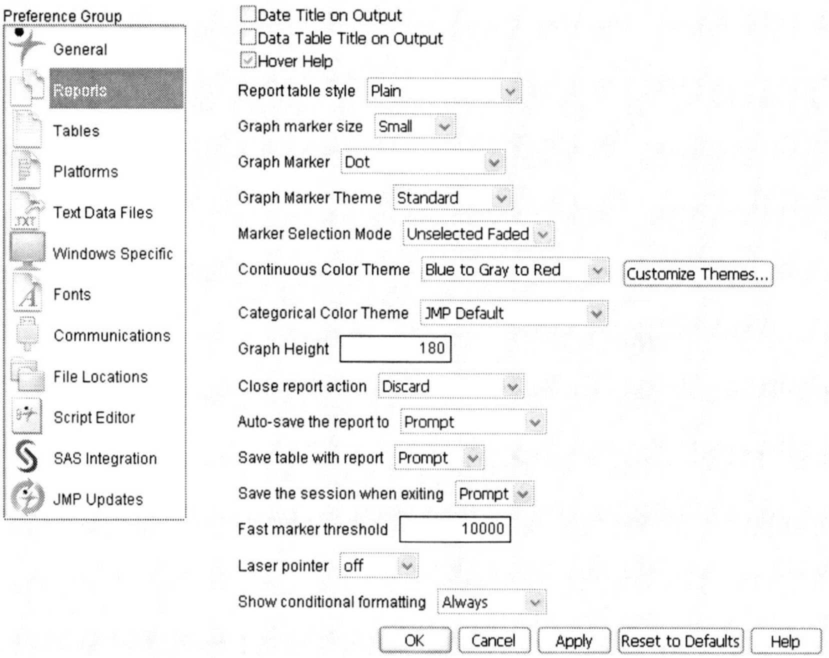

Table C.2 Preferences on the Reports Page

Preference	Description
Date Title on Output	Select this option to display the date and time the analysis occurred in your report windows. This option is cleared by default.
Data Table Title on Output	Select this option to display the name of the data table and notes, if there are any, at the top of the report. This option is cleared by default.
Hover Help	On some numeric output, JMP provides tooltip-style help when you circle the mouse over a result. Select this option to see hover help where it is offered. This option is selected by default.
Report table style	Select a style for your report tables: **Plain**, **Bordered**, **Embossed**, **Beveled**, or **Heading Separator**. **Plain** is the default value.

Table C.2 Preferences on the Reports Page *(Continued)*

Preference	Description
Graph marker size	Select a default size for the markers in graphs: **Dot**, **Small**, **Medium**, **Large**, **XL**, **XXL**, or **XXXL**.
Graph Marker	Select a default marker shape for the markers in graphs.
Graph Marker Theme	Select a default theme for the markers in graphs when you mark by row or column.
Marker Selection Mode	Select the default formatting for selected markers. See "Marker Selection Modes," p. 195 in the "Output Reports" chapter for details.
Continuous Color Theme	Select a default color theme for continuous data.
Customize Themes	Lets you customize color themes defined in JMP. See "Create Color Themes," p. 139 in the "Properties and Characteristics of Data" chapter for details.
Categorical Color Theme	Select a default color theme for categorical data.
Graph Height	Enter the default graph height, measured in pixels.
Close report action	Select one of the options from the menu to determine what happens when you close a report: **Prompt** When you close a report, you are prompted to save it. **Discard** When you close a report, you are not prompted to save it, and the report is not saved. **Save / Auto-save** When you close a report, it is automatically saved using the preference selected in the **Auto-save the report to** menu.
Auto-save the report to	Select one of the options from the menu to determine where a report is saved: **Prompt** When you close a report, you are prompted for a location to save it. **Data Table Script** When you close a report, it is saved as a script in the open data table. **Journal Script** When you close a report, it is saved to a journal window. **Log Window** When you close a report, its script is written to the log window.

Table C.2 Preferences on the Reports Page *(Continued)*

Preference	Description
Save table with report	Select one of the options from the menu to determine how a table is saved to a report: **Embed** When you save a report, the table is embedded into the report. Choose this option if you want to share the report with others. The data table that is stored in the report is reopened (unchanged) each time the report is opened. If you make changes to the table, they are not saved into the report unless you re-save the report. **Separate** When you save a report, the table is referenced in the report. **Prompt** When you save a report, you are prompted to specify how to save the report: either to embed the table within the report, or to add a reference to the table in the report. **Prompt** is selected by default.
Save the session when exiting	Select one of the options from the menu: **Always** When you quit JMP, the session is always automatically saved. **Never** When you quit JMP, the session is never saved. **Prompt** When you quit JMP, you are asked if you would like to save the session.
Fast Marker Threshold	When JMP refreshes a report window, it can draw markers on a plot at two different speeds: normal and fast. If JMP is in normal drawing mode, and the number of markers in a graph are more than the specified threshold number, JMP automatically switches to fast mode. Enter the number of markers that separate normal and fast mode. The default is 10,000.
Laser pointer	JMP has a built-in laser pointer that enables you to visually emphasize parts of a report. It is off by default. To turn it on, select a color for the laser pointer from the menu.

Table C.2 Preferences on the Reports Page *(Continued)*

Preference	Description
Show conditional formatting	Conditionally formats the color of the text that represents the values. Available for correlation values, p-values, and factor pattern values. See "Changing Numeric Formats and Field Widths," p. 188. Select one of the options from the menu: **Always** Conditional formatting is always applied to correlation values, p-values, and factor pattern values. **Screen Only** Conditional formatting is always applied to correlation values, p-values, and factor pattern values. However, the conditional formatting does not show when the report is printed. **Never** Conditional formatting is never applied to correlation values, p-values, and factor pattern values.

Tables

Table preferences customize JMP data tables, including formula handling, appearance, and compatibility with SAS data sets.

Figure C.4 Tables Preferences

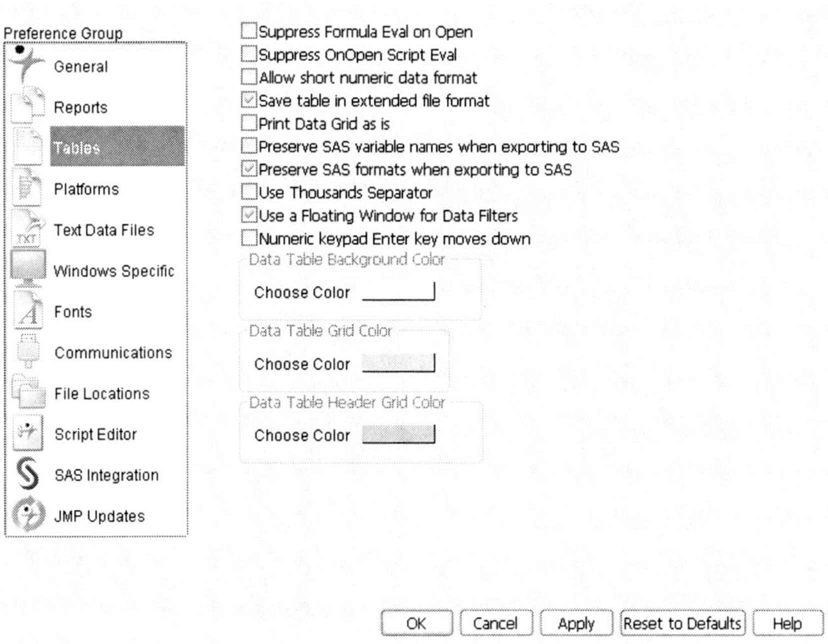

Table C.3 Preferences on the Tables Page

Preference	Description
Suppress Formula Eval On Open	Select this option to prevent columns formulas from being evaluated when you open data tables. Clear this option to allow formulas to be evaluated when you open data tables. This option is cleared by default.
Suppress OnOpen Script Eval	If you save a script to a data table and name it OnOpen, that script is automatically run whenever the data table is opened. Select this option to prevent any script named OnOpen in the data table from running when you open the data table. Clear this option to allow OnOpen scripts to run. This option is cleared by default.

Table C.3 Preferences on the Tables Page *(Continued)*

Preference	Description
Allow short numeric data format	JMP has the ability to store numeric data in as few as 8 bits (one byte). This option makes short-integer formats available to you when you select **Cols > Column Info** and assign a column a data type. When you use the correct short-integer format for your data, the numbers are not displayed differently, but the data table uses less disk space. See "Using the Short-Integer Format," p. 60, for details.
Save table in extended file format	Starting with JMP 8, data tables can contain a very large number of columns. To accomplish this feature, much of the internal data table coding was changed to support the large (that is, extended) file format. By default, this option is on. Clear this check box if you want to open your JMP data table in versions earlier than JMP 8.
Print Data Grid as is	Select this option to print the JMP data table as it appears on the screen. Clear this option to resize column widths to accommodate the content width. This option is cleared by default.
Preserve SAS variable names when exporting to SAS	Select this option to use variable names that are compliant with SAS when you export a JMP data table to a SAS data set. This option is cleared by default.
Preserve SAS formats when exporting to SAS	Select this option to use formats that are compliant with SAS when you export a JMP data table to a SAS data set. This option is selected by default.
Use Thousands Separator	Select this option to display numbers using the locale-appropriate thousands separator. This option is not selected by default.
Use a Floating Window for Data Filters	Select this option to float the Data Filter window on top of its associated data table. Clearing this option causes the Data Filter window to behave like any other window. This option is selected by default.
Numeric keypad Enter key moves down	(Windows only) When a data table cell is selected and you press the Enter (or Return) key on the keyboard, the next table cell down is selected. Pressing either the Tab key or the Enter key on the numeric keypad located to the right of the keyboard selects the next table cell to the right. Select this option to change the behavior of the Enter key on the numeric keypad to select the next table cell down instead of to the right. This option is cleared by default.
Data Table Background Color	Click the color box to select a background color for data tables.
Data Table Grid Color	Click the color box to select a color for the grid lines in a data table.

Table C.3 Preferences on the Tables Page *(Continued)*

Preference	Description
Data Table Header Grid Color	Click the color box to select a background color for the column names.

Platforms

Each analysis report has a variety of plot and table options that are shown by default. However, there might be additional options that you want to see each time you run a particular analysis. For example, a bivariate analysis shows a scatterplot by default, but you might also always want to see a linear fit each time. By selecting the **Platforms** option in the left panel of the Preferences window, you can set the default options for analyses. (Analyses are run by using *platforms*, such as the Bivariate platform. Thus, the name of this category is **Platforms.**)

Highlight an analysis name in the Platforms list. Its available options appear in the **Options** box with the defaults checked.

Figure C.5 Platforms Preferences

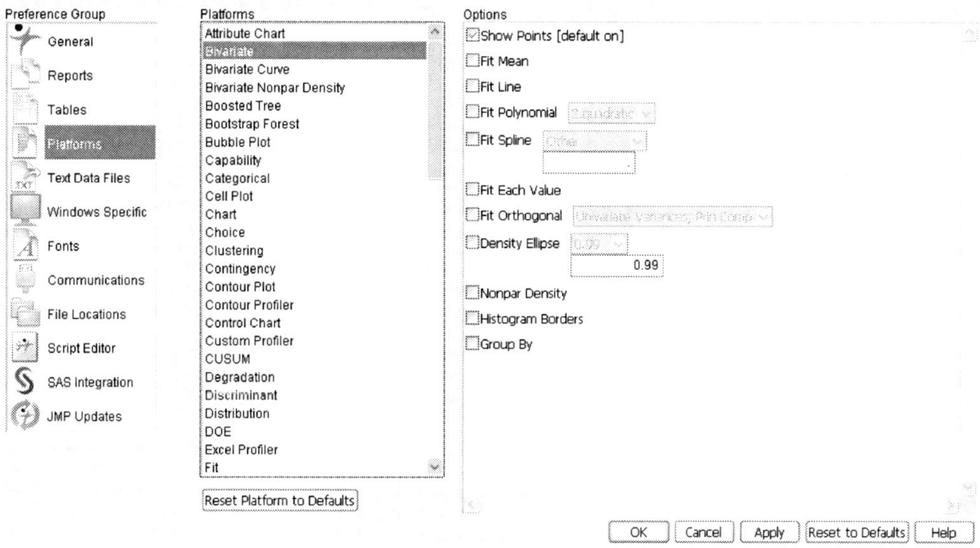

Graph Builder Preferences

There are a few preferences in Graph Builder that do not appear as options within JMP and cannot be scripted. Select **Platforms**, and then select **Graph Builder** from the list of platforms to see these preferences.

Table C.4 A Few Graph Builder Preferences

Continuous Points Limit	For continuous variables, this number indicates the number of rows at which the Points element is no longer shown initially. The default value is 1500.
Continuous Alternate	For continuous variables, indicates the element to display when there are more rows than the Points limit can handle. The default value is **None**, or you can select **Contour** (density contour). If **None** is selected, and the **Continuous Smoother** is turned off, Points appear when the limit is breached.
Continuous Smoother	For continuous variables, indicates whether the **Continuous Smoother** should appear.
Categorical Points Limit	For categorical variables, this number indicates the number of rows at which the Points element is no longer shown initially. The default value is 1500.
Categorical Alternate	For categorical variables, indicates the element to display when there are more rows than the Points limit can handle. The default value is **Box Plot**, or you can select from the following options: • **None** • **Line** • **Smoother** • **Box Plot** • **Bar** • **Histogram**
Jitter	Indicates whether jittering is on by default. Small spaces are displayed between the data points so that you can see each point more clearly. (Only applies to elements that support jittering.)

Text Data Files

Text Data File preferences customize the handling of importing and exporting text files.

Figure C.6 Text Data Files Preferences

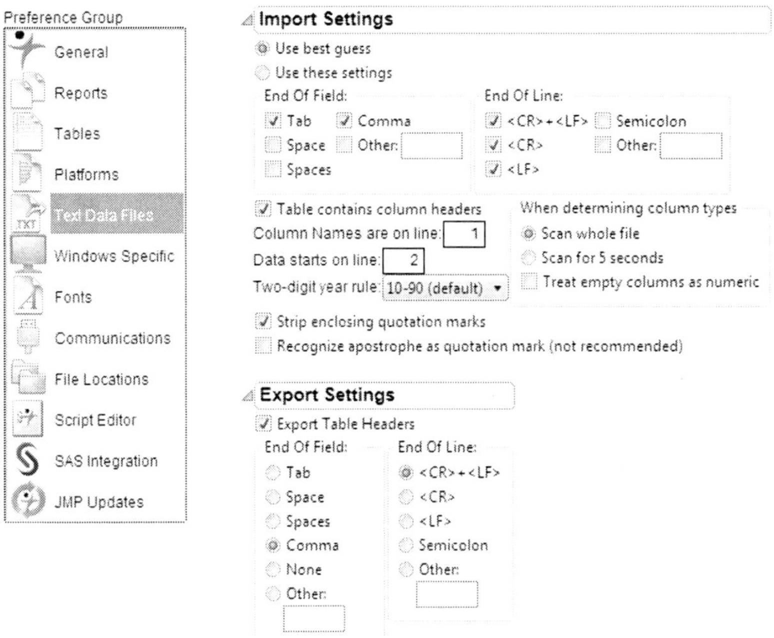

Table C.5 Preferences for Import Settings for Text Files

Preference	Description
Import Settings	Select the strategy JMP uses to open text files. The default selection is **Use these settings**. In that case you need to ensure that the settings reflect your text files. If you select **Use Best Guess**, JMP collects statistics in the text file on tabs, commas, blanks, and a few other characters and uses a rule-based system to decide what the file format might be. The rules try to make reasonable field widths and a reasonable number of fields per line. If your data format is too different from what the rules are designed to guess, JMP guesses incorrectly. In that case, either use the wizard or explicitly describe your data in these preference settings.
End Of Field	Select one or more characters to use as the delimiter that signifies the end of a field when importing text data. Select the **Other** option and enter a character to specify a delimiter that is not listed.

Table C.5 Preferences for Import Settings for Text Files *(Continued)*

Preference	Description
End Of Line	Select one or more characters to use as the delimiter that signifies the end of a line (row). Select the **Other** option and enter a character to specify a delimiter that is not listed. Note that if double-quotes are encountered when importing text data, the delimiter rules change to look for an end double-quote. Other text delimiters, including spaces, that are embedded within the quotes are ignored and treated as part of the text string.
Table contains column headers	Select this option if your text file contains columns names. If you select this option, enter the line number where the column names are located in the field next to **Column Names are on line**.
Column Names are on line	If you select the **Table contains column headers** option, enter the line number where the column names are located in this field.
Data starts on line	Enter the line number where the data starts in your text file.
When determining column types	Set how long JMP scans a text file to determine data types for the columns. The default value is **Scan whole file**. Note that the **Scan whole file** option can cause importing a text file to be slow for large files. When your text file contains columns of missing data, select **Treat empty columns as numeric** to import the columns as numeric rather than character. A period, unicode dot, *NaN*, or a blank string are possible missing value indicators. This option is deselected by default.
Two-digit year rule	Select the rule that you want to use to import dates that use two digits instead of four. For details about these rules, see Table 2.3 "The 10-90 Rule for Importing Two-Digit Years," p. 31.
Strip enclosing quotation marks	Select this option to remove quotation marks that enclose data in the text file.
Recognize apostrophe as quotation mark	Select this option to treat apostrophes as quotation marks. This option is off by default. This option is not recommended unless your data comes from a nonstandard source that uses apostrophes around data fields rather than quotation marks.

Table C.6 Preferences for Export Settings for Text Files

Preference	Description
Export Table Headers	Select this option to include column names when you save data tables as text files.
End Of Field	Select one or more characters to use as the delimiter signifying the end of a field when exporting text data. Select the **Other** option and enter a character to specify a delimiter that is not listed.
End Of Line	Select one or more characters to use as the delimiter that signifies the end of a line (row). Select the **Other** option and enter a character to specify a delimiter that is not listed.

Windows Specific

These preferences customize settings for machines running the Windows operating system, including graphics formats and the window background color.

Figure C.7 Windows Specific Preferences

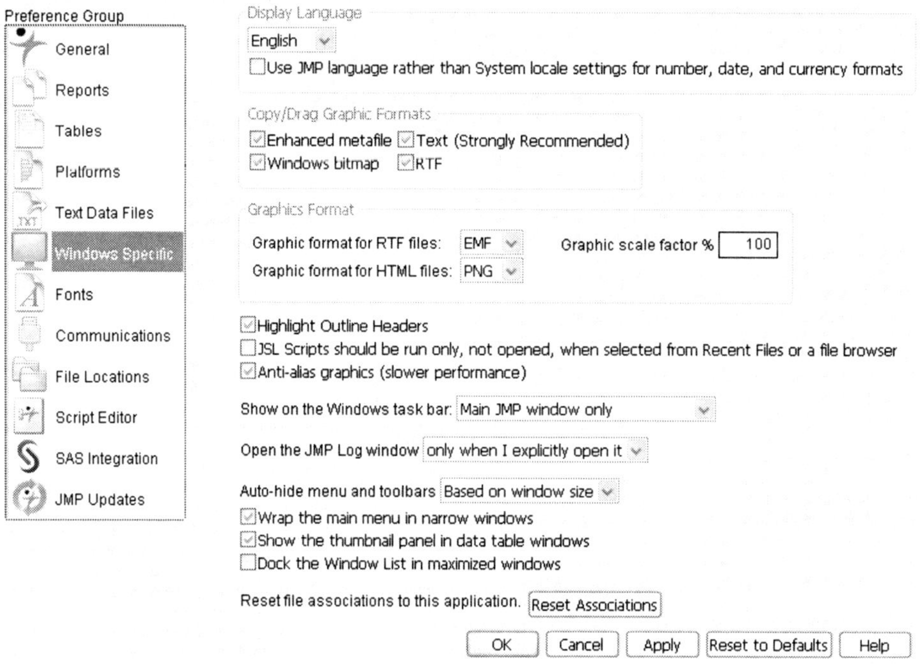

Table C.7 Windows Specific Preferences

Preference	Description
Display Language	Select the language in which you want to run JMP. The locale settings for your operating system normally determine settings for number, date, and currency formats. Select the option below the language menu to use that language to determine these formats instead. Text that appears in windows provided by the operating system (for example, **File > Open**), do not reflect changes in the language setting.
Copy/Drag Graphic Formats	Select one or more graphic formats to use when copying and pasting (or dragging and dropping) graphics from JMP into other applications.
Graphics Formats	Select the format to use for graphics when you save a JMP report as an RTF file or an HTML file. In the **Graphic Scale Factor %** box, enter the percentage at which you want graphics to appear in other applications. This feature might not work with all versions of your chosen application.

Table C.7 Windows Specific Preferences *(Continued)*

Preference	Description
Highlight Outline Headers	Select this option to give title bars in the data table and report windows a light shade of gray. This option is selected by default. Clear this option to give the title bars a darker shade of gray.
JSL Scripts should be run only, not opened, when selected from Recent Files or a file browser	Select this option to force all scripts to run when opened. If this is selected, the script window for the script is not opened. This option is off by default.
Anti-alias graphics (slower performance)	Select this option to use anti-aliased graphics. Anti-aliased graphics appear smoother, but take longer to draw. If you are running JMP on an older computer, clearing this option might enhance performance when drawing graphics. This option is selected by default. It is strongly recommended that you leave this option on.
Show on the Windows task bar	Select which JMP windows you would like displayed on the Windows task bar. • On Windows XP, the default selection is **Main JMP window only.** • On Windows Vista and Windows 7, the default selection is **All Windows.** • You can also choose to display the main JMP window and data table windows.
Open the JMP Log window	Select one of the options to open the log only when text (such as error messages) is added to it or every time you start JMP. The default setting displays the log only when you open it.
Auto-hide menu and toolbars	Select the rule used to determine when menus and toolbars are hidden. The default value is **Based on window size**.
Wrap the main menu in narrow windows	Select this option to wrap the menu to additional lines when the window is narrower is than the menu. This option is on by default.
Show the thumbnail panel in data table windows	Select this option to show the thumbnail panel of reports at the bottom of a data table. This option is on by default.
Dock the Window List in maximized windows	Select this option to automatically dock the Window List if you maximize your JMP windows. This option is off by default.
Reset file associations to this application	Click this button to associate all JMP file types with the JMP version that you are running.

Mac OS Settings

Mac OS Settings preferences customize settings for Macintosh machines, including the display language, graphics formats, and file associations.

Figure C.8 Mac OS Settings Preferences

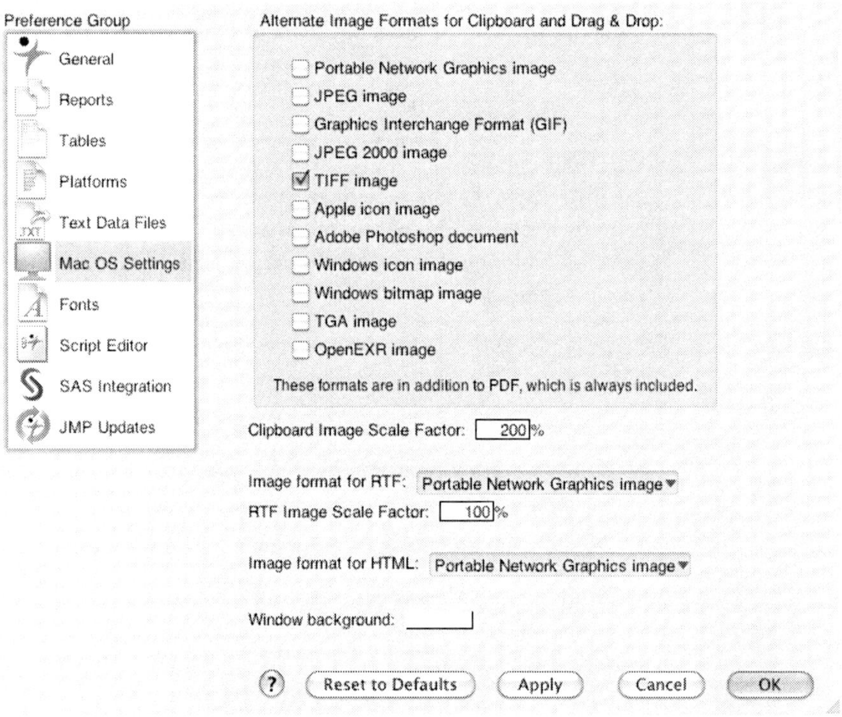

Table C.8 Macintosh OS Settings

Preference	Description
Alternate Image Formats for Clipboard and Drag & Drop	Select one or more graphic formats to use when copying and pasting (or dragging and dropping) graphics from JMP into other applications.
Clipboard Image Scale Factor	Enter the percentage at which you want graphics to appear in other applications.
Image format for RTF	Select the format to use for graphics when you save a JMP report as an RTF file.

Table C.8 Macintosh OS Settings *(Continued)*

Preference	Description
RTF Image Scale Factor	Enter the percentage at which you want graphics to appear in RTF documents.
Image format for HTML	Select the format to use for graphics when you save a JMP report as an HTML file.
Window background	Click the color box to select a background color for windows.

Fonts

Font preferences customize the appearance of reports, data tables, and scripts, including fonts, text size, and font style.

Figure C.9 Fonts Preferences

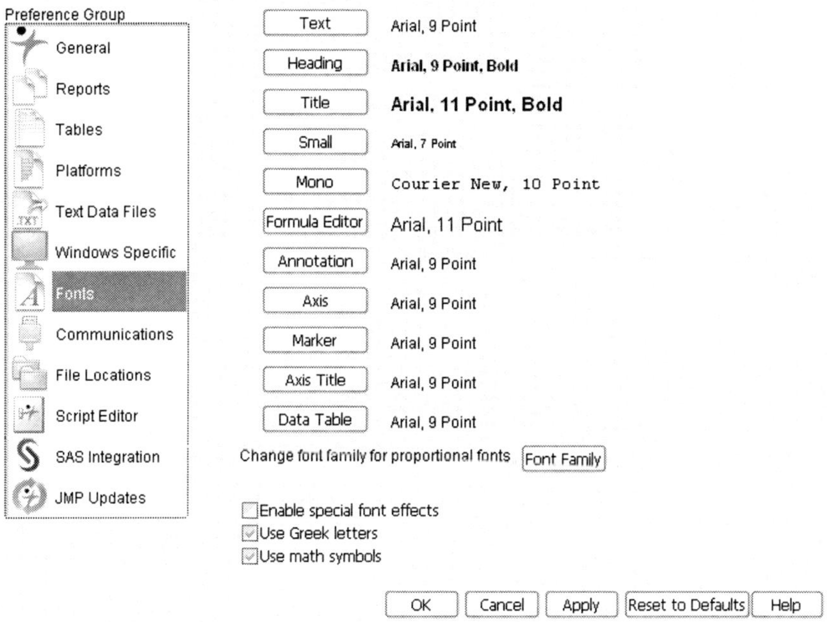

Table C.9 Preferences for Customizing Fonts

Preference	Description
Text	Sets the font for the text portion of a JMP analysis report.

Table C.9 Preferences for Customizing Fonts *(Continued)*

Preference	Description
Heading	Sets the font for the heading of columns in an analysis report and a data table.
Title	Sets the font for the title shown in all title bars.
Small	Sets the font for small text, which is used in the upper left corner of the data grid to show the number of columns and rows.
Mono	Sets the font used in the JMP Scripting Language (JSL) editor for script commands.
Formula Editor	Sets the font for the expressions entered into the formula editor.
Annotation	Sets the default font used in annotations.
Axis	Sets the font used for the axis tick labels.
Marker	Sets the font used for alphanumeric markers.
Axis Title	Sets the font for axis titles.
Data Table	Sets the font for displaying data in the data table.
Font Family	Select a proportional font. The font that you select is immediately applied to all settings except for **Mono**, which is used for scripts.
Enable special font effects	(Windows only) Select this option to enable setting **Underline** and **Strikeout** in the other font settings. This option is cleared by default.
Use Greek letters	Select this option to use Greek letters instead of spelling out Greek letters. (For example, π instead of pi.) This option is selected by default.
Use math symbols	Select this option to use math symbols instead of simple text representations of math symbols. (For example, ± instead of +/-) This option is selected by default.

Communications

(Windows only) Communications preferences customize settings for reading data from an external source. These settings only need to be specified if you are using an instrument to do so.

Figure C.10 Communications Preferences

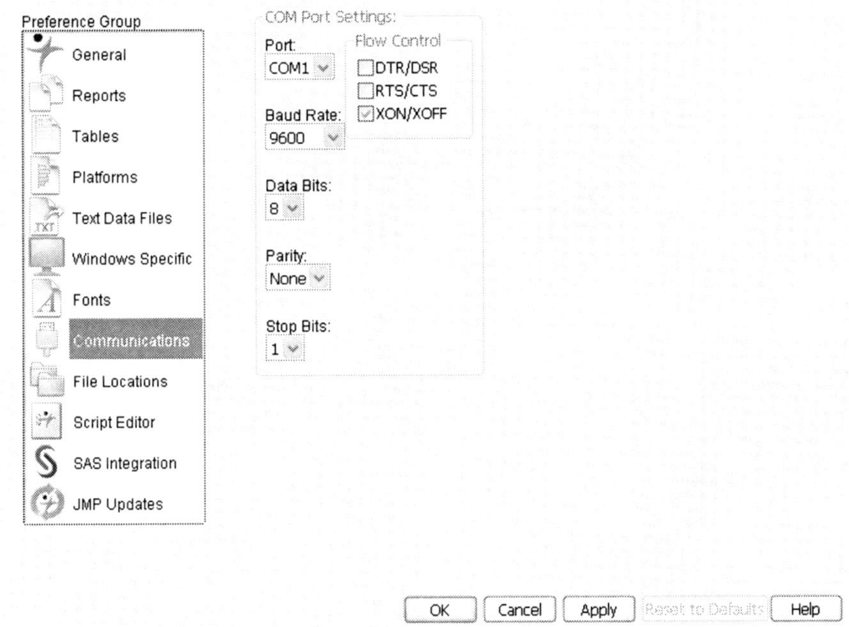

Table C.10 Preferences on the Communications Page

Preference	Description
Port	Select the port that your data source uses.
Baud Rate	Set the baud rate for your data source.
Data Bits	Set the number of data bits (7 or 8).
Parity	Set the parity bit. **None** is the default value.
Stop Bits	Set the stop bits (1 or 2).
Flow Control	Set the flow control. **XON/XOFF** is the default value.

File Locations

(Windows only) File Locations preferences set the default locations of JMP system files. Usually, the files can stay where JMP installs them, and you do not have to change anything here. However, if you do move files, such as Help files, elsewhere, you should update the location here.

Figure C.11 File Locations Preferences

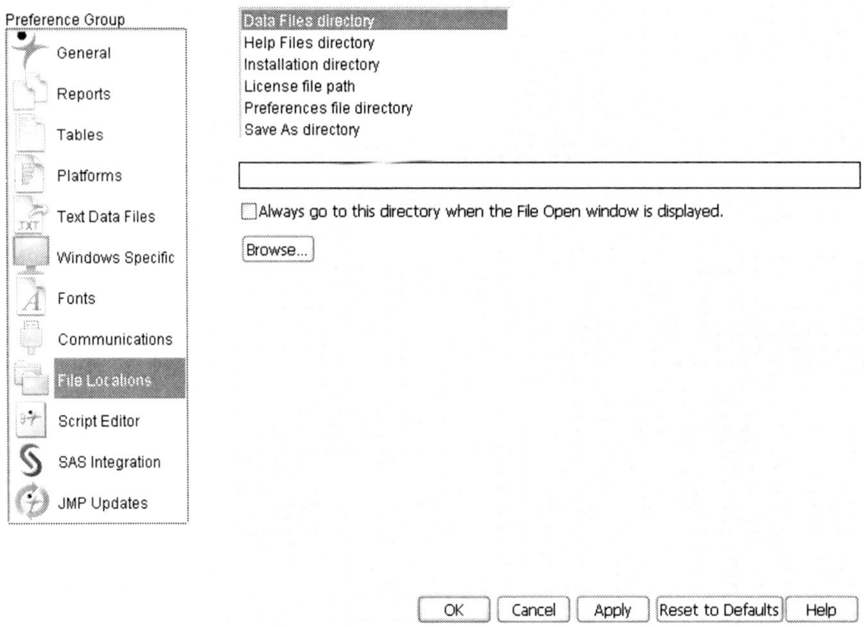

Table C.11 Preferences on the File Locations Page

Preference	Description
Data Files directory	Change the folder in which JMP looks for data files (for example, data tables).
Help Files directory	Change the folder in which JMP looks for Help files.
Installation directory	Change the JMP installation folder.
License file path	Change the folder in which JMP looks for your JMP License file.
Preferences file directory	Change the folder in which JMP looks for preference information.
Save As directory	Change the folder in which JMP saves data files when you select **File > Save As**.

Script Editor

Script Editor preferences customize the appearance of the JSL Script Editor, such as tab width, syntax coloring, and tooltips.

Figure C.12 Script Editor Preferences

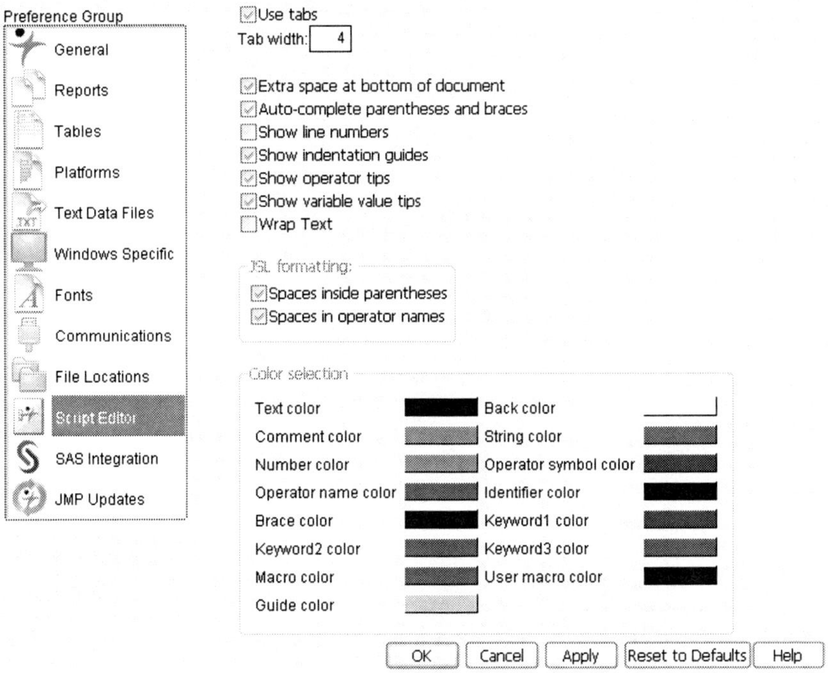

Table C.12 Preferences on the Script Editor Page

Preference	Description
Use tabs	Select this option to enable tabs in your scripts. This option is selected by default. Clear this option to replace any tab that you type with spaces.
Tab width	Enter how many spaces a tab should indent. If you have disabled tabs, any tab you type is replaced with the number of spaces specified. The default value is 4.
Extra space at bottom of document	Select this option to enable scrolling the last line of a script to the top of the script editor. This option is selected by default on Windows and deselected on Macintosh.

Table C.12 Preferences on the Script Editor Page *(Continued)*

Preference	Description
Auto-complete parentheses and braces	Select this option to enable the script editor to automatically add closing parentheses, square brackets, and curly braces when you type an opening one. This option is selected by default.
Show line numbers	Select this option to show the line numbers on the left side of the script editor. This option is cleared by default.
Show indentation guides	Select this option to see faint vertical lines that mark indention. This option is selected by default.
Show operator tips	Select this option to see tooltips for JSL operators. This option is selected by default.
Show variable value tips	Select this option to see tooltips for variable values. This option is selected by default.
Wrap Text	Select this option to always wrap text in the script editor. This option is off by default.
Spaces inside parentheses	Select this option to cause the script editor to add spaces between parentheses, brackets, and braces and their contents for automatically formatted scripts. This is on by default.
Spaces in operator names	Select this option to cause the script editor to add spaces between words within operator names. For example, turning on this option results in `New Window` instead of `NewWindow`. This option is selected by default.
Color Selection	To set your own color for any of the listed types, click the color box and select your color.

SAS Integration

SAS Integration preferences customize the default settings for working with SAS servers. For details about using the SAS Integration capabilities, see "External Data and Analytical Sources," p. 357.

Figure C.13 SAS Integration Preferences

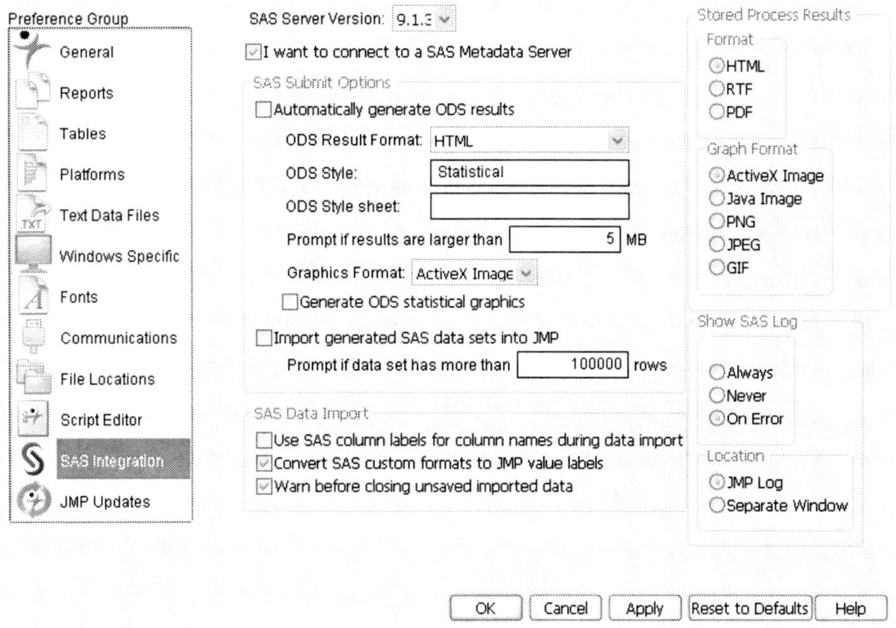

Table C.13 Preferences on the SAS Integration Page

Preference	Description
SAS Server Version	Select the default version for the SAS Server. This can also be changed in the Server Connections window. (See "To connect to a metadata server," p. 360.) This option does not apply when connecting to local SAS on Windows.
I want to connect to a SAS Metadata Server	Select this option to always connect to a SAS Metadata Server. This option is selected by default. Clear this option if you do not have a SAS metadata server available, and you connect directly to SAS workspace servers instead.
Automatically generate ODS results	Select this option to generate ODS results. This option is cleared by default.
ODS Result Format	Select the format for ODS reports. The default value is **HTML**.
ODS Style	Enter the style name for ODS reports. The default value is Statistical.
ODS Style sheet	Enter the style sheet name for ODS reports.

Table C.13 Preferences on the SAS Integration Page *(Continued)*

Preference	Description
Prompt if results are larger than __ MB	Enter the number of MB that triggers a prompt for you to continue or cancel the operation. The default value is 5.
Graphics Format	Select the format for graphics for ODS reports.
Generate ODS statistical graphics	Select this option to include statistical graphics in the ODS reports.
Import generated SAS data sets into JMP	Select this option to import any generated data sets into JMP automatically.
Prompt if data set has more than __ rows	Enter the number of rows that triggers a prompt for you to continue or cancel the operation. The default value is 100,000.
Use SAS column labels for column names during data import	Select this option to use the column labels in the SAS data set as the JMP data table column names when importing a SAS data set into a JMP data table.
Convert SAS custom formats to JMP value labels	Select this option to use the information in the SAS column formats to set JMP value labels when importing a SAS data set into a JMP data table.
Warn before closing unsaved imported data	Select this option to trigger a prompt to save any SAS data sets that you imported into JMP and have not saved before closing.
Stored Process Results: Format	Select a report format: **HTML**, **RTF**, or **PDF**.
Stored Process Results: Graph Format	Select a graph format: **ActiveX Image** (only available on a SAS server that runs on Windows), **Java Image**, **PNG**, **JPEG**, or **GIF**.
Show SAS Log	Select **Always**, **Never**, or **On Error** to set when the SAS log is displayed
Location	Select the location for SAS log information: within the JMP log window, or in a separate SAS log window.

JMP Updates

The JMP Update preferences show you what version of JMP you currently have, the last time it was updated, and whether and how often JMP checks automatically for updates.

Figure C.14 JMP Updates Preferences

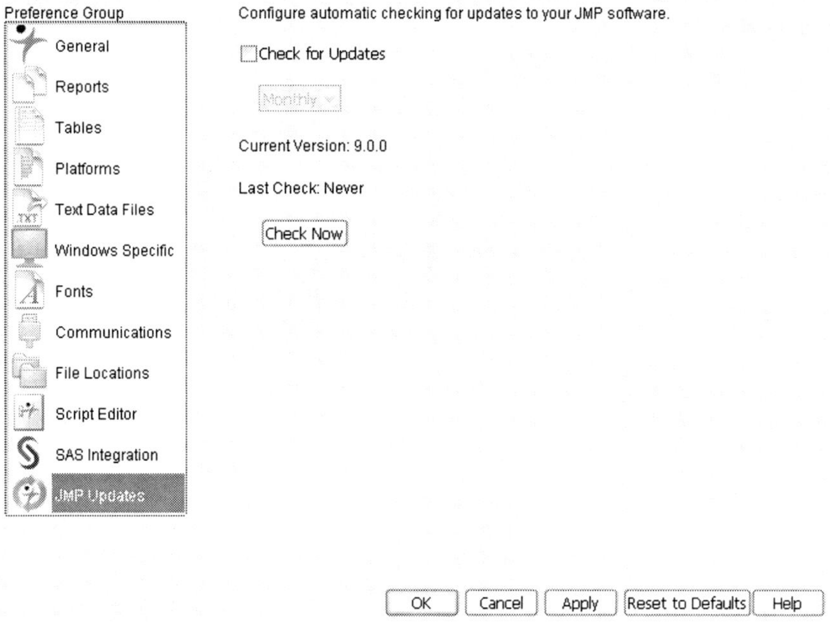

Table C.14 Preferences on the JMP Updates Page

Preference	Description
Check for Updates	Select this option to have JMP automatically check for updates. If this option is selected, specify how often the check is performed: **Daily**, **Weekly**, or **Monthly**.
Current Version	What version of JMP you are currently running.
Last Check	The last time you checked for an update.
Check Now	Click this button to perform an update check immediately.

Scripting-Only Preferences

The following preferences can be changed only by using a JSL script. See the *Scripting Guide* for information about scripting.

Table C.15 Scripting-Only Preferences

Preference	Syntax	Description
Disable Recent Files From JSL	Preferences(Disable Recent Files From JSL(0|1))	Sets the ability to run files on the JMP Recent Files list using the Main Menu command in JSL. Available only on Windows. 0 turns it off and 1 turns it on.
JMP 6 Scoping	Preferences(JMP 6 Scoping(0|1|)	Turns JMP 6 scoping on or off.
Warn On Scoping Conflict	Preferences(Warn On Scoping Conflict(0|1))	Turns the warning for possible scoping conflicts on or off.

Appendix D

Formula Functions Reference

A Description of Functions Available in JMP

For v10: Due to time constraints, screenshots of graphs could not be retaken and cataloged. Cross-references to those shots were removed from this chapter along with the figures themselves. Add them back for v10.

You can add functions to a formula. All of these functions are organized in the function browser, which groups collections of functions and features in lists organized both alphabetically (**Functions (all)**) and by topic (**Functions (grouped)**. This chapter gives a description of functions in the formula editor.

More information about functions is available in the following resources:

- *JSL Functions* describes all functions and their arguments, demonstrates how the functions work, and links to online Help. In JMP, select **Help > JSL Functions** to view this interactive resource.
- The *Scripting Guide* also provides the arguments for all JMP functions, not just those available in the formula editor. In JMP, select **Help > Books > Scripting Guide** to open a PDF of the *Scripting Guide*.

Figure D.1 Functions in the Formula Editor

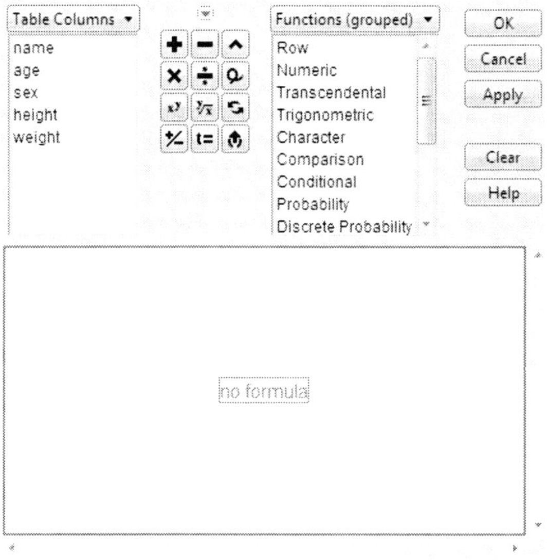

For instructions on how to create a formula that contains a function, see "Creating a Formula," p. 303.

Contents

Row Functions

Adding a row function to a formula lets you reference specific rows or cells within specific rows. You can also insert values based on an arithmetic sequence. See the *Scripting Guide* for details about syntax.

Sequence

Produces an arithmetic sequence of numbers across the rows in a data table, where the start value, ending limit, and increment are specified as arguments.

Count

Creates a list of values beginning with the **from** value and ending with the **to** value. The number of **steps** specifies the number of values in the list between and including the **from** and **to** values. Each value determined by the first three arguments of the count function occurs consecutively the number of times that you specify with the **times** argument. When the **to** value is reached, **Count** starts over at the **from** value.

Also, you can add the **times** argument with the insert button on the keyboard. This argument is one by default, but repeats the count process as many times as you specify, as illustrated by the Count4 column in the data table in Figure D.2. To add any argument to the **Count** function, highlight the argument preceding the one that you want to enter. Either type a comma or use the insert button on the formula editor keypad.

The columns in the data table below result from the following formulas:

- **Count (1, 9, 2)** gives Count 1
- **Count (1, 9, 3)** gives Count 2
- **Count (1, 9, 9)** gives Count 3
- **Count (1, 9, 3, 3)** gives Count 4

Figure D.2 Example of the Count Function

	count1	count2	count3	count4
1	1	1	1	1
2	9	5	2	1
3	1	9	3	1
4	9	1	4	5
5	1	5	5	5

The **Count** function is useful for generating a column of grid values. For example, the following formulas create a square grid of increment **NRow(). NRow()** is the **Row** function that gives the total number of rows in the data table) and axes that range from –5 to 5:

```
Count (-5, 5, Root(NRow()))
Count (-5, 5, Root(NRow(), Root(NRow())))
```

Lag

Returns the value of the first argument in the row defined by the current row less the second argument. The default **Lag** is one, which you can change to any number. The value returned for any lag that identifies a row number less than one is missing. Note that `Lag(X, n)` gives the same result as the subscripted notation, $X_{\text{Row( )-n}}$.

Dif

Returns the difference between the value of the first argument in the current row and its value in the row defined by the current row less the second argument. The default **Dif** is one, which you can change to any number. Note that `Dif(X, n)` gives the same result as $X_{\text{Row()}} - X_{\text{Row()-n}}$, or as $X_{\text{Row()}}$`-Lag(X, n)`.

Subscript

Enables you to use a column's value from a row other than the current row. After choosing **Subscript** from the list, enter a numeric expression into the subscript argument. Subscripts that evaluate to nonexistent row numbers produce missing values. Column names with no subscript refers to the current row. To remove a subscript, select the subscript and delete it. Then delete the missing box.

The formula $\text{Count}_{\text{Row()}} - \text{Count}_{\text{Row()-1}}$, where **Row()** is the row number as described below, uses subscripts to calculate the difference between each pair of values from the column named Count. This result is the same as that given by the **Dif()** function. When **Row()** is 1, the computation produces a missing value.

The formula below calculates a column called Fib, which contains the terms of the *Fibonacci* series (each value is the sum of the two preceding values in the calculated column).

$$\text{If}\begin{bmatrix} \text{Row()}<=2 \Rightarrow 1 \\ \text{else} \quad \Rightarrow Fib_{\text{Row()-1}} + Fib_{\text{Row()-2}} \end{bmatrix}$$

It shows the use of subscripts to do recursive calculations. A recursive formula includes the name of the calculated column, subscripted such that it references only previously evaluated rows (rows 1 through (i–1)). The calculation of the Fibonacci series shown includes a conditional expression and a comparison. See the sections "Conditional Functions," p. 489, and "Comparison Functions," p. 488, for details.

Row

Returns the current row number when an expression is evaluated for that row. You can use **Row()** in any expression, including column name subscripts. The default subscript of a column name is **Row()** unless otherwise specified.

NRow

Returns the total number of rows in the active data table.

Numeric Functions

You can create a formula that contains arithmetic operators that are commonly used in formulas. See the *Scripting Guide* for details about syntax.

Abs

Returns a positive number of the same magnitude as the value of its argument. For example, |5| and |–5| both result in 5.

Modulo

Returns the remainder when the second argument is divided into the first. For example, **Modulo(6, 5)** results in 1.

Ceiling

Returns the smallest integer greater than or equal to its argument. For example, **Ceiling(2.3)** results in 3, while **Ceiling(–2.3)** results in –2.

Floor

Returns the largest integer less than or equal to its argument. For example, **Floor(2.7)** results in 2, but **Floor(–0.5)** results in –1.

Round

Rounds the first argument to the number of decimal places given by the second argument. For example, **Round(3.554, 2)** rounds to 3.55 and **Round(3.555, 2)** rounds to 3.56.

Transcendental Functions

You can create a formula that supports transcendental functions, such as logarithmic functions for any base, functions for combinatorial calculations, the **Beta** function, and several gamma functions. See the *Scripting Guide* for details about syntax.

Exp

Raises *e* to the power that you specify. Thus, **Exp(1) = e**.

Log and Log10

Calculates the natural logarithm (base *e*). To change the default base, highlight the argument and type a comma or click the insert key on the keypad. The base appears and is editable. The **Log** argument can be any numeric expressions. The expression **Log(e)** evaluates as 1, and **Log2(32)** is 5. The **Log10** function calculates the logarithm of base 10 only.

Squish

Is an efficient computation of the function $1 / (1+e^{-x})$, where *x* is any numeric column, variable, or expression.

Root

Calculates the root of its argument as specified by the index. **Root** initially shows with an index of 2. To change the index, highlight the index argument and enter the value that you want.

Factorial

Returns the product of all numbers 1 through the argument that you specify. For example, **Factorial(5)** evaluates as 120.

N hooseK

Returns the number of *n* things taken *k* at a time (*n* select *k*) and is computed in the standard way using factorials, as $n! / (k!(n-k)!)$. For example, **NChooseK(5,2)** evaluates as 10.

Beta

Adds the two parameter **Beta** function and is written terms of the Gamma function as:

$$B(m, n)) = \frac{\Gamma(m)\Gamma(n)}{\Gamma(m+n)}$$

Gamma

Adds the **Gamma** function, denoted $\Gamma(i)$, and is defined as:

$$\Gamma(i) = \int_0^{\infty} (x^{i-1})(e^{-x})dx$$

Gamma with a single argument is the same as **Gamma(x, infinity)**. The optional second argument changes the upper integer from infinity to the value that you enter. Other interesting gamma function relationships are

- for any $\alpha > 1$, $\Gamma(\alpha) = (\alpha-1) \bullet \Gamma(\alpha-1)$
- for any positive integer, n, $\Gamma(n) = (n-1)!$
- $\Gamma(0.5)$ = the square root of π

LGamma

Is the natural log of the result of the gamma function evaluation. You get the same result using the **Log** (natural log) function with the **Gamma** function. However, the **LGamma** function computes more efficiently than do the **Log** (natural log) and the **Gamma** functions together. **NChooseK** is implemented using **LGamma** functions. The result is not always an exact integer. If the result is close to an integer, it is rounded up using the **Floor** function.

Digamma

The logarithmic derivative of the **Gamma** function.

Trigamma

The derivative of the **Digamma** function, or the logarithmic second derivative of the **Gamma** function.

Arrhenius

Calculates the non-specific component of the Arrhenius relationship that is then multiplied by the activation energy in the Arrhenius equation.

$$\frac{11605}{T + 273.15}$$

Arrhenius Inv

The inverse of the Arrhenius function:

$$\left(\frac{11605}{y}\right) - 273.15$$

Logit

Applies the logit transformation to the argument using

$$\text{logit}(x) = \log\frac{x}{1-x}$$

Scheffe Cubic

Is used in fitting certain models. **Scheffe Cubic (X1, X2)** is equivalent to **X1*X2*(X1-X2)**.

Trigonometric Functions

You can create a formula that supports transcendental functions, such as logarithmic functions for any base, functions for combinatorial calculations, the **Beta** function, and several gamma functions. See the *Scripting Guide* for details about syntax.

Sine, Cosine, Tangent

The **Sine** and **Cosine** functions calculate the sine and cosine of their respective arguments given in radians. For example, the expression **Sine(0)** evaluates as 0, and **Cosine(0)** evaluates as 1. The tangent function calculates the tangent of an argument given in radians. The expression **Tan(.25)** evaluates as 0.255342.

ArcSine, ArcCosine, ArcTangent

The **ArcSine** and **ArcCosine** functions return the inverse sine and inverse cosine of their respective arguments. The returned value is measured in radians. For example, both expressions **ArcSine(1)** and **ArcCosine(0)** evaluate as 1.57080. The **ArcTangent** function returns the inverse tangent of its argument. The returned value is measured in radians. The expression **ArcTangent(0.5)** evaluates as 0.46364.

SinH, CosH, TanH

The **SinH** and **CosH** functions return the hyperbolic sine and hyperbolic cosine of their respective arguments. The expression **SinH(1)** evaluates as 1.175201, and **CosH(0)** evaluates as 1.0. The **TanH** function returns the hyperbolic tangent of its argument. The expression **TanH(1)** evaluates as 0.761594.

ArcSinH, ArcCosH, ArcTanH

The **ArcSinH** and **ArcCosH** functions return the inverse hyperbolic sine and inverse hyperbolic cosine of their respective arguments. The expression **ArcSinH(1)** evaluates as 0.881374, and **ArcCosH(1)** is 0. The **ArcTanH** function returns the inverse hyperbolic tangent of its argument. The expression **ArcTanH(0.5)** evaluates as 0.549306.

Character Functions

You can create a formula that accepts character arguments or returns character strings and converts the data type of a value from numeric to character, or character to numeric. When you create these formulas, note that:

- Character functions can result in either character or numeric data. If you calculate a data type different from the one specified, the data type of the computed column is automatically changed to match the result.
- Arguments that are literal character strings must be enclosed in quotation marks.

See the *Scripting Guide* for details about syntax.

Char

Produces a character string that corresponds to the digits in its numeric argument. For example, **Char(1.123)** evaluates as 1.123. See the *Scripting Guide*, for details.

Concat ||

Concatenates character strings to produce a new string with the function's second character argument appended to the first. For example, **"Dr." || " " || name** produces a new string consisting of the title **Dr.** followed by a space and the contents of the *name* string. (See also "Concat Items," p. 484.)

Contains

Returns the numeric position within the first argument of the first instance of the second argument, if it exists. The second argument can contain one ore more characters. If the second argument does not exist, **Contains** returns a zero. For example, **Contains("Veronica Layman", "ay")** evaluates as 11. **Contains("Lillie Layman", "L")** evaluates as 1. The third argument is optional and is a numeric value that specifies the starting position. If *offset* is negative, **Contains** searches backward from *offset* from the end of the string.

Munger

Computes new character strings from existing strings by inserting or deleting characters. It can also produce substrings, calculate indices, and perform other tasks depending on how you specify its arguments. The **Munger** function treats uppercase and lowercase letters as different characters.

Text is a character expression. **Munger** applies the other three arguments to this string to compute a result.

Offset is a numeric expression indicating the starting position to search in the string. If **Offset** is greater than the position of the first instance of the find argument, the first instance is disregarded.

Find/Length is a character or numeric expression. Use a character string as search criterion, or use a positive integer to return that number of consecutive characters starting from the **Offset** position. If you specify a negative integer as the **Length** value, **Munger** returns all characters from the **Offset** through to the end of the string.

Replace (optional argument) can be a string or unspecified. If it is a string and the **Find/Offset** value is numeric, **Munger** replaces the search criterion with the **Replace string** to form the result. If the Find/Offset value is numeric and no string is specified, **Munger** calculates a substring. If the **Find/Length** value is a character string, **Munger** always returns the numeric offset, disregarding the **Replace** value if it exists. To insert the **Replace** argument, click any argument in the **Munger** function and then click the insert button. Use the delete key on your keyboard or the delete button ( ) on the formula editor keypad to remove the **Replace** argument.

Lowercase, Uppercase

The **Lowercase** function converts any uppercase character found in its argument to the equivalent lowercase character. For example, **Lowercase("VERONICA LAYMAN")** evaluates as veronica layman. The **Uppercase** function converts any lowercase character found in its argument to the equivalent uppercase character. For example, **Uppercase("Veronica Layman")** evaluates as VERONICA LAYMAN.

Length

Calculates the length of its argument. For example, **Length("Veronica")** evaluates as 8. If the argument is

- a string, length returns the number of characters;
- a list, length returns the number of items in the list;
- a blob (binary object), the number of bytes.

Num

Produces a numeric value that corresponds to its character string argument when the character string consists of numbers only. If a character string contains a non-numeric value, the result is a missing value. For example, **Num("1.123")** evaluates as 1.123.

Substr

Extracts the characters that are the portion of the first argument. Begins at the position given by the second argument, and ends based on the number of characters specified in the third argument. The first argument can be either a character column or a literal value. The starting argument and the length argument can be numbers of expressions that evaluate to numbers. For example, to show the first name only, **Substr("Veronica Layman", 10, 6)** starts at position 11 and reads through position 16, which yields **Layman.**

If *start* is negative, **Substr** searches backward from *start* from the end of the string. If *length* is negative or absent, **Substr** returns a string that begins with *start* and continues to the end of *s*.

Substr can also be used with lists.

Trim

Produces a new character string from its argument, removing any trailing blanks. For example, **Trim("john ")** evaluates as john.

Word

Extracts the nth word from a character string. One or more spaces define where each word begins and ends unless the optional **delimiters** argument is specified. For example, **Word(2, "Veronica Layman")** returns the word **Layman**.

To insert the **delimiters** argument, click on any argument in the **Word** function and then click the insert button on the formula editor keypad. Use the delete key on your keyboard or the delete button on the formula editor keypad to remove the **delimiters** argument. If you do not specify a delimiter, space is used as the delimiter. If you define the delimiter as an empty string, each character is treated as a separate word.

Most special characters act as single delimiters. You can enter any character or set of characters to act as a word delimiter. For example, to extract the last name in the following example, use a comma and blank together as the delimiting characters and ask for the first word. **Word(1, "Layman, Veronica", ", ")** returns the word **Layman**.

Words

Extracts the words from *text* according to the delimiters listed in the optional second argument. The default delimiter is space. For example, **Words("the quick brown fox")** returns {**"the","quick","brown","fox"**}.

If you include a second argument, any and all characters in that argument are taken to be delimiters. For example, **Words("Doe, Jane P.",", .")** returns {**"Doe","Jane","P"**}.

To insert the **delimiters** argument, click on any argument in the **Words** function and then click the insert button on the formula editor keypad. Use the delete key on your keyboard or the delete button on the formula editor keypad to remove the **delimiters** argument. If you do not specify a delimiter, white space is used as the delimiter. If you define the delimiter as an empty string, each character is treated as a separate word.

Left, Right

Returns a substring of the left-most or right-most *n* characters of the string *text*, respectively. Both functions also work with lists.

Starts With, Ends With

Returns 1 if *whole* begins or ends with *part*, respectively. Returns 0 otherwise. Both functions also work with lists.

Item

Is different than the **Word** function because of the way it treats word delimiters. If a delimiter is found multiple times, or you enter a delimiter with multiple characters, the **Word** function treats them as a single delimiter. The **Item** function uses each delimiter to define a new word position. To compare, suppose a name is of the form lastname, firstname. The delimiter is a comma followed by a blank, such as:

Item(2, "Layman, Veronica", ", ")

Word(2, "Layman, Veronica", ", ")

The **Item** function returns a missing value because it treats the comma and blank separately and finds nothing between them. The **Word** function treats the comma and blank as a single delimiter and finds **Veronica** as the second word.

If you do not specify a delimiter, white space (blank space) is used as the delimiter. If you define the delimiter as an empty string, each character is treated as a separate item.

Char to Hex, Hex, Hex to Char, Hex to Number

Converts between Hex and other formats.

Hex returns the hexadecimal representation of its argument. If the argument is character (in quotes), then the result is a character string twice as long containing the hexadecimal codes for the character values. For example, **Hex("A")** returns the string 41.

If the argument is numeric and "integer" is specified, the Hex function returns an 8-hexadecimal-character representation of the integer returned. For example, **Hex(12, "integer")** returns the string **0000000C**.

Hex to Char converts hexadecimals to characters. The resulting character string might not be valid display characters. All the characters must be in pairs, in the ranges 0-9,A-Z, and a-z. Blanks and commas are allowed and skipped.

Char to Hex converts characters to hexadecimals.

Hex to Number converts hexadecimals to numbers.

For details, see the *Scripting Guide* book.

Repeat

Creates a string that is the first argument repeated the number of times specified by the second argument. The first argument can be either a character literal, a character variable, or a character expression. For example, **Repeat("Katie", 3)** creates **KatieKatieKatie**.

A third argument applies when **Repeat** is used in a JSL script to repeat a matrix. When the first argument is a matrix, the second argument is the rowwise repeat and the third argument is the columnwise repeat.

Insert, Insert Into

Insert inserts a new item into the *list* or *expression* at the given position. If position is not given, it is inserted at the end.

Insert Into is the same as insert, but it inserts in place.

Remove, Remove From

Remove the character(s) at the indicated *position*. If *n* is omitted, the item at *position* is deleted. If position and n are omitted, the item at the end is removed. There are three possible arguments: the string, followed by the position, followed by the number of characters to be removed.

Remove From returns items removed in place. The function returns the removed item(s), but you do not have to assign them to anything. The first argument is a variable name, followed by the position, followed by the number of characters to be removed.

Shift, Shift Into

Shift shifts an item or *n* items from the front to the back of the *list* or *expression*. Shifts items from back to front if *n* is negative. **Shift Into** shifts items in place.

Reverse, Reverse Into

Reverse reverses the characters in the string. **Reverse Into** reverses the characters in place.

Concat Items

Concat Items converts a list of string expressions into one string, with each item separated by a delimiter. The delimiter is a blank, if unspecified.

Substitute, Substitute Into

The first argument is a string, the second is a pattern, and the third is a replacement string. **Substitute** finds all matches to the pattern in the string, and replaces them with the replacement string. **Substitute Into** does the same substitution in place.

Regex

The first argument is the source string that Regex searches for a match to the pattern. The second argument is the pattern, in the form of a regular expression. The Formula Editor prompts you for these two required arguments.

Tip: For more information about using regular expressions, search the Internet for *regular expression tutorial.*

By default, Regex performs a case-sensitive search and returns the parts of the source string that match the pattern that you specified (or returns MISSING if the match fails). There are two optional arguments that you can add. You can type a third argument—the *format*—that specifies the string to return. If you choose, you can use regular expressions to specify replacement text in the returned string. If you specify the third argument, you can also specify **IGNORECASE** so that Regex ignores capitalization when searching the source string for a match.

Table D.1 Regex Examples

Sample Regex function	String that is returned
`Regex( "@ q3 #", "([a-z])([0-9])" )`	q3 The function is case sensitive, so q3 matches but Q3 would not.
`Regex( "@ Q3 #", "([a-z])([0-9])", "\0",IGNORECASE)`	Q3 Although \0 is the default argument, it is required in this example so that IGNORECASE can be specified.
`Regex( "@ Q3 #", "([a-z])([0-9])", "\2\1",IGNORECASE)`	3Q

For more information and an example that you can run, select **Help > Indexes > JSL Operators** and select **Regex**.

Hex to Blob, Char to Blob, Blob to Char

Hex to Blob converts the hexadecimal to a blob (Binary Large Object).

Char to Blob converts the string to a blob. You can specify the encoding in an optional second argument. Supported encodings are: utf-8, utf-16le, utf-16be, us-ascii, iso-8859-1, and ascii~hex.

Blob to Char converts the blob to a string. You can specify the encoding in an optional second argument. Supported encodings are: utf-8, utf-16le, utf-16be, us-ascii, iso-8859-1, and ascii~hex.

Character Pattern Functions

These functions provide powerful pattern matching abilities. Pattern matching is a flexible method for searching and manipulating strings, and regular expressions are also supported. When you create these formulas, note that:

- First, you define a pattern with one more of the character patterns.
- Then, you use **Pat Match** to compare a string to the pattern.
- **Pat Match** returns True (1) if the pattern is found in the string, or it returns False (0) if the pattern was not found in the string.
- To use regular expressions instead of patterns, use **Regex Match**.

For complete details, see the *Scripting Guide*.

Pat Any

Constructs a pattern that matches a single character in the argument.

Pat Not Any

Constructs a pattern that matches a single character that is not in the argument.

Pat Break

Constructs a pattern that matches zero or more characters that are not in its argument; it stops or breaks on a character in its argument. It fails if a character in its argument is not found. In particular, it fails to match if it finds the end of the source string without finding a break character.

Pat Span

Constructs a pattern that matches one or more (not zero) occurrences of characters in its argument. It is greedy; it always matches the longest possible string. It fails rather than matching zero characters.

Pat String

Constructs a pattern that matches its string argument.

Pat Len

Constructs a pattern that matches n characters.

Pat Pos

Constructs patterns that match the null string if the current position is *int* from the left end of the string, and fail otherwise.

Pat R Pos

Constructs patterns that match the null string if the current position is *int* from the right end of the string, and fails otherwise.

Pat Tab

Constructs a pattern that matches forward to position *int* in the source string. It can match 0 or more characters. It fails if it would have to move backwards or beyond the end of the string.

Pat R Tab

Constructs a pattern that matches up to position n from the end of the string. It can match 0 or more characters. It fails if it would have to move backwards or beyond the end of the string.

Pat Test

Constructs a pattern that succeeds and matches the null string if *expr* is not zero and fails otherwise.

Pat At

Constructs a pattern that matches the null string and stores the current position in the source string into the specified JSL variable (*varName*). The assignment is immediate, and the variable can be used with `expr()` to affect the remainder of the match.

Pat Rem

Constructs a pattern that matches the remainder of the string. It is equivalent to patRTab(0).

Pat Arb

Constructs a pattern that matches an arbitrary string. Initially it matches the null string. It matches one additional character each time the pattern matcher backs into it.

Pat Succeed

Constructs a pattern that always succeeds, even when the matcher backs into it. It matches the null string.

Pat Fail

Constructs a pattern that fails whenever the matcher attempts to move forward through it. The matcher backs up and tries different alternatives. If and when there are no alternatives left, the match fails and `Pat Match` returns 0.

Pat Abort

Constructs a pattern that immediately cancels the pattern match. The matcher does not back up and retry any alternatives. Conditional assignments are not made. Immediate assignments that were already made are kept.

Pat Fence

Constructs a pattern that succeeds and matches the null string when the matcher moves forward through it, but fails when the matcher tries to back up through it. It is a one-way trap door that can be used to optimize some matches.

Pat Arb No

Constructs a pattern that matches zero or more copies of pattern.

Pat Repeat

Matches pattern between minimum and maximum times.

Pat Conditional

Saves the result of the pattern match, if it succeeds, to a variable named as the second argument (type) after the match is finished.

Pat Immediate

Saves the result of the pattern match to a variable named as the second argument (*varName*) immediately.

Pat Altern

Constructs a pattern that matches any one of the pattern arguments.

Pat Concat

Constructs a pattern that matches each pattern argument in turn.

Pat Regex

Constructs a pattern that matches the regular expression in the quoted string argument.

Pat Match

Pat Match executes a pattern match using the source in the first argument and the pattern in the second argument. The pattern must be constructed first, either inline or by assigning it to a JSL variable elsewhere. A third argument, if present, is the replacement text for the matched characters in the source argument (if the source argument is a variable). Pat Match returns true if the match succeeds. Additional arguments, in any order, are ANCHOR (match must begin at start of source), FULLSCAN (turn off some optimizations for special situations), and MATCHCASE (by default, A == a).

Pat Match returns true or false rather than a string, so Pat Match is somewhat difficult to use in a formula. You might find the Regex function (p. 484) easier to use when you are adding pattern-matching formulas in the Formula Editor.

Regex Match

Regex Match is similar to Pat Match. Regex Match executes a pattern match using the source in the first argument and the pattern in the second argument. Regex Match uses a regular expression for the second argument and returns a list of information about the result of the match.

A simpler function, Regex (p. 484), is also available. Regex returns a string value rather than a list, so Regex is usually easier to use in the Formula Editor than RegEx Match.

Comparison Functions

You can create a formula that compare the values of two arguments by using the comparison function. Each comparison relationship evaluates as *true* or *false* based on numeric magnitudes or character rankings. A true relationship evaluates as one, and false evaluates as zero.

Comparisons are useful when you include them in conditional expressions, but they can also stand alone as numeric expressions if neither term in comparison is missing. A relational symbol's arguments can be any two expressions. However, both arguments in a comparison function must be of the same data type. Also note that:

- JMP displays an error if you use a single "=" in a conditional where "==" is expected.
- The formula editor uses the International Utilities package when comparing character strings. This package contains different rankings for each international character set and takes diacritical marks into consideration.
- You should not use comparison operators to specifically compare to a missing value. Instead, use the **Is Missing** function to detect a missing value.

See the *Scripting Guide* for details about syntax.

< Less than

> Greater than

<= Less than or equal to

>= Greater than or equal to

== Equal to

!= Not equal to

a<b<=c *b* is greater than *a* and less than or equal to *c*

a<=b<c *b* is greater than or equal to *a* and less than *c*

Is Missing Returns a one (1) if the value of the argument for the current row is missing, and a zero if the value is not missing. The formula editor excludes missing numeric values from its statistical calculations.

Conditional Functions

You can include conditional expressions (called *conditionals* for short) in your formulas. These expressions let you build a sequence of clauses paired with result expressions. Constructing a sequence of clauses is the way you conditionally assign values to cells in a calculated column. Conditionals follow these rules:

- When no clause is true, the formula editor evaluates the result expression that accompanies the else clause.
- All result expressions in a conditional expression must evaluate to the same data type.
- A missing term matches any data type.
- By definition, expressions that evaluate as zero are false.
- If an expression evaluates as missing, no clauses are executed and missing is returned. All other numeric expressions are true.

See the *Scripting Guide* for details about syntax.

Use the insert and delete clause buttons on the formula editor panel to expand the expression. For maximum efficiency, list the most frequently evaluated clause and result pairs first in the sequence.

Note: Interpolate, **Step**, **For**, and **While** are most often used in conjunction with other commands to build a JSL script. You can use the formula editor to create and execute a script in that column, but this is not recommended because of dependencies and ambiguities that can result. Most often, scripts are stored as .JSL files, and can be saved with a data table as a table property. See "The Table Panel," p. 48, for details about table properties, and see the *Scripting Guide* for documentation of all scripting commands.

If

Shows a single **If** condition with a missing expression and a missing **then clause**. Highlight either **expr** or **then clause** and enter a value. For example, to calculate count as a percentage of total when total is not 0, enter the conditional expression (using columns called **count** and **total**) in Figure D.3.

Figure D.3 A Conditional Expression

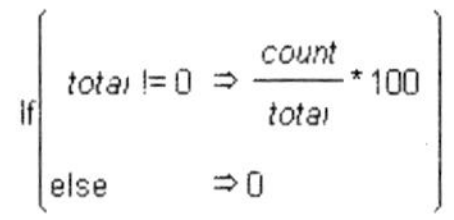

To add a new condition to the **If** conditional, highlight **then clause** and click the insert button ( ) on the formula editor keypad. Initially, this changes the existing **else** condition to an **expr** clause. Click the insert button again to add an **else clause**. Highlighting **then** or **else** and repetitively clicking the insert button changes the **else** to **expr** or adds a new **expr** clause.

To delete a clause, select the **then clause** above it and press the delete key on your keyboard or click the delete button ( ) on the formula editor keypad.

By definition, expressions that evaluate as zero are false. If an expression evaluates as missing, no clauses are executed and missing is returned. All other numeric expressions are true.

Match

Compares an expression to a list of clauses and returns the value of the resulting expression for the first matching clause encountered. You provide the matching expression only once and then give a match value for each clause.

After you select **Match** in the formula editor, a list appears with two options:

- Select **Add Match Arguments from Data**, and clauses that correspond to all of the levels in your data are added automatically. Alternatively, hold down the SHIFT key, select **Conditional**, and then select **Match**. In Figure D.4, the example on the left shows clauses that were added automatically.
- Select **Don't Add** so that you can add each clause individually. In Figure D.4, the example on the right shows an empty clause, which you fill with the missing expressions.

Figure D.4 Examples of Using the Match Function

automatically added Match arguments

empty Match argument

In an automatically filled argument, you should highlight **then clause**, and then enter an expression. In an empty argument, you highlight either **expr**, **value**, or **then clause**, and then enter an expression. (Or, if you highlight an expression and click **Match**, the formula editor creates a new **Match** conditional, with the original highlighted expression as **expr** and nothing for the **value** and **else clause.**) Also, keep in mind that:

- **Match** evaluates faster and uses less memory than an equivalent **If** because the variable is evaluated only once for each row in the data table. The **If** condition must evaluate the variable at each **If** clause for each row until a clause evaluates as true. See “Comparison Functions,” p. 488, for a comparison of **Match** and **If** conditionals.
- With **If** and **Match**, the formula editor searches down from the top of the sequence for the first true clause and evaluates the corresponding result expression. Subsequent true clauses are ignored.

In the following example, each value is assigned depending on the value of the *age* variable.

Figure D.5 An Example of Using the Match Function

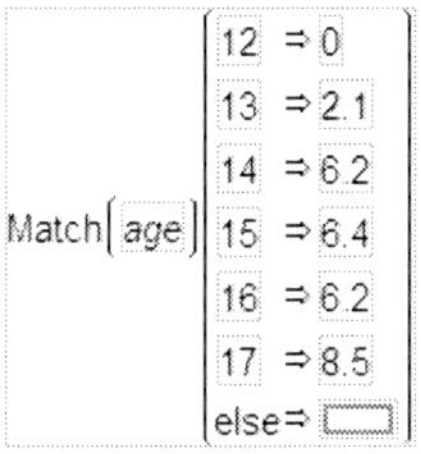

Note: **Match** ignores trailing spaces, while **If** does not.

Although `Match` returns missing for any missing values, you can also specifically match missing values.

Choose

Choose is a special case of **Match** in which the arguments of the condition are a sequence of integers starting at one. The value of **clause** replaces the match condition. An example of a **Choose** condition is shown in Figure D.6. With **Choose**, the formula editor goes directly to the correct choice clause and evaluates the result expression.

Figure D.6 Example of a **Choose** Condition

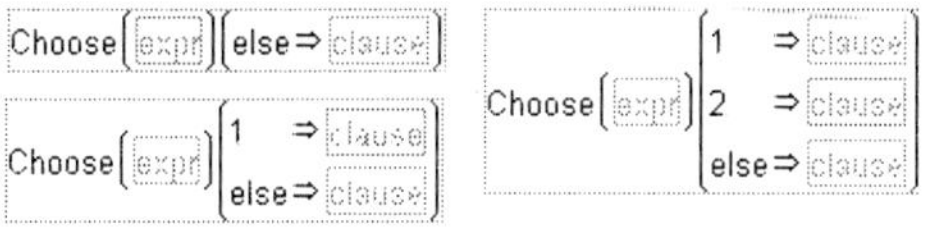

When you highlight an expression and click **Choose**, the formula editor creates a new conditional expression with one **clause**. Use the insert ( ) and delete ( ) buttons on the keypad to add new clauses or remove unwanted clauses, as described previously for the **If** conditional.

Choose evaluates the **choose** expression and goes immediately to the corresponding result expression to generate the returned value. With **Choose**, you provide a choosing expression that yields sequential integers starting at 1 only once, and then you give a choice for each integer in the sequence.

IfMax

Evaluates the first of each pair of arguments and returns the evaluation of the result expression (the second of each pair) associated with the maximum of the expressions. If more than one expression is the maximum, the first maximum is returned. If all expressions are missing and a final result is not specified, missing is returned. If all expressions are missing and a final result is specified, that final result is returned. The test expressions must evaluate to numeric values, but the result expressions can be anything.

IfMin

Evaluates the first of each pair of arguments and returns the evaluation of the result expression (the second of each pair) associated with the minimum of the expressions. If more than one expression is the minimum, the first minimum is returned. If all expressions are missing and a final result is not specified, missing is returned. If all expressions are missing and a final result is specified, that final result is returned. The test expressions must evaluate to numeric values, but the result expressions can be anything.

And &

Evaluates as 1 when both of its arguments are true. Otherwise it evaluates as 0. (See Figure D.9.) The formula in Figure D.7 labels Group 1 as drivers only if both comparisons are true.

Figure D.7 Creating an **And** Function

```
If ( sex == "M" & age > 13  ⇒ "Group 1"
   ( else                   ⇒ "Group 2"
```

Or |

Evaluates as 1 when either of its arguments is true. If both of its arguments are false, then the `Or` expression evaluates as 0. (See Figure D.9.) The formula in Figure D.8 assigns males and all participants over 13 to Group 1.

Figure D.8 Creating an **Or** Function

```
If ( sex == "M" | age > 13  ⇒ "Group 1"
   ( else                   ⇒ "Group 2"
```

The truth tables on the left in Figure D.9 illustrate the results of the **And** (&) and **Or** (|) functions when both arguments have nonmissing values that evaluate to true or false. The table on the right illustrates the result when either the left or right expression (call them **a** and **b**) or both have missing values.

Figure D.9 Evaluations of And and Or Expressions

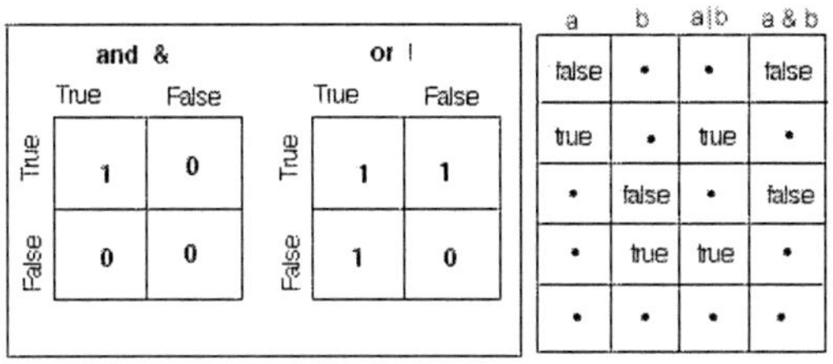

Not !

Evaluates as 1 when its argument is false. Otherwise, **Not** evaluates as **0**. When you apply the **Not** function, use parentheses where necessary to avoid ambiguity. For example, `!(weight==64)` can be either true or false (either 1 or 0), but `(!weight)==64` is always false (0) because `Not` can return only 0 or 1. Expressions such as `!(weight==64)` can also be entered as `weight != 64`.

Interpolate

Linearly interpolates the *y*-value between two points, *x1*, *y1* and *x2*, *y2* that corresponds to the arguments that you give. You can insert additional pairs of *x*, *y* arguments with the insert key. **Interpolate** finds the pair of *x*, *y* points that correspond to the *x*-value and completes the interpolation.

Step

Is like **Interpolate** except that it returns the *y*-value corresponding to the greatest *x*-value less than or equal to the *x* and *y* arguments. That is, it finds the corresponding *y* for a given *x* from a step function rather than a linear fit between points. Like **Interpolate**, you can have as many *x* and *y* argument pairs as you want.

Figure D.10 Example of Interpolate

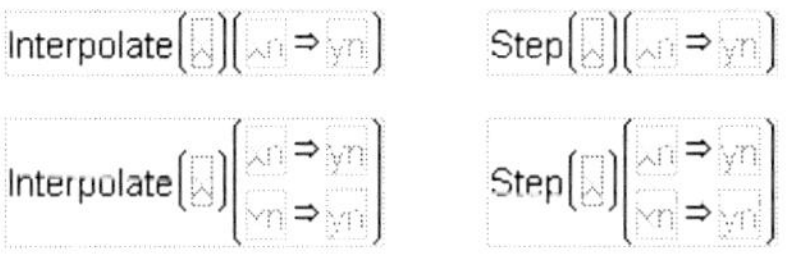

For

Repeats the statements in the **body** argument as long as the **while** condition is true. The **init** and **next** control the iterations.

While

Repeatedly tests the **expr** condition and executes the **body** until **expr** is no longer true.

Break, Continue

Break stops execution of a loop completely and continues to the statement following the loop. **Continue** ends the current iteration of a loop and begins the loop at the next iteration.

Both are used in **For**, **While**, and **For Each Row** loops.

Stop

Immediately stops a script that is running.

Probability Functions

You can create a formula that calculates probabilities and quantiles for statistical distributions like beta, Chi-square, F, gamma, normal, Student's t, Weibull distributions, Tukey HSD, and so on. See the *Scripting Guide* for details about syntax.

Beta Density

Requires three arguments: quantile argument and the shape parameters alpha and beta. A threshold parameter (θ) and a scale parameter ($\sigma > 0$) are additional arguments. It returns the value of the beta probability density function (pdf) for the given arguments. The beta density is useful for modeling the probabilistic behavior of random variables such as proportions constrained to fall in the interval [0, 1].

Beta Distribution

The beta distribution has two shape parameters: $\alpha > 0$ and $\beta > 0$. A threshold parameter (θ) and a scale parameter (σ) are additional arguments, where $\theta \leq x \leq \theta + \sigma$. The default value for θ is 0. The default value for σ is 1.

The beta distribution function is the inverse of the beta quantile function.

Beta Quantile

Accepts a probability argument, *p*, and shape and scale parameters, $\alpha > 0$ and $\beta > 0$. It returns the pth quantile from the standard beta distribution. The beta quantile function is the inverse of the beta distribution function.

ChiSquare Density

Accepts a quantile argument from the range of values for the Chi-squared distribution, a degrees of freedom argument, and an optional noncentrality parameter. It returns the value of the Chi-squared density function (pdf) for the arguments.

ChiSquare Distribution

Accepts a response argument (range of *x* values) and three parameter arguments: a quantile, a degrees of freedom, and a noncentrality parameter. It returns the probability that an observation from the Chi-squared distribution with the specified noncentrality parameter and degrees of freedom is less than or equal to the given quantile. For example, the expression **ChiSquare Distribution(11.264, 5)** returns the probability that an observation from the Chi-squared distribution centered at 0 with 5 degrees of freedom is less than or equal to 11.264. The expression evaluates as 0.95361.

Furthermore, the **ChiSquare Distribution** function accepts integer and noninteger degrees of freedom. It is centered at 0 by default. The **ChiSquare Distribution** function is the inverse of the **ChiSquare Quantile** function.

ChiSquare Quantile

Accepts three arguments: a probability *p*, a degrees of freedom, and a noncentrality parameter. It returns the pth quantile from the Chi-squared distribution with the specified noncentrality parameter and degrees of freedom. For example, the expression **ChiSquare Quantile(.95, 3.5, 4.5)** returns the 95% quantile from the Chi-squared distribution centered at 4.5 with 3.5 degrees of freedom. The expression evaluates as 17.50458.

The **ChiSquare Quantile** function accepts integer and noninteger degrees of freedom. It is centered at 0 by default. The **ChiSquare Quantile** function is the inverse of the **ChiSquare Distribution** function.

Dunnett P Value

Returns the *p*-value from Dunnett's multiple comparison test.

Dunnett Quantile

Returns the quantile needed in Dunnett's multiple comparison tests.

F Density

Accepts a quantile argument from the range of values for the *F*-distribution, numerator and denominator degrees of freedom arguments, and an optional noncentrality parameter. It returns the value of the *F*-density function (pdf) for the arguments.

F Distribution

Accepts four arguments: a quantile, a numerator and denominator degrees of freedom, and a noncentrality parameter. It returns the probability that an observation from the *F*-distribution with the specified noncentrality parameter and degrees of freedom is less than or equal to the given quantile. For example, the expression **F Distribution(3.32, 2, 3)** returns the probability that an observation from the central *F*-distribution with 2 degrees of freedom in the numerator and 3 degrees of freedom in the denominator is less than or equal to 3.32. The expression evaluates as 0.82639.

The *F*-distribution function accepts integer and noninteger degrees of freedom. By default, the non-central parameter is set to 0. The *F*-distribution function is the inverse of the **F Quantile** function.

F Quantile

Accepts four arguments: a probability *p*, a numerator and denominator degrees of freedom, and a noncentrality parameter. It returns the pth quantile from the *F*-distribution with the specified noncentrality parameter and degrees of freedom. For example, the expression **F Quantile(0.95, 2, 10, 0)** returns the 95% quantile from the *F*-distribution centered at 0 with 2 degrees of freedom in the numerator and 10 degrees of freedom in the denominator. The expression evaluates as 4.1028.

The **F Quantile** function accepts integer and noninteger degrees of freedom. By default, the non-central parameter is set to 0. The **F Quantile** function is the inverse of the **F Distribution** function.

Frechet Density

Returns the density at *x* of a Fréchet distribution with location *mu* and scale *sigma*.

Frechet Distribution

Returns the probability that a Fréchet distribution with location *mu* and scale *sigma* is less than *x*.

Frechet Quantile

Returns the quantile associated with a cumulative probability *p* for a Fréchet distribution with location *mu* and scale *sigma*.

Gamma Density

Requires a quantile argument. Also accepts an optional shape parameter, which must be greater than zero and defaults to 1. A scale parameter *b*, which must be greater than zero and defaults to 1 is optional. And a threshold parameter, which must be in the range $-\infty < \theta < +\infty$ and defaults to zero is optional.

Gamma Distribution

Is based on the standard gamma function, and accepts a single argument with a quantile value. The shape, scale, and threshold parameters are optional, with defaults as described previously in the discussion of the **Gamma Density** function. It returns the probability that an observation from a standard gamma distribution is less than or equal to the specified *x*. The **Gamma Distribution** function is the inverse of **Gamma Quantile** function.

Gamma Quantile

Accepts a probability argument *p*, and returns the pth quantile from the standard gamma distribution with the shape parameter that you specify. The **Gamma Quantile** function is the inverse of the **Gamma Distribution** function.

LEV Density

Returns the density at *x* of the largest extreme value distribution with location *mu* and scale *sigma*.

LEV Distribution

Returns the probability that the largest extreme value distribution with location *mu* and scale *sigma* is less than *x*.

LEV Quantile

Returns the quantile associated with a cumulative probability *p* of the largest extreme value distribution with location *mu* and scale *sigma*.

Logistic Density

Returns the density at *x* of a logistic distribution with location *mu* and scale *sigma*.

Logistic Distribution

Returns the probability that the logistic distribution with location *mu* and scale *sigma* is less than *x*.

Logistic Quantile

Returns the quantile associated with a cumulative probability *p* of the logistic distribution with location *mu* and scale *sigma*.

Loglogistic Density

Returns the density at *x* of the loglogistic distribution with location *mu* and scale *sigma*.

Loglogistic Distribution

Returns the probability that the loglogistic distribution with location *mu* and scale *sigma* is less than *x*.

Loglogistic Quantile

Returns the quantile associated with a cumulative probability *p* of the loglogistic distribution with location *mu* and scale *sigma*.

Lognormal Density

Returns the density at *x* of the lognormal distribution with location *mu* and scale *sigma*.

Lognormal Distribution

Returns the probability at *x* of the lognormal distribution with location *mu* and scale *sigma*.

Lognormal Quantile

Returns the quantile associated with a cumulative probability *p* of a lognormal distribution with location *mu* and scale *sigma*.

Normal Density

Accepts a quantile argument from the range of values for the standard normal distribution. It returns the value of the standard normal probability density function (pdf) for the argument. For example, you can create a column of quantile values (*x*) with the formula **count(-3, 3, nrow())**. A second column is computed as **Normal Density(X)** to generate density values. Then select **Graph > Overlay** to plot the normal density by *x*.

Normal Distribution

Accepts a quantile argument from the range of values for the standard normal distribution with mean 0 and standard deviation 1. It returns the probability that an observation from the standard normal distribution is less than or equal to the specified quantile. For example, the expression **Normal Distribution(1.96)** returns

0.975, the probability that an observation from the standard normal distribution is less than or equal to the 1.96^{th} quantile. Also, you can specify mean and standard deviation parameters to obtain probabilities from nonstandard normal distributions. The **Normal Distribution** function is the inverse of the **Normal Quantile** function.

Normal Quantile (Probit)

Accepts a probability argument *p*, and returns the pth quantile from the standard normal distribution. For example, the expression **Normal Quantile(0.975)** returns the 97.5% quantile from the standard normal distribution, which evaluates as 1.96. Also, you can specify parameter values for the mean and standard deviation to obtain quantiles from nonstandard normal distributions. The **Normal Quantile** function is the inverse of the **Normal Distribution** function.

Normal Biv Distribution

Computes the probability that an observation is less than or equal to (*x*,*y*) with correlation coefficient *r* where the observation is marginally normally distributed. You can specify the mean and standard deviation for the X and Y coordinates of the observation. The default values are 0 for both means and 1 for both standard deviations.

GLog Density

Returns the density or pdf at a particular quantile *q* of a generalized logarithm distribution with location *mu*, scale *sigma*, and shape *lambda*. When the shape parameter is equal to zero, the distribution reduces to a Lognormal(*mu*, *sigma*).

GLog Distribution

Returns the probability or cdf that a generalized logarithm distributed random variable is less than *q*. When the shape parameter is equal to zero, the distribution reduces to a Lognormal(*mu*, *sigma*).

GLog Quantile

Returns the quantile, the value for which the probability is *p* that a random value would be lower. When the shape parameter is equal to zero, the distribution reduces to a Lognormal(*mu*, *sigma*).

SEV Density

Returns the density at *x* of the smallest extreme distribution with location *mu* and scale *sigma*.

SEV Distribution

Returns the probability that the smallest extreme distribution with location *mu* and scale *sigma* is less than *x*.

SEV Quantile

Returns the quantile associated with a cumulative probability *p* of the smallest extreme distribution with location *mu* and scale *sigma*.

t Density

Accepts a quantile argument from the range of values for the *t*-distribution, a degrees of freedom argument, and an optional noncentrality parameter. It returns the value of the *t*-density function (pdf) for the arguments. To compare a *t*-density with 5 df with a standard normal distribution, you can create a column of quantile values (x) with the formula **count(-3, 3, nrow())**. A second column is computed as **t Density(X)**. A third column is computed as **Normal Density(X)**. Then select **Graph > Overlay** to plot the *t*-density and the normal density by x. You will see that the *t*-density has slightly more spread than the normal.

t Distribution

Accepts three arguments: a quantile, a degrees of freedom, and a noncentrality parameter. It returns the probability that an observation from the Student's *t*-distribution with the specified noncentrality parameter and degrees of freedom is less than or equal to the given quantile. For example, the expression **t Distribution(.9, 5)** returns the probability that an observation from the Student's *t*-distribution centered at 0 with 5 degrees of freedom is less than or equal to 0.9. The expression is evaluated as 0.79531. *t*-distribution accepts integer and noninteger degrees of freedom. It is centered at 0 by default, but you can enter a value for the noncentrality parameter. The **t Quantile** function is the inverse of the **t Distribution** function.

t Quantile

Accepts three arguments: a probability p, a degrees of freedom, and a noncentrality parameter. It returns the pth quantile from the Student's *t*-distribution with the specified noncentrality parameter and degrees of freedom. For example, the expression **Student's t Quantile(.95, 2.5)** returns the 95% quantile from the Student's *t*-distribution centered at 0 with 2.5 degrees of freedom. The expression evaluates as 2.558219. The **t Quantile** function is the inverse of the **t Distribution** function. This function also accepts integer and noninteger degrees of freedom. It is centered at 0 by default, but you have the option to enter a value for the noncentrality parameter. The **t Distribution** function is the inverse of the **t Quantile** function.

Weibull Density

Accepts a quantile argument from a range of values for the Weibull distribution. It returns the value of the Weibull probability density function (pdf). This function is the probability that an observation from a Weibull distribution is less than or equal to the specified quantile argument.

Weibull Distribution

Uses an argument with a quantile valued, an optional value for the scale parameter α and an optional shape parameter β. It returns the probability that an observation is less than or equal to the specified x for Weibull distribution with the shape and scale parameters that you specified. The **Weibull Distribution** function is the inverse of **Weibull Quantile** function.

The Weibull distribution has different shapes depending on the values of α (a scale parameter that affects the x direction) and β (a shape parameter). It often provides a good model for estimating the length of life, especially for mechanical devices and in biology. The two-parameter Weibull is the same as the three-parameter Weibull with a threshold of zero.

The Weibull distribution has two parameters, $\alpha > 0$ and $\beta > 0$. When $\beta = 1$ the pdf reduces to the exponential distribution (with $\gamma = 1 / \beta$). The exponential distribution is a special case of both the gamma and Weibull distributions. However, there are Weibull distributions that are not exponential distributions.

Weibull Quantile

Accepts a probability argument *p*, and returns the pth quantile from the Weibull distribution with the shape and scale parameters that you specify. The **Weibull Quantile** function is the inverse of the **Weibull Distribution** function.

Johnson Su Distribution

Returns the probability that a Johnson Su-distributed random variable is less than *x*.

Johnson Su Quantile

Returns the quantile whose value for which the probability is *p* that a random value would be lower.

Johnson Su Density

Returns the density at *x* of a Johnson Su distribution.

Johnson Sb Distribution

Returns the probability that a Johnson Sb-distributed random variable is less than *x*.

Johnson Sb Quantile

Returns the quantile whose value for which the probability is *p* that a random value would be lower.

Johnson Sb Density

Returns the density at *x* of a Johnson Sb distribution.

Johnson Sl Distribution

Returns the probability that a Johnson Sl-distributed random variable is less than *x*.

Johnson Sl Quantile

Returns the quantile whose value for which the probability is *p* that a random value would be lower.

Johnson Sl Density

Returns the density at *x* of a Johnson Sl distribution.

Tukey HSD Quantile

Accepts a probability argument *1-alpha*, and returns the $1\text{-}alpha^{th}$ quantile from Tukey's HSD test for the parameters that you specify. The *alpha* argument is the significance level that you want. *nGroups* is the number of groups in a study. *dfe* is the error degrees of freedom (based on the total study sample). This is the quantile used to calculate least significant difference in Tukey's multiple comparisons test.

Tukey HSD P Quantile

Returns the *p*-value from Tukey's HSD multiple comparisons test.

F Power and F Sample Size

The **F Power** function calculates the power from a given situation that involves an F-test or t-test, and the **F Sample Size** function computes the sample size. The arguments are the values you specify for computation of a prospective power analysis. (These functions perform the same computations as if you selected **DOE > Sample Size and Power**. See the *Design of Experiments* for a discussion of power and sample size.) The arguments include:

- **alpha** The significance level you are willing to tolerate (often 0.05).
- **dfh** The hypothesis degrees of freedom. It is one (1) for a t-test.
- **dfm** The model degrees of freedom (such that $dfe = n - dfm$).
- **SquaredSize** The squared effect size scaled by the error variance, which is used for making the noncentrality argument for the F-distribution. For this argument, use squared size = Δ^2/σ^2 where σ^2 is the error variance. That is, use:

$$\Delta^2 = (\bar{x} - \mu)^2 \text{ for a one-sample } t\text{-test}$$

$$\Delta^2 = \frac{(\bar{x}_1 - \bar{x}_2)^2}{4} \quad \text{for a two-sample } t\text{-test}$$

$$\Delta^2 = \sqrt{\sum_{i=1}^{k} \frac{(\bar{x}_i - \bar{x})^2}{k}} \quad \text{for a } k\text{-sample } F\text{-test}$$

- **n** (found only in the **F Power** function) The total number of observations (runs, experimental units, or samples) you expect to have. Power (in the **F Sample Size** function) is the probability that you want to have of declaring a significant result.

Discrete Probability Functions

Gamma Poisson Probability

Returns the probability or pmf that a gamma-Poisson distributed random variable is equal to x. In general, the gamma Poisson functions accept arguments that are the mean parameter *lambda*, the overdispersion parameter *sigma*, and the count of interest x. When the overdispersion is equal to one, the Gamma Poisson reduces to a Poisson(*lambda*) distribution.

Gamma Poisson Distribution

Returns the probability that a gamma-Poisson distributed random variable is less than or equal to x. In general, the gamma Poisson functions accept arguments that are the mean parameter *lambda*, the overdispersion parameter *sigma*, and the count of interest x.

Gamma Poisson Quantile

Returns the smallest integer quantile for which the cumulative probability of the Gamma Poisson (*lambda*, *sigma*) distribution is larger than or equal to p.

Binomial Distribution

Returns the probability that an observation from a binomial distribution with parameters p and n is less than or equal to k. In general, the binomial functions accept arguments that are the probability of success p (the event of interest), the number of trials n, and the number of successes k.

Binomial Probability

Computes the probability that a random variable from a binomial distribution is equal to k. In general, the binomial functions accept arguments that are the probability of success p (the event of interest), the number of trials n, and the number of successes k.

Binomial Quantile

Returns the smallest integer quantile for which the cumulative probability of the Binomial (p, n) distribution is larger than or equal to the specified probability.

Neg Binomial Distribution

Returns the probability that a negative binomially distributed random variable is less than or equal to k, where the probability of success is p, and the number of successes is n.

Neg Binomial Probability

Returns the probability that a negative binomially distributed random variable is equal to k, where the probability of success is p, and the number of successes is n.

Beta Binomial Distribution

Returns the probability or pmf that a beta binomially distributed random variable is less than or equal to x. In general, the beta binomial functions accept arguments that are the probability of success p (the event of interest), the overdispersion parameter *delta*, and the number of trials n. When the overdispersion parameter for the beta binomial is zero, the distribution reduces to a binomial(p, n).

Beta Binomial Probability

Returns the probability or cmf that a beta binomially distributed random variable is equal to x. When the overdispersion parameter for the beta binomial is zero, the distribution reduces to a binomial(p, n).

Beta Binomial Quantile

Returns the smallest integer quantile for which the cumulative probability of the Beta Binomial (p, n, *delta*) distribution is larger than or equal to the specified probability. When the overdispersion parameter for the beta binomial is zero, the distribution reduces to a binomial(p, n).

Hypergeometric Distribution

Computes the probability that a random variable from a hypergeometric distribution is less than or equal to x. The hypergeometric distribution models the total number of successes in a fixed sample drawn without replacement from a finite population. The hypergeometric functions accept as arguments the size of the population *N*, the total number of items with the desired characteristic in the population, *K*, the number of samples drawn *n*, and the number of successes in the sample *x*.

Hypergeometric Probability

Computes the probability that a random variable from a hypergeometric distribution is equal to *x*.

Poisson Distribution

Computes the probability that a random variable from a Poisson distribution with mean lambda is less than or equal to the count of interest. In general, Poisson functions accept an argument that is the count of interest, and lambda, the mean parameter.

Poisson Probability

Computes the probability that a random variable from a Poisson distribution with mean lambda is equal to the count of interest.

Poisson Quantile

Returns the smallest integer quantile for which the cumulative probability of the Poisson (*lambda*) distribution is larger than or equal to *p*.

Statistical Functions

There are two types of **Statistical** functions you can use in a formula:

- The functions with names that have the prefix **Col**. These functions compute statistics for a column of numbers or expressions involving columns.
- The **Mean**, **Std Dev**, **Number**, **Sum**, **Quantile**, **Maximum**, **Minimum**, and **N Missing** functions. These functions evaluate across columns or arguments. The statistic is computed for each row across the series of arguments. You can use the insert key () on the on-screen keypad, or type a comma to add arguments to the functions that accept multiple arguments. When there are multiple contiguous arguments, select the function and the first argument, and then Shift-click the last argument in the group. These functions then automatically show with the complete list.

See the *Scripting Guide* for details about syntax.

Col Mean

Calculates the mean (or arithmetic average) of the numeric values identified by its argument. The formula **Col Mean(age)** calculates the average of all nonmissing values in the **age** column.

Col Std Dev

Measures the spread around the mean of the distribution identified by its argument. In the normal distribution, about 68% of the distribution is within one standard deviation of the mean. 95% of the distribution is within two standard deviations of the mean. 99% of the distribution is within three standard deviations of the mean.

Col Number

Counts the number of nonmissing values in the column you specify. A missing numeric value occurs when a cell has no assigned value or is the result of an invalid operation (such as division by zero). Missing values show on the spreadsheet as a missing value mark (•). Missing character values are null character strings. In formulas for row state columns, an excluded row state characteristic is treated as a missing value.

Col N Missing

Counts the number of missing values in the column you specify. A missing numeric value occurs when a cell has no assigned value or is the result of an invalid operation (such as division by zero). Missing values show in the data grid with a missing value character (•). Missing character values are null character strings.

Col Sum

Computes the sum of the values in its numeric argument. Missing values are ignored.

Col Minimum and Col Maximum

Takes the minimum of its numeric arguments. **Col Minimum** ignores missing values. **Col Maximum** takes the maximum of a numeric column argument and ignores missing values.

Col Quantile

Computes the value at which a specific percentage of the values is less than or equal to that value. For example, the value calculated as the 50% quantile, also called the median, is greater than or equal to 50% of the data. Half of the data values are less than the 50th quantile.

The **Col Quantile** function's quantile argument represents the quantile percentage divided by 100. The 25% quantile, also called the lower quartile, corresponds to $p = 0.25$, and the 75% quantile, called the upper quartile, corresponds to $p = 0.75$.

The formula editor computes a quantile for a column of n nonmissing values by arranging the values in ascending order. The subscripts of the sorted column values, $y_1, y_2,...,y_n$, represent the ranks in ascending order.

The pth quantile value is calculated using the formula $p(n + 1)$, where p is the percent value and n is the total number of nonmissing values. If $p(n+1)$ is an integer, then the quantile value is $y_{p(n+1)}$. If $p(n + 1)$ is not an integer, then the value is interpolated by assigning the integer part of the result to i, assigning the fractional part to f, and applying the formula $(1 - f)y_i + (f)y_{i+1}$.

For example, suppose a column has values 2, 4, 6, 8, 10, 12, 14, 16, 18, and 20. The 50% quantile is calculated as `0.5(10 + 1) = 5.5`.

Because the result is fractional, the 50% quantile value is interpolated as

$$(1 - 0.5) \times 10 + (0.5) \times 12 = (0.5)10 + (0.5)12 = 6 + 5 = 11$$

The following are example **ColQuantile** formulas:

- **ColQuantile(age, 1)** Calculates the maximum age.
- **ColQuantile(age, 0.75)** Calculates the upper quartile age.

- **ColQuantile(age, 0.5)** Calculates the median age.
- **ColQuantile(age, 0.25)** Calculates the lower quartile age.
- **ColQuantile(age, 0)** Calculates the minimum age.

The **ColQuantile** argument can be any expression that evaluates to a value between (and including) 0 and 1. For example, the first formula in Figure D.11 calculates quantile values of age in ascending order for each row. The column then contains the interpolated values of **age** in ascending order in the calculated column. The second formula lists the interpolated values of age in descending order.

Figure D.11 Examples of the Quantile Function

$$\text{Col Quantile}\left[age, \frac{(\text{Row}()-1)}{(\text{NRow}()-1)}\right] \qquad \text{Col Quantile}\left[age, \frac{(\text{NRow}()-\text{Row}())}{(\text{NRow}()-1)}\right]$$

Col Rank

Ranks each row's value, from 1 for the lowest value to the number of columns for the highest value. Ties are broken arbitrarily.

Col Standardize

Performs the usual standardization on its numeric expression. For each row *i*, **Col Standardize(height)** is **(HeightRow()–Col Mean(Height))/Col Std Dev(Height)**.

Mean

Calculates the arithmetic average of the nonmissing arguments that you specify. The arguments can be constants, numbers, or expressions. The **Mean** function initially shows with a single argument. You add arguments with the insert button ( ) on the formula editor keypad or by typing a comma.

Std Dev

Computes standard deviation of the list of arguments that you specify. The arguments can be constants, numbers, or expressions. The **Std Dev** function initially shows with a single argument. You add arguments by clicking the insert button ( ) on the formula editor keypad or by typing a comma.

Number

Counts the number of nonmissing values in the list of arguments that you specify.

Sum

Returns the sum of the arguments.

Quantile

Calculates the quantile given by its first argument for all the following arguments given.

Summation (Σ)

Evaluates for an explicit range of values in a column, as given by the summation indices. This behavior is different from all other statistical functions (except **Product**), which *always* evaluate *on every row.* The **Summation** function uses the summation notation shown in Figure D.12. To calculate a sum, replace the missing *body* term with an expression containing the index variable i, or an index variable you assign. **Summation** repeatedly evaluates the expression for $i = 1$, $i = 2$, through $i =$ **NRow()** and then adds the nonmissing results together to determine the final result.

You can replace **NRow()**, the number of rows in the active spreadsheet, and the index constant, i, with any expression appropriate for your formula. For example, the summation formula in Figure D.12 computes the total for each row of all revenue values for rows 1 through the current row number, filling the calculated column with the cumulative totals of the revenue column.

Figure D.12 Example of the **Summation** function

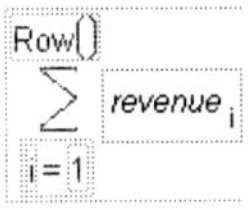

Product (Π)

Evaluates for an explicit range of values in a column, as given by the summation indices, as opposed to all other statistical functions (except **Summation**), which *always* evaluate *on every row.* **Product** uses the notation shown in the formulas on the right in Figure D.13. To calculate a product, replace the missing body term with an expression containing the index variable j. **Product** repeatedly evaluates the expression for $i = 1$, $i = 2$, through $i = n$ and multiplies the nonmissing results together to determine the final result.

You can replace **NRow()**, the number of rows in the active spreadsheet and the index constant, i, with any expression appropriate for your formula.

For example, the expression second product example in Figure D.13 calculates $i!$ (each row number's factorial).

Figure D.13 Examples of the **Product** Function

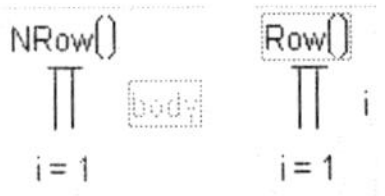

Minimum and Maximum

Return the minimum and maximum value, respectively, from the list of nonmissing arguments that you specify.

N Missing

Counts the number of missing values in the list of arguments that you specify.

Desirability

Are smooth piecewise functions that are crafted to fit the control points. The minimize and maximize functions are three-part piecewise smooth functions that have exponential tails and a cubic middle.

The target function is a piecewise function. It is a scale multiple of a normal density on either side of the target (with different curves on each side), which is also piecewise smooth and fit to the control points.

Random Functions

You can create formulas that generate real numbers by effectively "rolling the dice" within the constraints of the specified distribution. Each time you click **Apply** in the formula editor window, these functions produce a new set of random numbers.

Note: Random numbers are generated using the Mersenne-Twister technique. This technique has a period length of 2^{19937}-1. For details about the generators, see Matsumoto and Nishimura (1998). The new generators are verified to pass all the DIEHARD tests as documented in Marshalled (1996).

See the *Scripting Guide* for details about syntax.

Random Uniform

Generates random numbers uniformly between 0 and 1. This means that any number between 0 and 1 is as likely to be generated as any other. The result is an approximately even distribution. You can shift the distribution and change its range with constants. For example, `5 + Random Uniform()*20` generates uniform random numbers between 5 and 25.

Random Normal

Generates random numbers that approximate a normal distribution with a mean of 0 and standard deviation of 1 if no arguments are used, or with the mean and standard deviation entered as arguments. The normal distribution is bell shaped and symmetrical. You can also modify the `Random Normal` function with constants if no arguments are entered to give a normal distribution with specific mean and standard deviation. For example, the formula `Random Normal()*5 + 30` generates a random normal variable with a mean of 30 and a standard deviation of 5.

Random Exp

Generates a single parameter exponential distribution for the distribution parameter lambda=1. You can modify the exponential function to use a different lambda.

For example, `Random Exp()/.1` generates an exponential distribution for lambda=0.1. The exponential distribution is often used to model simple failure time data, where lambda is the failure rate.

Random Gamma

Gives a gamma distribution for the parameter, **alpha**, you enter as the function argument. The gamma distribution describes the time until the kth occurrence of an event. The gamma distribution can also have a scale parameter, **beta**. A gamma variate with shape parameter alpha and scale beta can be generated with the formula **beta*Random Gamma(alpha)**. If **2*alpha** is an integer, a Chi-squared variate with **2*alpha** degrees of freedom is generated with the formula **2*Random Gamma(alpha)**.

Random Beta

Generates a pseudo-random number distributed Beta(alpha, beta).

Random Cauchy

Generates a Cauchy distribution with location parameter 0 and scale parameter 1. The Cauchy distribution is bell shaped and symmetric but has heavier tails than the normal distribution. A Cauchy variate with location parameter alpha and scale parameter beta can be generated with the formula `alpha+beta*Random Cauchy()`.

Random Johnson Su

Returns a random number from the Johnson Su distribution.

Random Johnson Sb

Returns a random number from the Johnson Sb distribution.

Random Johnson Sl

Returns a random number from the Johnson Sl distribution.

Random Triangular

Generates a triangular distribution of numbers between 0 and 1, with the midpoint that you enter as the function argument. You can add a constant to the function to shift the distribution and multiply to change its span.

Random Integer

Generates a uniform distribution of integers between 1 and the argument you enter as **n1**, if nothing is entered for **n2**. If you enter both **n1** and **n2 (n1<n2)**, **Random Integer** generates a uniform distribution of the integers between and including **n1** and **n2**.

Random Binomial

Generates random numbers from a binomial distribution with parameters that you enter as function arguments. The first argument is n, the number of trials in a binomial experiment. The second argument is p, the probability that the event of interest occurs. When n is 1, the binomial function generates a distribution of Bernoulli trials. For example, $n = 1$ and $p = 0.5$, give the distribution of tossing a fair coin. The mean of the binomial distribution is np, and variance is $np(1 - p)$.

Random Negative Binomial

Generates a negative binomial distribution for the parameters that you enter as function arguments. The first parameter is the number of successes of interest (r) and the second argument is the probability of success (p). The random variable of interest is the number of failures that precede the ith success. In contrast to the binomial variate, where the number of trials is fixed and the number of successes is variable, the negative binomial variate is for a fixed number of successes and a random number of trials. The mean of the negative binomial distribution is $(r(1-p))/p$ and the variance is $(r(1-p))/p^2$.

Random Beta Binomial

Returns random numbers from the beta binomial distribution for n trials with probability p and correlation or overdispersion *delta*.

Random Frechet

Returns a random number from a Fréchet distribution with the location *mu* and scale *sigma*.

Random Geometric

Returns random numbers from the geometric distribution with the parameter you enter as the function argument. The parameter, p, is the probability that a specific event occurs at any one trial. The number of trials until a specific event occurs for the first time is described by the geometric distribution. The mean of the geometric distribution is $(1\text{-}p)/p$, and the variance is $(1-p)/p^2$.

Random Poisson

Generates a Poisson variate based on the value of the parameter, lambda, you enter as the function argument. Lambda is often the expected number of events occurring per unit time or unit of area. Lambda is both the mean and the variance of the Poisson distribution.

Random Gamma Poisson

Returns random numbers from the gamma Poisson distribution with parameters *lambda* and *sigma*.

Random Weibull

Returns a random number from a Weibull distribution.

Random Logistic

Returns a random number from a logistic distribution with the location *mu* and scale *sigma*.

Random Loglogistic

Returns a random number from a loglogistic distribution with the location *mu* and scale *sigma*.

Random Lognormal

Returns a Lognormal-distributed random number with location parameter *mu* and scale *sigma*.

Random GLog

Returns random numbers from the generalized logarithm distribution with parameters *mu*, *sigma*, and *lambda*. When *lambda* is equal to zero, the function returns a lognormal(*mu*, *sigma*).

Random Reset

Restarts the random number sequences with a seed that you specify.

Random LEV

Returns a random number from an LEV distribution with the specified location *mu* and scale *sigma*.

Random SEV

Returns a random number from the smallest extreme distribution with the specified location *mu* and scale *sigma*.

Col Shuffle

Selects a row number at random from the current data table. Each row number is selected only once. When **Col Shuffle** is used as a subscript, it returns a value selected at random from the column that serves as its argument. Each value from the original column is assigned only once as **Col Shuffle**'s result.

For example, to identify a 50% random sample without replacement, use the formula in Figure D.14.

Figure D.14 Formula Identifying 50% Random Sample

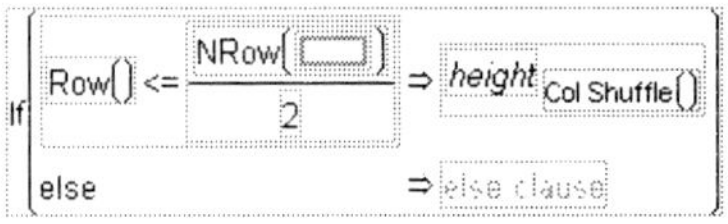

The formula in Figure D.14 selects half the values (*n*/2) from the column *x* and assigns them to the first half of the rows in the computed column. The remaining rows of the computed column fill with missing values.

Resample Freq

Generates a random selection with replacement frequency counts, suitable for use in bootstrapping. For example, it supports a second **Freq Column** argument, enabling it to do bootstrap samples relating to a pre-existing frequency column specified in the second argument. `Resample Freq()` generates a 100% resample. `ResampleFreq(rate)` generates a `rate` frequency sample. `Resample(rate, column)` generates a sample that is calculated by the rate multiplied by the sum of the specified column.

Date Time Functions

JMP stores dates and times in numeric columns using the Macintosh standard of the *number of seconds since January 1, 1904*. When a column has date values, you can assign a date format to that column by double-clicking a column name and selecting **Date** or **Time** from the **Format** menu. See "Numeric Format Options," p. 61, for details.

See the *Scripting Guide* for details about syntax.

In Minutes, In Hours, In Days, In Weeks, In Years

Converts from the units of the function name to the equivalent number of seconds for the argument. The argument must be a number or numeric expression. For example, **In Minutes(2)** yields **120**, and **In Years(1)** yields **31,557,600** (60 seconds * 60 minutes * 24 hours * 365.25 days).

Date DMY, Date MDY

Accepts numeric expressions for day, month, and year and return the associated JMP date. For example, **Date DMY (20, 3, 1991)** and **Date MDY(3, 20, 1991)** evaluate to **2,752,272,000**.

Today

Returns the number of seconds between January 1, 1904 and the current date. For example, at midnight on March 20, 1991 (a Wednesday), the **Today** function returns **2752272000** (2,752,272,000 seconds) and continues counting. If you evaluate the **Today** function later in the day, it reflects the additional seconds.

Day, Month, Year,

Returns the day of the month, the month (as a number from 1 to 12), a four-digit year, respectively. The argument for these functions is interpreted as a JMP date. For example, on March 20, 1991:

- **Day(2752272000)** returns the number **20**.
- **Month(2752272000)** returns the number **3**.
- **Year(2752272000)** returns the number **1991**.

Quarter

Returns the annual quarter of a *datetime* value as an integer 1-4.

Hour, Minute, Second

Returns the hour, the minute, and the seconds of a date-time value, respectively. The argument for these functions is interpreted as a JMP date. For example, on March 20, 1991:

- **Hour(2752572649)** returns the number **11**.
- **Minute(2752572649)** returns the number **30**.
- **Second(2752572649)** returns the number **49**.

Day of Week, Day of Year, Week of Year, Time of Day

The argument for these functions is a JMP date. **Day Of Week** returns a number from 1 to 7, where 1 represents Sunday. **Day Of Year** returns the number of days from the beginning of the year. **Week Of Year** returns a number from 1 to 52. **Time Of Day** returns a number from 0 to 86399 (time of day in seconds). For example, on Wednesday, March 20, 1991:

- **Day Of Week(2752272000)** returns the number **4**.
- **Day Of Year(2752272000)** returns the number **79**.
- **Week Of Year(2752272000)** returns the number **12**.
- **Time Of Day(2752272000)** returns the number **0**.

Informat

The argument for the **Informat** function is a date character string. For example, **Informat("03/20/1991")** returns the appropriate JMP date value, **2752272000**. JMP can read all the date formats except for **Abbrev Date** and **Long Date**.

Abbrev Date, Long Date, Short Date

The argument for these date functions is a JMP date. They return character strings that are the formatted representation of the argument. For example:

- **Abbrev Date(2752272000)** returns **Wed, Mar 20,1991**.
- **Long Date(2752272000)** returns **Wednesday, March 20, 1991**.
- **Short Date(2752272000)** returns **3/20/91**.

Format

The first argument in the **Format** function is a JMP date. This function returns the character string representation of the date by the date format that you specify in the second argument, which is a quoted string. If you apply this formula to a numeric column, JMP automatically changes the column's data type to character.

You can also supply a column for the first argument and leave the rest blank. The result is the formatted value of the column reference. This can be used to extract value labels of a column when the value labels are turned off.

MDYHMS

The argument of **MDYHMS** is a JMP date. This function shows all date and time fields, appending zeros as time fields if no time information is present. This is useful if a date column is formulated such that not all date information is displayed. The **MDYHMS** function can be used to see all available date and time information.

Date Increment

Adds 1 or more intervals to a starting datetime value. For example, **Date Increment(Today(), "Day", 3)** adds three days to the current date. **Date Increment(Today(), "Year", 3)** adds 3 years to the current date.

Date Difference

Returns the date difference in intervals of two datetime values. For example, following formula returns **207.890243055556**, the difference between `01Jan2010:00:00:00` and `27Jul2010:21:21:57`:

```
Date Difference(01Jan2010:00:00:00, 27Jul2010:21:21:57, "Day", "fractional");
```

Row State Functions

There are six characteristics that rows in a data table can have: selected, hidden, excluded, labeled, colored, and marked. If you give rows one or more of these characteristics and then create row state data table columns, you can then create a formula that computes and saves row state conditions. (See "Assigning Characteristics to Rows and Columns," p. 135, and "Using Row State Columns," p. 173, for more details.)This formula processes row state data just as it would process character and numeric data.

See the *Scripting Guide* for details about syntax.

Note: A row can be assigned any combination of row states; a row state column can have multiple row states as a value.

Table D.2 describes the type of argument each Row State function requires and what each returns.

Table D.2 Row State Functions

Function Name	Argument Type Required	What the Function Returns (Your Column Data Type Should be This Type)
Row State	none	row state of current row
As Row State	numeric	all row states of current row
Combine States	multiple row state arguments	multiple row state assignments
Excluded State	positive integer or zero	row state-excluded or not excluded
Hidden State	positive integer or zero	row state-hidden or not hidden
Labeled State	positive integer or zero	row state-labeled or not labeled
Color State	integer or color name or {*red, green, blue*}	row state color
Marker State	integer or character	row state marker
Selected State	positive integer or zero	row state-selected or not selected
Hue State	integer	row state hue
Shade State	integer 1-5	row state intensity
Excluded	Row State() or row state column	numeric 0 (not excluded) or 1 (excluded)
Hidden	Row State() or row state column	integer 0 (not hidden) or 1 (hidden)
Labeled	Row State() or row state column	integer 0 (not labeled) or 1 (labeled)
Color Of	Row State() or row state column	color map integer
Marker Of	Row State() or row state column	marker map integer

Table D.2 Row State Functions *(Continued)*

Function Name	Argument Type Required	What the Function Returns (Your Column Data Type Should be This Type)
Selected	**Row State()** or row state column	integer 0 (not selected) or 1 (selected)

Row State

Returns the active row state condition of the current row as true or false. You can use this function to conveniently write conditional clauses that depend on the status of the current row. For example, Figure D.15 assigns a 1 to rows that are currently selected and labeled and 0 otherwise.

Figure D.15 Row State

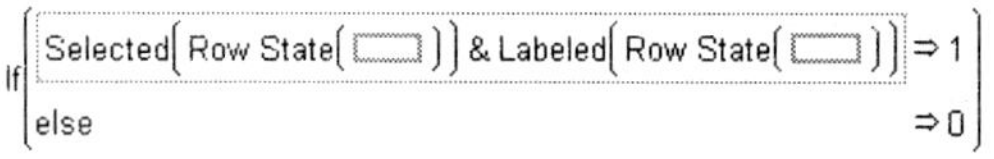

As Row State

Converts a numeric argument to a row state or set of row state conditions. Row states are stored internally in JMP as a 16-bit number, with each bit assigned to represent one of the possible row states as illustrated in Figure D.3. For example, the binary representation of **1327** is **0000010100101111. As Row State(1327)** would therefore set the row state as selected, excluded, hidden, labeled, with marker 2 and color 10.

Table D.3 Row States Stored as 16-Bit Numbers: Each Bit Represents a Row State

Bit	Row State
0	Not selected (0) or Selected (1)
1	Unexcluded (0) or Excluded (1)
2	Unhidden (0) or Hidden (1)
3	Unlabeled (0) or Labeled (1)
4-7	Marker
8-14	Color

Combine States

Generates a row state combination with two or more arguments. Use the insert button (^) on the formula editor keypad or type a comma to add arguments to the **Combine States** function. The currently selected expression becomes the first argument when you select **Combine States**. Replace each argument with an expression that evaluates to a row state. This formula:

```
Combine States(
   Selected State(Modulo(Row(),2),
```

```
Labeled State(Modulo(Row()+1,2))
```

alternately labels or selects each row in the calculated row state column. The **Selected State** and **Labeled State** functions are defined later in this section. Use the insert (⌃) and delete (⌫)buttons on the formula editor keypad to add more arguments or remove unwanted arguments.

If you include conflicting row states in a combination, the results are unpredictable.

Excluded State

Interprets a numeric argument as true or false. When an argument evaluates as true, the **Excluded State** function assigns the excluded condition as the value of the column for that row. For example, **Excluded State(Modulo(Row(),2))** assigns the excluded row state as the value of the row state column for each odd numbered row.

Hidden State

Assigns the hidden row state condition when its argument is greater than zero. If the argument is zero, the value in the column for that row is not hidden.

Labeled State

Gives the labeled row state condition when its argument is greater than zero. If the argument is zero the row value in the column for that row is not labeled.

Color State

Returns the color from the JMP color map that corresponds to its integer argument. JMP colors are numbered 0 through 84. Zero maps to black.

Marker State

Returns markers from the JMP marker map that correspond to its integer argument. JMP markers are numbered 0 through 16. The formula **Marker State(Row())** assigns all the row state markers in a repeating sequence determined by the current row number to the calculated row state column. A row state column can have multiple row states as a value.

Selected State

Gives the selected row state condition when its argument is greater than zero. If the argument is zero, the value in the column for that row is not selected.

Hue State

Returns the color from the JMP hue map that corresponds to its integer argument. JMP hues are numbered 0 through 11 but larger integers are treated as *modulo* 12. The **Hue State** function does not map to black, gray, or white. A hue of zero maps to red and hue of 11 maps to magenta. The formula on the left in Figure D.16 assigns row state colors in a chromatic spread based on the value of *z*. The **Hue State** function used with a row state data type column.

Shade State

Assigns five shade levels to a color or hue. A shade of –2 is darkest and shade of +2 is lightest. A shade of zero is a pure color. The formula on the right in Figure D.16 assigns shade values based on the value of *z*.

Figure D.16 Examples of Hue and Shade Functions

$$\text{Hue State}\left(\frac{(z - \text{Col Quantile}(z, 0))}{(\text{Col Quantile}(z, 1) - \text{Col Quantile}(z, 0))} * 12\right) \quad \text{Shade State}\left(\frac{(z - \text{Col Quantile}(z, 0))}{(\text{Col Quantile}(z, 1) - \text{Col Quantile}(z, 0))} * (5 - 2)\right)$$

To assign all shades of all the colors in the colors palette, you need to use the **Hue State** and **Shade State** assignments together. The formula in Figure D.17 uses the **Combine States** function described at the beginning of this section. The first argument in the **Combine States** function is the **Hue State** formula shown previously, and the second argument is the **Shade State** formula. In addition, the **Marker State** function with an argument of 2 assigns the **X** marker to each row, and the **Selected State** function with an argument of 1 selects each row.

Figure D.17 Combine States Example For Using Both Hue State and Row State

$$\text{Combine States}\left(\begin{array}{l}\text{Hue State}\left(\frac{(z - \text{Col Quantile}(z, 0))}{(\text{Col Quantile}(z, 1) - \text{Col Quantile}(z, 0))} * 12\right), \\ \text{Shade State}\left(\frac{(z - \text{Col Quantile}(z, 0))}{(\text{Col Quantile}(z, 1) - \text{Col Quantile}(z, 0))} * (5 - 2)\right), \\ \text{Marker State}(2), \\ \text{Selected State}(1)\end{array}\right)$$

Excluded, Hidden, Labeled, and Selected

Accepts a row state expression argument (row state column or row state constant) that evaluates as either 1 or 0 (true or false). These characteristics are inactive by default. Often, the **Row()** function is the argument, which detects the active row state condition of each row. For example, in Figure D.18, the formula assigns 99 whenever a row is actively selected, and 0 otherwise. Note that this formula is used in a column that has a numeric data type.

Figure D.18 Example of a Formula Using the **Selected** Function

$$\text{If}\left(\begin{array}{ll}\text{Selected}(\text{Row State}(\square)) & \Rightarrow 99 \\ \text{else} & \Rightarrow 0\end{array}\right)$$

The example in Figure D.19 assigns row state conditions to a row state column. The formula for the row state column (in the columns called **row state col**) checks to determine whether the active row state is either **Hidden** or **Excluded**, and if so, assigns the **Labeled** row state.

Figure D.19 Calculate Row State Information in a Row State Column

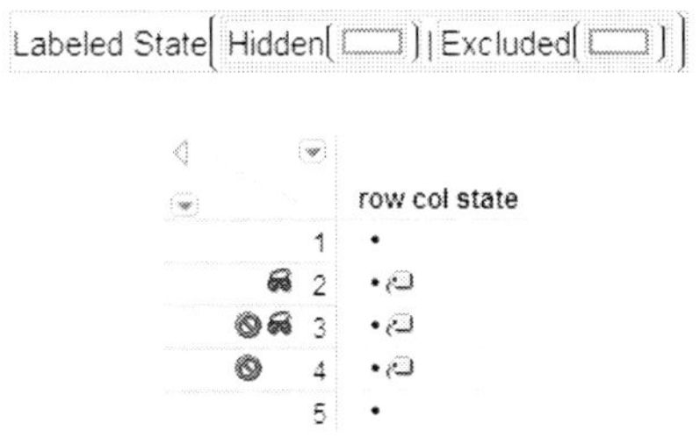

Color Of

Accepts any row state expression or column, or the **Row State()** function as its argument. Returns a number from the JMP color map that corresponds to the active color state, or zero if there is no assigned color.

Marker Of

Accepts any row state expression or column, or the **Row State()** function as its argument. Returns a number from the JMP marker map that corresponds to the active marker or zero if there is no assigned marker.

Assignment Functions

Assignment functions work in place. That is, the result returned by the operation (on the right of the operator) is stored in the argument on the left of the operator and replaces its current value.

Assignment statements are most often used in conjunction with other commands to build a JSL script. You can use the formula editor to create and execute a script in that column, but this is not recommended because of dependencies and ambiguities that can result. Most often, scripts are stored as .jsl files, and can be saved with a data table. See "Creating Scripts," p. 84.

See the *Scripting Guide* for details about syntax.

Note: The first argument of an **assignment** function must be capable of being assigned. This means you cannot have an assignment such as 3+=4, because 3 is a constant value that cannot be reassigned. You must first create a variable (a table variable or local variable) whose value is 3. (See "Adding Table Variables," p. 81, for details about table variables. See "Referencing Columns and Table Variables," p. 304 in the "Formula Editor" chapter, for details about local variables). Then use that variable as the left-hand argument of the assignment function.

= (assign) Puts the value of *b* into *a*. For example, (*a*=*b*).

+= (add to) Adds the value of *b* to *a* and puts the result back into *a*. For example, *a*+=*b*.

-= (subtract to) Subtracts the value of *b* and puts the result back into *a*. For example, *a*–=*b*.

***= (multiply to)** Multiplies *b* with *a* and puts the result back into *a*. For example, *a**=*b*.

/= (divide to) Divides *b* into *a* and puts the result back into *a*. For example, *a*/=*b*.

++ (post increment) Adds one (1) to *a*, in place, so that *a*++. For example, if the initial value of *a* is 4, the expression *a*++ changes a to 5.

-- (post decrement) Subtracts one (1) from *a*, in place, so that *a*– – . For example, if the initial value of *a* is 4, the expression *a*– – changes a to 3.

Parametric Model Functions

This category is a short cut to create three parametric models that are linear functions of set of window-selected columns.

Linear Model, Interactions Model, Full Quadratic Model

Selecting each of these opens a column selection box that lets you select one or more columns to be included in the model. The function then creates and populates the chosen model.

Finance Functions

Lets you create formulas to calculate principal payments, interest rate, rate of return, and so on.

Double Declining Balance

Returns the depreciation of an asset for a specified period of time. The function uses the double-declining balance method or some other depreciation factor.

Future Value

Returns the future value of an investment that is based on periodic, constant payments and a constant interest rate.

Interest Payment

Returns the interest payment for a given period for an investment that is based on periodic, constant payments and a constant interest rate.

Interest Rate

Returns the interest rate per period of an annuity.

Internal Rate of Return

Returns the internal rate of return for a series of cash flows in the *values* argument.

Modified Internal Rate of Return

Returns the modified internal rate of return for a series of periodic cash flows. The cost of investment and the interest received on reinvested cash is included.

Net Present Value

Returns the net present value of an investment by using a discount rate and a series of future payments (negative values) and income (positive values).

Number of Periods

Returns the number of periods for an investment that is based on periodic, constant payments and a constant interest rate.

Payment

Returns the payment for a loan that is based on constant payments and a constant interest rate.

Present Value

Returns the present value of an investment.

Principal Payment

Returns the payment on the principal for a given period for an investment that is based on periodic, constant payments and a constant interest rate.

Straight Line Depreciation

Returns the straight-line depreciation of an asset for one period.

Sum Of Years Digits Depreciation

Returns the sum-of-years' digits depreciation of an asset for a specified period.

Index

Using JMP

Symbols

Numerics

A

E

F

G

H

I

J

K

L

N

O

P

Q

R

S

U

V

W-Z

CPSIA information can be obtained at www.ICGtesting.com
Printed in the USA
269345BV00003BB/55/P

9 781607 645993